THE CYCLE OF COSMIC CATASTROPHES

Flood, Fire, and Famine in the History of Civilization

Richard Firestone, Allen West, and Simon Warwick-Smith

Bear & Company
Rochester, Vermont

Bear & Company
One Park Street
Rochester, Vermont 05767
www.BearandCompanyBooks.com

Bear & Company is a division of Inner Traditions International

Library of Congress Cataloging-in-Publication Data
Firestone, Richard B.
  The cycle of cosmic catastrophes : flood, fire, and famine in the history of civilization / Richard Firestone, Allen West, and Simon Warwick-Smith.
      p. cm.
  Includes index.
  Summary: "Newly discovered scientific proof validating the legends and myths of ancient floods, fires, and weather extremes"—Provided by publisher.
  ISBN-13: 978-1-59143-061-2
  ISBN-10: 1-59143-061-5
  1. Natural disasters—History. 2. Catastrophes (Geology) I. West, Allen. II. Warwick-Smith, Simon. III. Title.

GB5014.F57 2006
904'.5—dc22
                                                            2006007248

Printed and bound in the United States by Lake Book Manufacturing

10 9 8 7 6 5 4 3 2 1

Text design and layout by Priscilla Baker

This book was typeset in Sabon, with Avenir and Century Gothic used as display typefaces

CONTENTS

ACKNOWLEDGMENTS

First, the authors want to acknowledge the contributions of Dr. William Topping, who uncovered the first crucial pieces of scientific evidence and realized that a cosmic event of outstanding significance had occurred. Bill's work was partially funded by a grant from the National Science Foundation (#9986999), which made possible the initial scientific discoveries.

Some of the people who made significant contributions to Bill's discoveries and to the authors' work are listed below in roughly chronological order, as of the publication date, and other contributors are mentioned throughout the book. All provided crucial information, research, samples, or access to key sites. Their listing here, however, does not imply that they endorse any of the authors' theories or conclusions.

They are as follows: Denise Henry and Terrence Rettig of the National Science Foundation; Dr. Henry T. Wright and Dr. William Farrand, University of Michigan; Donald B. Simons, principal investigator, Gainey Paleo-Indian site; Dr. Reginald M. Ronningen and Paul Rossi, National Superconducting Cyclotron Laboratory; Philip Simpson and Henry Griffin, University of Michigan, Phoenix Memorial Laboratory; Dr. Dennis Stanford, National Museum of Natural History, Smithsonian; Dr. Zsolt Revay, Dr. Tamas Belgya, and Dr. Gabor Molnar (deceased), Budapest Neutron Centre; Dr. Alan R. Smith, physicist, Lawrence Berkeley National Laboratory; Dr. C. Vance Haynes Jr., University of Arizona; Dr. Everett Lindsey, University of Arizona; Jane Pike-Childress, BLM, Arizona; Dr. John Montgomery and Joanne Dickenson, Eastern New Mexico University, Blackwater Draw Site; Maria and Anton Chobot; John Issa, Canada Fossils, Ltd.; Dr. Matthew Boyd, Lakehead University; Dr. Brian Kooyman, University of Calgary; William Baldwin and Cheves Leland; Dr. Ted Bunch, formerly of NASA, and Dr. Jim Wittke, both of Northern Arizona University; Dr. Luann Becker, University of California, Santa Barbara; Dr. Robert Poreda, University of

Rochester; Dr. James Kennett, University of California, Santa Barbara; Dr. Han Kloosterman, the Netherlands.

Karen Misuraca, of Warwick Associates, offered invaluable manuscript suggestions and proofing. Last, we thank the team members at Bear & Company who put the story of these discoveries in your hands: Rob Meadows, Jon Graham, Jeanie Levitan, Anne Dillon, Peri Champine, and others.

NOTES ABOUT THE TEXT

Measurements: Most scientists use the metric system, but most Americans do not. In most cases, we provide measurements both ways.

Dating: For the ancient times covered in this book, most scientists use two dating systems, calendar dates and radiocarbon dates, which are based on the decay of carbon-14. They are not identical, so for consistency, we have converted most dates to calendar dates, unless noted.

Style: Each of the three authors wrote overlapping parts of this story. For simplicity, we decided to refer to the writers as "I" throughout the book, except in rare cases.

INTRODUCTION

Plato wrote about the catastrophic destruction of Atlantis, which occurred in a day and a night about 11,600 years ago. The Bible vividly describes torrential rains and an immense flood in which most of humanity perished. Native Americans have many rich stories about an enormous cataclysm involving worldwide fires and flooding. Altogether, the myth and folklore of as many as fifty different cultures around the planet tell of similar global devastations, during which humanity went through a trial by fire and flood.

Are the cultural legends based on facts, or are they fables? Did a major calamity actually happen? Lacking hard evidence, scientists have dismissed the old tales of epic natural disasters, and even though popular writers have speculated about such catastrophes, convincing proof has been elusive. Until now, no one has discovered and documented decisive evidence of a specific event that caused mass extinction.

You are about to embark on an expedition to find the truth. Along the way, across the United States, you will meet archaeologists, paleontologists, geologists, and other scientists at their excavation sites and in their laboratories. You will witness their electrifying discoveries, from solving the mystery of thousands of Paleo-Indian hunters who suddenly vanished to explaining the radioactive bones of "Eloise," one of the last mammoths to walk the planet.

In this book, we intend to prove that a cosmic chain of events began 41,000 years ago and culminated in a major global catastrophe 28,000 years later. We refer to that culmination period of 13,000 years ago as simply the "Event." During our search, we three authors, who are trained researchers who have worked as government scientists, are joined in our quest for truth by a wide variety of other experts, including a retired NASA section chief and others who are researchers at well-known universities.

Through a series of unusual and unpredictable on-site investigations, we turn up clues to a violent calamity that doomed millions of animals to extinction, caused widespread major mutations, decimated the human race, and set the stage for the rise of modern civilization.

Now, for the first time, this work presents tangible explanations for the following:

- *Mass extinctions.* What caused the sudden annihilation of Ice Age mammoths, mastodons, saber-toothed tigers, and 40 million other animals, along with much of the human race?
- *Fire and flood stories.* Is there any hard evidence to corroborate the megafloods and cataclysms recorded in the Bible, Plato's story of Atlantis, Native American lore, and the myths and literature of other cultures?
- *Ice Age warming.* What caused the sudden end of the last Ice Age, when Greenland's temperatures suddenly shot up an amazing seventeen degrees within a few hundred years and the great continental ice sheets began to melt away?
- *The Carolina Bays.* What was responsible for the formation of hundreds of thousands of puzzling miles-long depressions on the Atlantic Coast that look just like Martian impact craters?
- *The "black mat."* What is the mysterious layer of black sediment found spanning North America that rests directly over the bones of the last mammoths to walk the Earth?
- *Black glass.* How does this unique and unusual material provide a crucial clue to the nature of the Event?

THE SIXTH EXTINCTION

The Earth has been shaken by five major extinctions in the last 500 million years. In the most destructive of these events, about 90 percent of life on Earth disappeared, and even in the mildest of the extinctions, hundreds of millions of living things vanished. Look around you and imagine that 90 percent of every species you see suddenly disappeared, including trees, fleas, flowers, dogs, frogs, fish, cats, bats, and, of course, your family and friends.

We are now living in the period of what is called the "sixth extinction," which some scientists believe will be equally devastating. In our time, the extinction of modern species and the gradual worsening of our climate and environment appear to be due primarily to the actions of human beings.

This is what we are facing today:

- *Overpopulation.* Every day, more than one in six of the world's 6 billion people do not have enough to eat.
- *Unpredictable climate.* Weather is more unstable and extreme; tornadoes and hurricanes are more destructive; and global temperatures are rising.
- *Natural disasters.* Most of history's costliest earthquakes and tsunamis have occurred in the last 100 years.
- *Environment.* Humans are destroying forests, wetlands, and grasslands, which are crucial to environmental balance worldwide.
- *The nuclear threat.* Terrorism and nuclear proliferation are on the rise.
- *Warfare.* Recent and current wars reflect frequent struggles over increasingly scarce land, oil, water, and other natural resources.

This litany of dangers confronting our planet may be familiar to you, yet there is a hidden side to the story that we reveal in this book. *Our current species extinction, overpopulation, and environmental degradation were in fact kicked off 41,000 years ago by the Event, and they continue today—they have a cosmic cause, not just a man-made one.*

The Event nearly wiped out the entire human race at that time, and after our extensive research, important answers emerged that affect us today.

THE STORY OF THE EVENT

Our evidence indicates that a colossal cosmic catastrophe culminated 13,000 years ago, and *Cycle of Cosmic Catastrophes* is about how we uncovered that evidence.

Part 1 ("The Search") describes the hunt. We started with just a few compelling clues and used them to unravel the mystery of an awesome cosmic event.

In Part 2 ("The Main Event"), we set a dramatic scene for you, with a descriptive chronological overview of what happened.

In Part 3 ("The Evidence"), we lay out the evidence and explain how it all fits into the bigger picture. Along the way, we hear from some ancient storytellers through their cultural legends, which we maintain are the accounts of those who survived the disaster. While we do not purport their stories to be accurate in all details, we believe them to be valid recordings of subjective impressions of actual ancient events.

Are you ready to begin the search for answers?

THE SEARCH

1
THE SEARCH FOR CLUES

PREPARING FOR THE HUNT

Looking at the man with the shotgun, Dr. Bill Topping, an archaeologist, held up a shotgun shell and cautioned, "We need to hit it dead center." The other man nodded, as his colleagues Bob and Donna Miller watched.

Bill was there to conduct a low-tech experiment to test his theory for explaining the sudden disappearance 13,000 years ago of the Paleo-Indians, the mammoths, and dozens of other large Ice Age species that are referred to as megafauna, meaning that individual members weighed more than about 100 pounds (45 kg). Many scientists agree that the extinctions began and ended abruptly as the Ice Age ended.

Since useful scientific research can be done in both high-tech and low-tech ways, Bill planned to use, first, a $200 shotgun and then, second, a $200 million cyclotron, or "atom smasher." Even though the cost and level of sophistication of the two are vastly different, each would provide equally vital clues for solving the mystery of the Event.

Bill's associate Ray DeMott took the shell and flipped it around in his hand to check it one last time. The two had removed the birdshot, replaced it with fine metallic grains, and sealed the shell with wadding. Satisfied, Ray chambered the shell into the .410 shotgun, slid the bolt shut, and locked it with a sharp click. Next, he shouldered the shotgun and shifted it until it was comfortable, taking careful aim at their highly unusual target—a soup bone.

Bill had purchased the thick bone to substitute as a bone from an extinct mammoth. He had attached it to a plywood target, along with a thick piece of quartz glass to substitute for a Paleo-Indian spear point. The goal was to find out what high-speed tiny metallic particles would do to the simulated mammoth bone and the glass. Would they bounce off harmlessly—or would they penetrate with potentially lethal results? Bill hoped the results would reveal clues to explain the mysterious extinctions of 13,000 years ago. This was a key test for his theory.

Ray closed his left eye and squinted down the barrel with his right eye, lining up the two sights with the soup bone. Holding his breath to steady his aim, he slowly squeezed the trigger. The loud blast kicked up the gun barrel, and the air filled with smoke.

Walking forward quickly as the air cleared, the men were eager to check the results. Nothing. The target had been too far away.

They would have to load another shell and try again, but that's what experimental science is all about. If you fail once, twice, or a dozen times, you just keep trying. A colleague once summed up the scientific method by saying, "The best scientist gets it right the *last* time."

BEGINNING THE SEARCH FOR ANSWERS

The unlikely string of events that led to that shotgun experiment began by accident several years earlier while Bill Topping worked on his doctorate in archaeology. Since he lived in Michigan, a state with many archaeological sites dating from the Paleo-Indian era about 13,000 years ago, Bill's research led him to explore some long-standing mysteries about ancient Ice Age people and the animals they hunted. Scientists do not agree on why they vanished, and Bill was interested in finding out more.

Throughout this book, we use the designation "Paleo-Indian" interchangeably with "Clovis." These terms refer to the Ice Age people who scientists believe came from Asia to Alaska to spread from the western plains of North America across the rest of the continent. Those new arrivals found plenty of large animals to hunt—mammoths, mastodons, and bison, which are the familiar buffalo of Western movies. Most of them are now extinct.

Clovis-era spear points, or Clovis points, were first discovered at the Blackwater Draw site near Clovis, New Mexico, in the 1920s, so the Clovis name is now synonymous with those Paleo-Indian lancelike flint points, which have very distinctive flutes that were created when long, single flakes were removed from each side of a spear point's base. Those flutes are the hallmarks of a Clovis point, distinguishing it from nearly all other types of stone spear points (fig. 1.1).

Fig. 1.1. Classic Clovis point with a long flute at the base. *Source: The Arizona State Museum collection*

WHERE DID THE CLOVIS PEOPLE GO?

The Clovis people appeared suddenly in the New World and stayed on the scene for just a few hundred years, and then their distinctive culture vanished as suddenly as did the megafauna around 13,000 years ago. Some archaeologists believe that the Clovis culture did not actually disappear; it simply underwent radical changes after the sudden extinction of the mammoths and other large animals of prey. However, that theory does not explain why the archaeological record is sparse or empty over large areas of North America where there is almost no sign of Paleo-Indian descendants for hundreds of years after Clovis times.

MILLIONS OF MISSING ANIMALS

The greatest puzzle is why the mammoths, mastodons, American camels, American horses, and saber-toothed tigers all became extinct almost simultaneously. According to Paul Martin in the book *Quaternary Extinctions* (1984), far more species of large North American land mammals died out in the last part of the Ice Age than had gone extinct in the previous 3.5 million years. As the graph of the information in his article makes clear (fig. 1.2), something extraordinary must have happened at the end of the last Ice Age—something that had not occurred for millions of years.

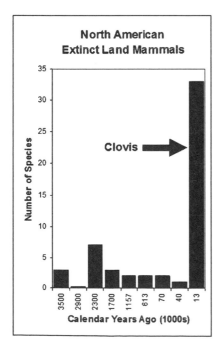

Fig. 1.2. Clovis-era extinctions compared to those of the previous 3.5 million years. *Data from Martin (1984)*

Martin believes that the Clovis hunters are to blame, and that they hunted the animals to extinction. He calls this the overkill theory, one of the three most popular theories on the extinctions. However, one problem with that theory is that the extinctions occurred at nearly the same time across all the Northern Hemisphere and parts of South America, as shown in figure 1.3. Mead and Meltzer (1984) conclude that those extinctions could have been *finished* by 12,900 years ago, and so, to our way of thinking, could be construed as *having occurred* at precisely the same time. But it is hard to conceive that Ice Age hunters, who had been hunting animals in Asia and Europe for tens of thousands of years with stone spears, could suddenly force millions of rabbits, deer, hyenas, bison, saber-toothed tigers, bears, mammoths, camels, horses, and oxen into extinction. Undoubtedly, their hunting had an effect on the populations of those animals, but something else must have been at work.

Another proposal, called the chill theory, suggests that sudden climate change at the end of the Ice Age exterminated all the large animals, in spite of the fact that those animals had been through millions of years of previous seesawing climatic change and had survived.

The third hypothesis is the ill theory, which proposes that epidemics or plagues killed off the megafauna, even though there is no conclusive evidence that such a thing happened. Together, these three theories make up what has been termed the "ill, chill, or overkill" way of explaining what happened to the large animals. We will look at all of them in detail later.

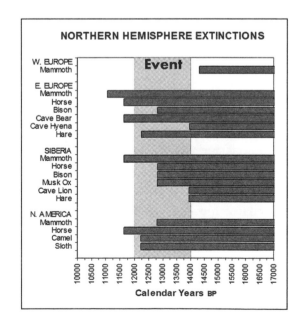

Fig. 1.3. The most likely extinction peaks for Northern Hemisphere megafauna. Note that, after having been around Asia, North America, and Europe for millions of years, many species went extinct within just a short time. *Interpreted from data in McDonald (1984), Mead and Meltzer (1984), and Vereshchagin and Baryshnikov (1984)*

None of the theories offers an adequate explanation, nor is there general agreement among scientists about what happened. Researchers just do not know the answer, although many believe in a combination of the three theories. With that in mind, Bill set out to find some evidence to clarify the mystery of the missing people and animals.

A RADIOCARBON PUZZLE

The first clue he found is related to radiocarbon dating. Radiocarbon, or ^{14}C, is usually formed when a cosmic ray, traveling at the nearly the speed of light, strikes a nitrogen atom (^{14}N) in the atmosphere, transforming it into a radioactive form of carbon, called ^{14}C. Scientists use the rate of radioactive decay of those atoms to estimate the age of old carbon-based objects such as charcoal, wood, seeds, and bones, which may be associated with harder-to-date items such as pottery and stone tools.

Although this technique is reliable in most cases, radiocarbon researchers have discovered a major flaw in the theory. At one time, scientists thought the radiocarbon in the atmosphere remained constant, but to their surprise, they discovered that radiocarbon levels varied considerably over thousands of years—a fact that makes radiocarbon dating substantially less reliable for those times when atmospheric radiocarbon changed dramatically.

At its worst, the effect is so dramatic that scientists refer to it as a "radiocarbon plateau," or a reversal, meaning that the ^{14}C dates are in reverse order, with younger dates seeming to predate older ones. During such a reversal, radiocarbon dates can be off by many thousands of years, as you can see from the graph of IntCal04 dates near 18,000 and near 13,000 years ago (fig. 1.4), just the time when the Clovis people and the giant animals were disappearing. Bill suspected that this was no coincidence.

The only way that this anomaly could occur is if something added a lot of radiocarbon to the atmosphere. The question is, what could have done that? Bill knew that there are only a few causes of large and rapid increases in the radiocarbon pool on Earth. First, the cosmic radiation rate can increase from something like a supernova, the explosion of a massive star that is much larger than the sun, when the pulse of radiation hitting Earth's atmosphere creates a surge of radiocarbon. Intense radiation from a huge solar flare can also cause an increase, although not as great.

Second, since Earth's magnetic field and atmosphere shield the planet from cosmic rays, if the magnetic field strength declines or the atmosphere

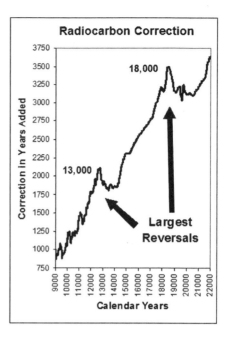

Fig. 1.4. The ¹⁴C plateau problem, or reversal. The sharp peaks at 13,000 and 18,000 indicate that something added ¹⁴C to the environment. *Data from Reimer et al. (2004)*

gets thinner, more rays can get through, thus producing more radiocarbon. There were widely measured magnetic field anomalies, called excursions, about 41,000 and 34,000 years ago. At those times, Earth's magnetic field wavered and almost reversed, meaning that the north magnetic pole nearly flipped down to become the South Pole. At such times of field instability, radiocarbon levels increase.

Third, radiocarbon can increase during impacts by comets and asteroids, which can contain elevated levels of radiocarbon. ·

Two of these three causes—radiation and impacts—have a direct cosmic connection, and both can affect the third one, the fluctuation of Earth's magnetic field. Knowing this, Bill believed it highly likely that a cosmic event was involved.

Other scientists have proposed a connection between extinctions and cosmic events: supernovae by Clark (1977), Brackenridge (1981), and Dar (1998); solar flares and weakness of Earth's magnetic field by Wdowczyk and Wolfendale (1977), Alvarez (1980), and Zook (1980); and cosmic impacts by Clube and Napier (1984), along with Hoyle and Wickramasinghe (2001). Some other scientists, however, ignored or rejected their theories, so they have remained controversial. With that in mind, Bill knew he would need solid evidence.

RADIOCARBON SAYS CLOVIS
INDIANS STILL ALIVE!

Exploring the radiocarbon issue further, Bill discovered that some Paleo-Indian radiocarbon dates were laughably wrong. For example, dates from the Paleo-Indian sites at Leavitt and Gainey, in Michigan, came from layers that scientists knew were 13,000 years old; and yet the radiocarbon date came back suggesting that, inexplicably, the long-vanished Ice Age Indians were still hunting extinct camels when the Egyptian pharaohs were building the Temple of Karnak 2,800 years ago.

Another 13,000-year-old site, at Thedford, Ontario, Canada, seemed to show that the long dead Indians miraculously came back to life and lived up until about the time of Jesus. In addition, the most astounding Clovis-era site of all was at Grant Lake in Nunavut Province in northern Canada, where the long gone Ice Age Paleo-Indians had apparently been hunting mammoths during the time of the Battle of Gettysburg in the U.S. Civil War!

Clearly, these dates and others are impossibly wrong, although many others are correct. This indicated to Bill one of two things: either something had dumped extra radiocarbon onto the planet at that time, selectively resetting the radiocarbon clocks in some areas and not others; or some terrestrial process had produced the erroneous radiocarbon dates.

While assembling these data, Bill realized that the most incorrect dates clustered near Canada and the Great Lakes, whereas those farther to the south tended to be more correct. This puzzling fact would turn out to be an important clue to the Event.

PEPPERED FLINT

Intrigued by the excess ^{14}C at Paleo-Indian sites, Bill reasoned that there might be other clues to the mystery at those sites. Contacting Don Simons, the lead investigator who excavated the Butler and Gainey Paleo-Indian sites in Michigan, Bill arranged to test some flakes of chert (a flintlike rock) that were on the excavated surface at Gainey where the Paleo-Indians had discarded them as they chipped new points.

In Don's basement, Bill slid a flake from the Gainey site under Don's microscope and focused at ten-power (10×) magnification. Squinting, he noticed an odd pattern, and increased the magnification. At 50×, the pattern became clearer, allowing Bill to determine that the chert was covered with tiny black dots resembling fine pepper. He was intrigued but mystified.

Don, who was nearby, thought something was wrong with the microscope and came over. Bill increased the magnification to 100× and stared at the black dots. Now he was able to see them very clearly. With a dawning sense of discovery, he leaned back in astonishment.

"Bullets!" he said to Don. "Looks like tiny black bullets. Must be iron. Could be micrometeorites or something like that. Thousands of them! Maybe millions!" Later tests would confirm that they were high in iron and other rare metals.

The implications staggered them both. Millions of tiny iron balls had smashed into that Paleo-Indian flake. Had they hit the Indians as well? And how about the megafauna—could the tiny black bullets have caused those extinctions? Bill's mind raced with possibilities.

Bill pulled the flake off the microscope stage, flipped it over, and checked it under the microscope. The dots were only on one side. That fit, since if that flake of chert had been lying on the ground at an active Paleo-Indian campsite when the metal balls hit it, they could have come only from space and would embed themselves on only one side.

Everything was falling into place, but Bill knew that one sample was not enough proof. Pulling out several boxes of flakes, he checked dozens more, and while not all of them had the black dots, some did, and they had the dots on only one side.

When he finished, Bill thanked Don for his help and said, "You know, Don, these particles must have come in *very* fast. And probably very hot, too." They were deep in the chert, and there were tiny raised rims around some of the craters, just as you would see with a meteorite crater.

Later, Bill conferred with Dr. Henry Wright of the University of Michigan about the results. Henry was on Bill's Ph.D. dissertation committee, and Bill's ideas about a cosmic catastrophe intrigued Henry. As a good scientist, he was open to new ideas, while insisting on solid, repeatable evidence. Henry traveled to Don's basement to verify the tracks in the Gainey artifacts, and then arranged to analyze chert samples at Leavitt, another Paleo-Indian site near Gainey. He and Bill went to the Leavitt Museum, put flakes under the microscope, and saw the identical pits and tracks as at Gainey. The proof was mounting.

Energized by his discovery, Bill had his first tangible evidence for a kind of catastrophe that might have caused the extinction of the megafauna. The Event that bombarded the chert was most likely cosmic. He knew that the embedded particles came from an enormous blast, such as a solar flare, a supernova, or an asteroid or comet impact. Now, he had to amass more evidence to solve the mystery.

ETCHING THE CHERT WITH ACID

Bill decided to etch the surface of the chert with hydrofluoric acid (HF) to expose the embedded particles. HF is very dangerous and must be handled with great care, so Bill worked with an eye shield and long rubber gloves while he carefully processed the chert.

With the anticipation of discovery, he slid the first flake under the lens of the microscope, wondering what he would find. What popped into view was far beyond his expectations. The chert was riddled with small holes, and at the bottom of most of them he saw what looked like small, roughened balls of rusty iron. Most were incredibly small, about ten times narrower than a human hair, and each was at the bottom of a small crater.

Bill had collected Paleo-Indian chert flakes from a number of other sites from Canada, to New Mexico, to Indiana and Illinois, and as he recorded the depth and number of the pits, he noticed a clear-cut pattern. As with the radiocarbon dates, the depths of the pits decreased as the distance increased away from the Great Lakes (see fig. 1.5). The New Mexico sample had fewer and much shallower pits than the ones from Gainey. Once again, this suggested that the Event had occurred in the area around the Great Lakes.

Then he noticed an important clue. In some pieces of chert, the impacts had formed craters so deep that they made tracks that looked like steep funnels or even tubes (see fig. 1.6), and in any given piece of chert, all the tracks were parallel, indicating that they had come from the same direction. Furthermore, at the sites closest to and in Michigan, these tracks were nearly vertical; on the flakes found farther away from the Great Lakes, the tracks were at an angle. This too suggested that the particles radiated out from an overhead explosion or an impact centered on the Great Lakes or central Canada. An overhead explosion? Bill was surprised and puzzled.

He did not have a lot of information to go on. However, it was clear that if a supernova or flare had occurred, it would have been so far away from Earth that the tracks in New Mexico should be angled the same as those in Michigan. Since that was not the case, it was difficult to attribute metal balls and tracks to a supernova or a solar flare. Perhaps a meteor had exploded in the atmosphere over Michigan.

This reminded Bill of Tunguska, Russia, in 1908, when a cosmic object, most likely a house-sized meteorite or comet fragment, exploded several miles above the Earth. An immense fireball and shock wave flattened the forest in a radial pattern under the explosion and touched off huge forest fires. Incredibly, the blast zone of such a seemingly small object spread over about 1,000 square miles, an area twice as large as either of America's two largest cities, New York City and Los Angeles. Most important for Bill, he remembered that the Tunguska explosion had peppered the ground with millions of tiny particles, some of which researchers found embedded in tree trunks facing the blast and not on the opposite side of the trees. Had a Tunguska-like event happened over the Great Lakes?

EXPERIMENTS TO DUPLICATE THE TRACKS

Now that he had found the tracks on the chert, Bill set out to understand how they had formed. He had not given up on the solar flare or supernova ideas, so he wanted to duplicate how each might affect a piece of Paleo-Indian chert. He knew that both a flare and a supernova could produce a wide range of particles, from smaller than an atom, to a single atom, on up to much larger clusters of many atoms, all of which could cause what he saw in the flint. With that in mind, he contacted the National Superconducting Cyclotron Laboratory at Michigan State University, where Dr. Reginald (Reg) Ronningen agreed to run a test for Bill.

Traveling to East Lansing, Michigan, Bill carried several pieces of Paleo-Indian chert flakes for the tests. After meeting Reg, Bill handed him the samples to use as targets for the cyclotron, which, in simplified terms, is a huge atomic shotgun, except that the "bullets" travel blindingly fast at an incredible 20 percent of the speed of light. With that "shotgun," they

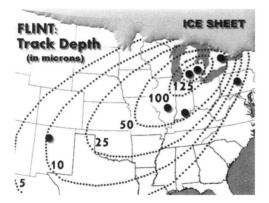

Fig. 1.5. Depth of craters in many Paleo-Indian chert samples. The depth decreases away from the Great Lakes, suggesting that the area was near the center of the Event.

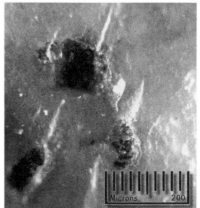

Fig. 1.6. The acid etching reveals a distinctive entry track with each particle. The direction of travel was from the upper right. The velocity was at least 1,000 miles per hour. *Source: Bill Topping*

planned to shoot the chert with atoms of iron to try to duplicate the tracks that Bill had found.

As they entered the protective shell of the control room and the test got under way, Bill knew there would be no shotgunlike blasts with puffs of smoke. At least, he hoped not, since that would mean serious trouble—no more experiment and, maybe, no more researchers. He thought he might hear the steadily rising whine of expensive high-tech machinery, yet there were no sounds except the faint steady drone of the air conditioner.

Later, Reg confirmed this in an e-mail: "During the irradiations, we were in the cyclotron control room. Even if we had stood by the sample (which we are not allowed to do for radiation protection purposes) there would have been no smell, visible signals, or audible sounds during this irradiation." This was a peculiar silence for having unleashed such awesome power.

When the test was over, they retrieved the chert. Paul Rossi, the radiation officer, checked it with a Geiger counter to make sure it was not radioactive, and then handed it to Bill, who flipped it over in his hand. Seeing nothing, he was a little disappointed, although he already knew that if the test worked, the tracks most likely would be too small to see without a high-powered microscope. Returning home, he slid the first sample into view under the microscope, and there they were—tiny craters in the chert (see fig. 1.7). They were almost identical to what he had seen in the Paleo-Indian flakes.

With a quickening pulse, he realized that something else matched, too. Some of the craters looked like trapezoids or diamonds. He knew that this was because of the angles at which crystals form in the chert, which is a variety of quartz. The cyclotron atoms had blasted away part of the crystal lattice, leaving a distinctive diamond-shaped hole. In the chert flakes, the Event had left identical holes 13,000 years ago. A thrill came over him as he considered the new evidence. Maybe high-speed iron particles caused some of the holes in the Paleo-Indian chert. If so, maybe a supernova or a giant solar flare was the culprit after all.

CONTINUING THE SHOTGUN TEST

Bill and his associate Ray then decided to replicate their original low-tech test, which had utilized the shotgun. After loading the magnetic grains into several more shotgun shells, Ray took aim at the wallet-sized piece of thick glass. The first time, from about ten feet away, the black sand peppered the glass, causing small pits in the surface. Only a few penetrated and stuck; most just bounced off.

They tried it several more times at different ranges and then walked over to check the glass. The blasts had pitted the glass more than the last time. Although the particles that stuck looked very similar to the ones in the Gainey chert, fewer particles had lodged in the glass than in the Gainey flint. Clearly, the "bullets" traveled much faster during the Event.

UNDERSTANDING THE EVENT

We drew several conclusions from the tests. First, to penetrate as deeply as Bill saw in the Gainey chert, the particles had to move faster than the shotgun's muzzle velocity of 700 miles per hour, most likely *much* faster. Based on the depth of penetration, we concluded that the particles might have been moving as fast as 3,000 miles per hour.

Second, we knew that most explosions cause intense heat, as does radiation from a solar flare. If either or both had occurred 13,000 years ago, they might have heated the chert and made it easier for particles to penetrate deeper. There is no evidence of anything having been melted at that time, yet there is one bit of evidence that Bill noticed: many of the Paleo-Indian flakes were reddish colored, whereas the parent chert source was not. Scientists know that heat and neutron radiation will cause chert to redden, so perhaps there was a connection. Both supernovae and solar flares give off neutrons.

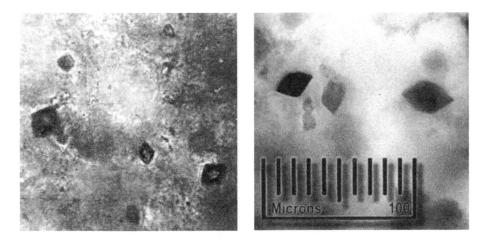

Fig. 1.7. At left, cyclotron photo shows impact craters caused by high-velocity iron ions. Notice the distinctive diamond shape. *Source: Reg Ronningen, National Superconducting Cyclotron Laboratory.* On the right, acid-etched Gainey flakes show almost identical diamond-shaped impact craters. *Source: Bill Topping*

NEW OR CORROBORATING CLUES TO THE MYSTERY

- An extraordinary "Event" injected a surge of radiocarbon into the atmosphere.
- It initiated a chain reaction that caused widespread extinctions of large animals.
- It produced high-velocity, high-density metallic grains that left tiny craters in chert.
- It affected the Great Lakes region more than areas farther to the south.
- The Event probably had a cosmic origin, such as a supernova, solar flare, and/or cosmic impact.

KEY DATES FOR THE EVENT

41,000 years ago
—Massive increase in global radiocarbon
—Magnetic field almost reverses
—Millions of animal species go extinct in Australia

⬇

34,000 years ago
—Large increase in global radiocarbon
—Magnetic field almost reverses again

⬇

16,000 years ago
—Northern ice sheets in rapid meltdown

⬇

13,000 years ago
—Sudden increase in global radiocarbon
—Magnetic field wavers again
—Millions of animal species go extinct in the North
—Clovis-era cultures disappear
—Clovis flint peppered with iron pellets

Fig. 1.8. The most important Event-related dates in calendar years before the present day (BP).

2
"BULLETS" IN THE DIRT

RADIOACTIVE CLOVIS ARTIFACTS

The needle of the radiation meter pegged off the scale. *This stuff is radioactive,* Bill thought, as he pulled away the meter from the Gainey chert. *It's really "hot"!*

Earlier, as Bill pondered what might have caused the Event, he had an unusual idea. Whether a supernova, a flare, or an asteroid impact caused the Event, each might have produced low levels of radioactive materials. Were the Paleo-Indian flakes radioactive? Using his radiation detector, he analyzed several flakes from different sites, and there it was—radioactivity.

To confirm it, he arranged to have the chert retested at McMaster University in Ontario, Canada, and by Herman Rao, with the Nuclear Technology Service in Atlanta, Georgia. When he got the results, he was surprised at how radioactive some of the chert samples were. Various pieces had significantly elevated levels of uranium and plutonium. In chert from the Taylor Paleo-Indian site, the uranium averaged more than 30 percent higher than normal, and the Gainey and Leavitt chert showed substantially enriched radioactive thorium.

In addition, both the Gainey and Taylor chert showed levels of radioactive plutonium that were far above normal. In fact, there should have been almost none, since plutonium is extraordinarily rare in chert, and yet there it was. Usually, high plutonium is found only around atomic-bomb test sites or nuclear reactors. At first Bill thought that was what it was—modern contamination from atomic-bomb testing. He changed his mind when he saw the results from cesium-137, a direct by-product of atomic testing; the testing found no radioactive cesium in the chert samples. Since atomic-bomb fallout usually contains both plutonium and cesium, he concluded that the radioactive elements might have come from a source other than bomb testing.

Bill reasoned that if the radioactivity was in the flint, then it should be in the sediment too, and it should be in the Clovis-era level at all the sites. To test that idea, Bill went to a location near his family home in Michigan that he calls the Woods Site. In spite of the absence of Clovis artifacts there, the Clovis-era level is easy to identify on top of the glacial gravel deposits. After digging a trench with a backhoe, he took thirty-four separate samples of sediment every few inches (5 cm) down the trench wall. He planned to use his radiation meter to analyze each sample.

When Bill returned home, his testing confirmed the chert results. With growing excitement, he looked at the radioactivity readings, which showed a major and clearly defined peak in the Clovis level (fig. 2.1). The amount of radioactivity was nearly 2,000 percent higher than most of the levels around it, and much higher than at the surface where modern fallout would concentrate. It seemed highly likely that the Event had contributed those radioactive elements to our planet. These new data were a surprising twist. He needed more information, along with some expert help; he is an archaeologist, not a nuclear physicist.

FINDING HELP

In his search for help solving the radioactivity puzzle, Bill contacted me (Rick) at Lawrence Berkeley National Laboratory (LBNL) at about this time. As a nuclear physicist with an interest in supernovae, I have a background that complements Bill's training as an archaeologist. Over the next few years, the other authors (Allen and Simon) joined in as the project grew.

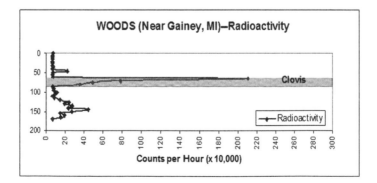

Fig. 2.1. The Woods Site profile shows a major peak of radioactivity in the Clovis level. Above and below it, there is almost none.

Bill discussed with me the implications of the research and sent me photos of the tracks and the other evidence he had found. I saw right away that he was on to something.

Although supernovae have been known for hundreds of years, astronomers observed the first new one with modern sophisticated equipment not long ago, in 1987. While scientists do not understand all the processes yet, they accept that supernovae can either directly or indirectly produce the effects that Bill was seeing. They do so in several ways. First, when a nearby supernova explodes, it bathes our planet in intense radiation. This produces powerful effects, such as increased radiocarbon along with mutations and extinctions. Second, the matter of the star is propelled outward into space at a slower yet still very high rate in a giant expanding shell of star-stuff. When that shell of debris hits our planet, it causes various effects, such as the tracks in the flint. In addition, the force of the debris wave can cause the sun to produce massive solar flares, and it can knock asteroids and comets out of orbit to collide with the Earth. In short, a supernova can create cosmic chaos.

In addition, supernovae influence our everyday life far more than most people know. For example, as you are reading this, you are looking at the book's paper and ink, and, most likely, you are holding the book with one or two hands. All the things involved—the paper, the ink, your hands, your eyes, and your brain—contain atoms that formed in the distant past in a supernova explosion. And not just a few atoms—*all* of them except hydrogen, including all the iron in your blood and the calcium in your bones. Supernovae gave birth to the essential parts of nearly every atom in our solar system and in the entire universe.

You already know that supernovae created all the iron atoms in your bloodstream, and one probably changed your blood type, as well, which is some variation of O, A, and B, the only blood types among humans. DNA research scientists know that there was only type-O blood prior to about 100,000 to 40,000 years ago. That means that if your blood type is A or B now, that blood did not exist tens of thousands of years ago. So what happened—where did A and B come from? The answer is mutation. The new blood types arose suddenly in humanity's family tree due to an instantaneous mutation. Now, the most common source of mutations is radiation, and one common source of intense radiation is a supernova. If you are type A or B, a supernova probably created your blood type not too long ago. Quite likely, you are a "blood brother" or "blood sister" to a giant supernova, but more about that later.

LOOKING FOR MORE RADIOACTIVITY

Since supernovae create radioactivity, as do solar flares, Bill's radioactivity measurements intrigued me, so I arranged to have the chert samples tested by Al Smith at the Low Background Counting Facility at LBNL. Al's lab is located far beneath the massive concrete and earth of a California dam to shield it from the constant radioactive noise of the universe, allowing him to obtain precise measurements. One of the things we wanted him to look for was a particular isotope of potassium, called potassium-40 or ^{40}K, which is a classic signature of a supernova. When supernova radiation bombards normal potassium, or ^{39}K, it turns a small amount into ^{40}K, and wherever we go on Earth, the ratio is nearly the same. However, if we could find elevated levels of ^{40}K in the Clovis layer and not above or below it, that would be powerful evidence that a supernova or the debris from a supernova arrived on Earth during the Clovis era.

When the results came back, the ^{40}K was there. In the chert from the Leavitt and Butler sites, the ^{40}K was 150 percent of normal values, and at Gainey, it was nearly 200 percent of normal, a surprisingly high amount. To find out whether this was just a natural radioactive oddity, we acid-etched a piece of the same Gainey chert, removing the outer surface. If the ^{40}K was a natural part of the chert, it would show up throughout the sample, and the test should show the same levels both before and after etching. It did not. The results were normal, indicating that something had deposited radioactive ^{40}K onto the flint surface. These results pointed to a supernova and its related effects, yet they were far from conclusive. We had a few tantalizing clues with many things that did not fit together, and we needed more evidence. At that point, all that we knew was that high radioactivity showed up at the time when the Clovis people vanished and the megafauna went extinct.

RADIOCARBON AND SUPERNOVAE

I was also interested in the radiocarbon anomalies that Bill had found, and not necessarily only in the samples dated at 13,000 years ago. In looking over the IntCal04 and Cariaco combined graph, which extends back to about 54,000 years ago, I saw an even more interesting time.

Beginning about 44,000 years ago, there is a steep and sudden increase in radiocarbon, as you can see in figure 2.2. The change was so extreme that by 40,000 years ago, it threw the ^{14}C correlation off by an astounding 8,000 years. This means that for radiocarbon dates of 44,000 years ago, there is almost no need for correction; a 44,000-year radiocarbon date for a piece of charcoal equals 44,000 calendar years.

However, 4,000 years closer to today, at 40,000 calendar years ago, the researchers find an 8,000-year mismatch. Instead of showing that a piece of charcoal is 40,000 years old, radiocarbon testing mistakenly says that it is 32,000 years. Such variations make it extraordinarily difficult to get accurate dates. However, the problem is not that the radiocarbon process is wrong. The test process itself is measuring the amount of ^{14}C correctly; the problem is that a lot more radiocarbon suddenly appeared on our planet about 44,000 to 41,000 years ago. How did it get there?

Only a major cosmic event of some kind could account for the dramatic rise, but that creates a puzzle. The rise began about 44,000 to 40,000 years ago, or more than 30,000 years before the extinction in Clovis times. How could this event have caused the extinction 30,000 years later? Maybe there was a connection, but we could not yet answer this question with the limited evidence we had. However, we knew there was good evidence embedded in the chert, so we thought there might be more evidence in the sediment at each Paleo-Indian site, notably at Gainey, Michigan.

HEADING TO GAINEY

As Bill drove from Baldwin, Michigan, his hometown, toward the Gainey Paleo-Indian site near Grand Blanc, the traffic was light and the driving was easy. However, if he had been at the same spot 13,000 years ago, millions of tons of ice would have crushed him. Back then, the Laurentide ice sheet was about 2,000 feet thick at that spot, and maybe more.

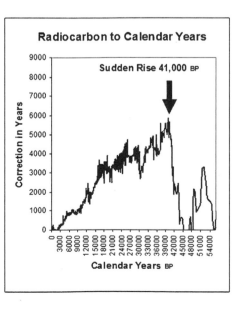

Fig. 2.2. About 41,000 calendar years ago, the ^{14}C in Earth's atmosphere increased dramatically. Since then, it has been mostly in a gradual decline. *Sources: Reimer et al. (2004) and Hughen et al. (2000)*

For tens of thousands of years, an immense glacier covered most of Canada, spilling over into Lake Superior and across much of northern Michigan. Then suddenly, around 16,000 to 13,000 years ago, the ice sheet retreated from Lake Superior and began to pull back hundreds of miles into Canada, ending the Ice Age. That was not surprising by itself, since ice ages end rather suddenly. There have been nine or ten of them in the last million years. All were approximately 100,000 years long, and all ended abruptly.

After an ice age ends, the period of mild climate before the next ice age is what scientists call an interglacial, which is typically 10,000 to 30,000 years long. We now live in an interglacial period, as shown in figure 2.3.

The last ice age ended about 13,000 years ago, and if this interglacial is like the last one, our civilization is on the brink of descending into another bitterly cold ice age. It is ironic that, in a time with many headlines about global warming, there is another, equally dangerous possibility—global freezing. It has happened like clockwork every 100,000 years for the last million years, and many scientists expect that the countdown to the Big Chill *will* continue; the ice will return.

While there is still a lot of debate about what causes ice ages to end suddenly, the most accepted explanation for that retreat 13,000 years ago concerns long-term cyclical changes of the Earth's rotation and its orbit around the sun. These are called Milankovitch cycles, named after the scientist who discovered them, but there is a growing debate about the extent of the effect on Earth's climate caused by these cycles. Some scien-

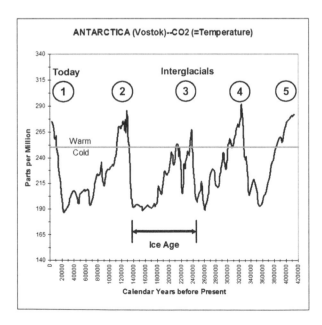

Fig. 2.3. Ice ages from Vostok, Antarctica, ice core CO_2 data. Covering the last 420,000 years, the graph shows four regular ice ages, each about 100,000 years long. Currently, Earth is in a brief interglacial, as you can see from the five sharp peaks. After each one, the cold returns. *Data from Petit, NCDC.*

tists believe the effect is too small to account for the ice ages. Whatever the cause, we think there was another major contributing factor this last time: the sudden occurrence of the Event.

A LINE IN THE SAND

Evidence for a connection between the sudden end of the last ice age and the Event comes from research done in Michigan by two scientists in the late 1950s.

Ronald Mason discovered that archaeologists found Paleo-Indian points only in the southern half of Michigan. George Quimby, who had been mapping the occurrence of extinct megafaunal skeletons, came across Mason's map of point locations and could hardly believe his eyes. Mason's map looked just like his own. Both the megafauna and Paleo-Indian points existed only in southern Michigan below a distinct line that spread from Lake Michigan to Lake Huron, a line that became known as the Mason-Quimby line (fig. 2.4). Other scientists have determined that the Mason-Quimby line is where the edge of the ice sheet stood about 13,000 years ago, which makes sense, since no animals or humans could live there.

If we put all this information together, the implications of that line are staggering. At the time when the glacier stood at that exact point along the Mason-Quimby line, three things happened. First, the large-mammal extinction occurred. No more mammoths or mastodons roamed the forests and fields of Michigan; they disappeared suddenly. Second, until that time, the ice sheet had alternately retreated and advanced near its maximum limit below the Great Lakes. After that moment in time, the ice sheet retreated

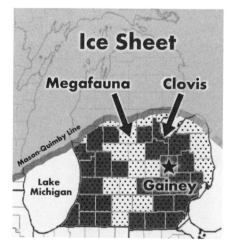

Fig. 2.4. Dotted areas are Michigan counties with extinct megafauna finds. Dark areas represent both megafauna and Paleo-Indian finds. The Mason-Quimby line is where the glacier stood during the Event. Gainey is located at the star. *After Martin (1984)*

rapidly into Canada, and it never returned deep into Michigan. The Ice Age was over. Third, the local Paleo-Indian cultures mysteriously vanished. No one ever has found any Clovis-era campfire sites or spear points that reliably date later than that time.

The date of the Gainey site matches the date of the Mason-Quimby line, and Bill found the site's flint artifacts peppered with high-velocity iron particles. Is that evidence of the extinctions? Did the bombardment end the Ice Age? Bill thought it likely, although he still needed hard evidence.

CLOVIS HUNTERS

As he drove south on Highway 23 toward the Gainey site, a large Paleo-Indian campsite that covers about forty acres, Bill tried to imagine what one of the Gainey people might have seen.

If an Indian hunter had peered off to the north, he would have seen the jagged face of a blue-white ice sheet glistening in the brilliant sun. The frigid cliffs were about 2,000 feet tall, forming a huge wall of ice tall enough to change the direction of the jet stream and create its own local climate, with fierce, frigid winds roaring down and out from the ice edge. This made warm clothing a necessity, and the Paleo-Indians probably dressed much like the modern Inuit, with multiple layers of animal fur and skins.

Complete Paleo-Indian skeletons have not been found, only several partial ones, so we do not know what they looked like. Most scientists believe that they came to North America from Asia, and so they probably looked like today's Indian tribes. No one has found widespread evidence in Asia, however, of artifacts similar to the Clovis- and Gainey-style spear points. The unique toolmaking technology suddenly appeared in the New World along with the recently arrived Paleo-Indians. That odd fact is contrary to most migrations, in which the new people carry their old ways with them.

Because of the toolmaking anomaly, some scientists looked elsewhere to find similar technology that predates Clovis, and they did find it in Europe. The technology for making Clovis points is very close to that used by the Solutrean people, who lived in ancient France up until about Clovis times. If researchers can confirm that theory, it means that the Paleo-Indians in Michigan are descendants of the French rather than the Asians.

For warmth, the Indians picked the Gainey site because the low hill would have faced the western sun and the local forests were full of firewood, a necessity for survival. Another crucial need was water, which they found nearby in the dense tangle of fast-moving glacial streams. The rivers would have contained enough fish and water plants for food for a low-density population.

The forests around Gainey were sparse by today's standards, containing widely spaced trees such as spruce and pine, oaks, and some Douglas fir. In addition, there would have been pecan trees for nuts, and hickory trees, which presumably allowed the Gainey people to make tasty hickory-smoked barbecued mammoth ribs.

Among the trees, the Paleo-Indians would have found tough, wiry, cold-resistant grasses, like sedge, which they used for starting campfires, for insulating their hide-covered shelters, and for braiding into ropes and hunting nets. Hardy herbs grew among the trees and along the rivers, providing medicines, food, and fibers for weaving.

Altogether, evidence indicates that the Clovis-era people were skilled in making use of their environment, and since they were seminomadic, they must have had a well-developed oral tradition to pass down to successive generations. This is vital to our story, since we believe that some of their oral traditions have survived to tell us about the Event. While they most likely had a complex language (all organized hunting societies do), there is no evidence that they had a written language, nor much artwork. They may have enjoyed jewelry and adornments just as modern people do, even though their possessions were limited to what they could carry on their backs or on the backs of domesticated dogs; the horse was not yet domesticated.

These ancient people came to the edge of the ice sheet in Michigan

Fig. 2.5. The Paleo-Indians shaped bones and ivory into tools, yet archaeologists have found very little decorative carving. Even so, it seems likely that, like this modern Inuit, they enjoyed carving designs into animal tusks. *Source: Library of Congress*

because many animals traveled there for the food and water. The nutritious grasses supported herds of mammoth, bison, deer, stag moose, and caribou. Mastodons, cousins to the mammoths, browsed the forests, and giant beavers built equally giant dams along the rivers. The Paleo-Indians quite likely hunted all those fauna, although their hunting methods are unclear. Since many of the larger animals, such as mammoths and bison, were "well armed" and dangerous, it is likely the Indians preferred to trap or corner the animals in relatively safe ways, although there was no truly safe way for them to hunt an enraged ten-ton mammoth. Eventually, some inspired hunter, weary of being chased by mad animals, invented the *atlatl*, an advanced spear-thrower for hurling spears at dangerous animals from a distance.

SEEING GAINEY FOR THE FIRST TIME

Bill flipped on his turn signal, changed lanes, and got off the freeway south of Grand Blanc. He planned to meet Don Simons, the chief excavator of the Gainey Paleo-Indian site, and get a tour of the site. As he drove down the off-ramp, what he saw looked nothing at all like the Gainey of Paleo-Indian times. The Ice Age landscape was long gone. Instead of mammoth-hide tents held up by head-high tusks and rib bones, he passed compact farmhouses and 7-Elevens. Ahead of him, he saw green, neatly plowed rectangular fields bordered by tall windrows of trees. The Gainey excavation site blended into what once had been a well-tended green field.

Along the dusty dirt roads near the Gainey gate, Bill saw farmhouses, many of them with cats to keep the field mice in check. The cats likely hunted in the tree lines that he drove past, but today's cats are a pale reflection of their cousins who hunted those woods around Gainey long ago. Twenty times larger than a house cat and the equal of any African lion, saber-toothed tigers once roamed that same piece of ground, and they didn't eat mice. Those tigers snacked on giant bison and mammoth calves.

Before long, Bill turned in at the gate where Don Simons waited by his car. Don waved, and after saying hello, they entered the site. Bill felt a surge of excitement—he had read and heard so much about Gainey, one of the earliest Paleo-Indian sites in North America. What Don and the other excavators had found there was so distinctive that scientists came to accept the style of spear points as the "Gainey style," just as the Blackwater Draw site produced the "Clovis style." The two point styles were different enough that some archaeologists believe their makers lived at different times. Other archaeologists propose that the differences are due just to distance, and that the Clovis and the Gainey peoples were contemporaries, living around 13,000 years ago. Our evidence supports the latter position that both groups lived at the same time and were merely different tribes or family groups derived from the same rootstock.

When they stopped, Don got out and explained some of the history of the site to Bill. "The farmer who owned this place found the first points after he plowed the field," Don said. "He knew I was interested in things like that, so he called it to my attention. Since I live only about a mile away, it was easy to come out and look around. I found a few nice points the first few times. Then I came back again, finding more, and I just kept coming back. I've been at it a lot of years now." Bill had seen some of the amazing array of artifacts Don had uncovered. There are thousands of them.

"Once we realized what we had," Don said, "we decided to work the site on a larger scale. See the piles of dirt over near those trees?" he asked, pointing across the field. "That's what we had to remove from the site to get to the occupation surface." Bill could see that most of the excavation was roughly a foot deep, so Don had removed many tons of sand to get to the Paleo-Indian living levels.

As Bill walked out across the site, he noticed that the sediment was the consistency of very fine flour with a yellowish tint to it. Occasionally, a light gust of wind stirred up small clouds of yellow dust that whirled around them. Don said, "Sometimes it was rough working out here. This stuff gets in your nose, mouth, and ears in no time." At the end of the day, they were often dusted up like Pillsbury flour-factory workers.

COLLECTING YELLOW DUST FROM GAINEY

Bill was at Gainey precisely because of that yellow dust. After he found the metallic particles embedded in the chert from Gainey, he reasoned that the same particles might also be in the sediment. He went there that day with six large, heavy plastic bags in which to put the samples. With Don's help, he filled the bags and loaded them into his car. From many locations around the excavation, he took samples of the old surface, the same one the Paleo-Indians had walked on 13,000 years ago.

Starting at the embankment at the edge of the excavation area, he dug down until he exposed about three feet of the strata well below the Paleo-Indian level. Then he removed sediment samples about every two inches and bagged them.

After all the back-straining, hot work, he said good-bye and thanks to

Don and set off for home, feeling as if his car trunk was filled with six bags of gold doubloons.

DIGGING THROUGH THE DIRT

At home, Bill eagerly unloaded the bags and prepared to search for magnetic grains hiding in the dirt. Bill knew that if there were any in the sediment, they would be far too small to pick out with tweezers or similar tools. He needed a very powerful magnet. He had already bought one made of neodymium, which is one of the best. Manufacturers usually call them rare-earth magnets or simply supermagnets. They are powerful and mildly dangerous, because they are forcefully attracted to anything with iron in it.

Bill prepared the sediment for separation and set up the supplies to run the experiment. (See appendix A for the technical details if you would like to do this yourself.) He sifted through the fine yellow flourlike sediment, and as the dust flowed over the edge of the magnet, he looked for anything unusual, but the magnet was too dusty to see.

Then he held the magnet up to the light. Peering closely, he saw a coating of fine yellow dust all over the magnet. Gently blowing it away, he saw the magnet's edge clearly for the first time. Looking like bits of ground pepper (fig. 2.6), tiny black magnetic grains glittered in the light!

Fascinated, Bill realized that these looked just like the particles that he had seen stuck in the chert from the Gainey and other Paleo-Indian sites. But were they the same? There was only one way to be sure. He had to see if there were more particles in the Clovis level than there were above and below it. If they peaked in the Clovis layer, that would be the decisive fac-

Fig. 2.6. Gainey magnetic grains stuck to the edge of the supermagnet.

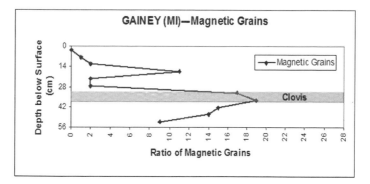

Fig. 2.7. Distribution of magnetic grains at Gainey. The highest concentrations were at about 32 to 38 centimeters below the surface in the Clovis-era sediment. There were nearly 2,000 percent more grains there than at the 2-centimeter depth.

tor. It would mean that somehow billions of these magnetic grains suddenly entered the Clovis world 13,000 years ago.

Rapidly, Bill checked samples from all the other levels that he had collected at Gainey. Yellow dust filled the air as he hastily checked bag after bag with the magnet. Straightening up, with Gainey dust covering his forearms and hands, he had the answer. There were *far* more grains in the Clovis samples than in the nearby levels, as many as 2,000 percent more, as shown in figure 2.7.

He thought, *There's the smoking gun!* And not just figuratively. If those things came in fast enough 13,000 years ago to blow craters in flint, then they were traveling faster than any smoking gun could shoot a bullet.

TINY BALL BEARINGS

Wanting to see the grains up close, Bill cleaned some and put them on a slide under the microscope. Focusing, he saw a jumble of somewhat rounded, rough grains. He surmised that the slide held a lot of natural magnetite, since it weathers out of granite and other igneous rocks, such as basalt. The high level in the Clovis sample suggested that there was something unusual going on with the sediment. Or was it just natural? Without a chemical analysis of the grains, he could not tell whether the particles were extraterrestrial. That would come later.

Disappointed, he scanned through more of the grains on the slide, and what he saw made his heart beat a little faster. He saw spherules—beautiful, perfectly round, highly polished microspherules (see fig. 2.8). He had seen

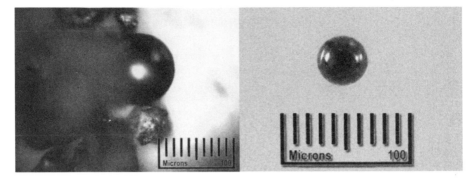

Fig. 2.8. Gainey magnetic spherules. Other grains almost hide the biggest spherule (left), which is about three times the size of the small one (right). This is about the min–max range of the spherules.

photos of them before, yet he had never seen one under a microscope. He knew that if they were at Gainey, they could come from only one direction—up! Spherules fall from two likely sources: the first is volcanoes, and he ruled those out, as there are no volcanoes near Michigan; and furthermore, volcanic spherules are typically composed of nonmagnetic glass, unlike the ones he had that stuck to his magnet. That left only one answer. The spherules were of cosmic origin. They had come from space, most likely from a meteorite impact or shower.

He set the microscope for higher magnification and saw not one but dozens of spherules. Nearly all were extraordinarily small. Fifty of them, side by side, would be about as wide as one human hair. *If there are that many in one view*, he realized, *there are tens of thousands of them here.*

So now he knew there was a very large amount of spherules in the Clovis layer, but that was not enough. Were they also in the nearby layers in the same quantities? That was a crucial test, since if he found lots there too, then most likely they were not extraterrestrial.

Quickly, he checked the other layers, making a comparative count of the number. What he found was that the Clovis layer had about 3,000 percent more microspherules than the nearby layers, compelling evidence supporting a cosmic event (see fig. 2.9).

Curious about the amount, Bill went to the Internet to check out microspherule quantities, and what he found surprised him. When scientists find them, they usually find only a few dozens or hundreds at a time. In one case, the Army recovered 181 from a site in Antarctica. In a larger find, scientists discovered 6,000 in a very old layer in North America. Bill thought he had found many more than that.

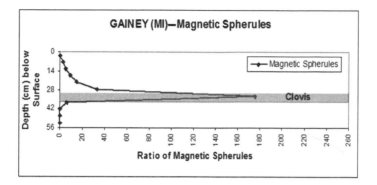

Fig. 2.9. Gainey magnetic spherules. Most layers show almost no spherules. By far the most are in the Clovis layer.

He estimated the number of spherules in the sixty pounds of sediment from Gainey. He did not believe his figures the first time, so he calculated again, with the same result. In the bags on his floor, there were more than 100,000 spherules! He most likely had the largest collection of microspherules ever found on the planet.

Astounded, he stared at the six bags of Gainey sediment with its huge load of cosmic spherules, another fact to support the case that something extraordinary happened 13,000 years ago. The evidence clearly indicated that it was cosmic.

With the discovery of the spherules, the Event became a more powerful phenomenon. Bill imagined what it must have been like on the Ice Age plains of Michigan when literally trillions of these tiny lethal "ball bearings" descended at hypervelocity onto the Clovis people and the herds of mammoth, caribou, and bison. Involuntarily, he shivered a little, realizing that another cloud of spherules could be hurtling through space toward us at this very moment. Could we do anything about it? He did not know.

A SUPERMAGNET AND A FILE CABINET

Bill explained to me how he found the grains and spherules with his supermagnet, and one day I checked some Clovis sediment with my own supermagnet. Bill had cautioned me that supermagnets are superstrong, and that they can catch you by surprise. As I reached over to put the magnet down on the workbench, it flew out of my hand and clanged loudly into the metal file cabinet a few inches away, sticking firmly to the side. Startled, I stared at the magnet as if some invisible spirit had just yanked it away from me.

I gripped the magnet to pull it off and was amazed when I could not do it. It would not move. I yanked even harder, and to my surprise, the mostly empty file cabinet slid toward me with a grinding screech. Still the magnet did not budge. I struggled with it repeatedly, and each time I yanked, the cabinet just slid noisily forward.

About that time, the door opened and my wife and dog came in, curious about the screeching noises in the garage. What they found was a man tugging a file cabinet around the shop by a small piece of metal.

"What's going on?" my wife asked in an even voice. She was used to my unusual projects.

"Working," I answered testily.

"You say working, but what you're really doing is dragging a file cabinet. Right?"

"Very funny," I responded. "That magnet is stuck." I pointed and shook my finger at the thing. "It's a supermagnet, and it apparently takes superstrength to pull one off a cabinet."

Adopting the same tone of voice she sometimes uses with the grand-children, she said, "Awww. You mean you can't get that little old piece of metal unstuck from that big old cabinet?"

Then she laughed. She thought I was joking. My icy stare told her I wasn't.

"Don't you have something else to do?" I inquired.

Ignoring my comment, she said, "Well, why don't you just pry it off?"

Great! I thought to myself. *I'm getting technical solutions from someone whose favorite technical solution is to ask ME!*

I started to make a more forceful "go away" suggestion, then real-ized that I was close to starting an argument. Staying quiet, I turned back to the magnet problem when suddenly it hit me. *She's right! Pry it off.*

I looked at her with a sheepish smile and said, "Thanks! I think that will work." Smiling as if she had just invented quantum physics, she and the dog went back into the house.

Firmly gripping the magnet one more time, I slid it slowly to the edge of the cabinet until half of it was off the edge. Then, slowly and with surprisingly little effort, I pried it up and off.

With that incident burned vividly into my memory, I consciously kept the magnet away from the file cabinets after that, although those are not the only hazards. There are many metal things around a house and shop, and supermagnets happily jump on all of them. Since I first bought mine, I have inadvertently stuck it to kitchen sinks, forks, shovels, knives, desk lamps, refrigerators, wrenches, trash cans, several cars, belt buckles, and a backhoe. If you decide to get a supermagnet of your own, you probably will add more things to your own list, but don't say I didn't warn you. Supermagnets seem to have a mind of their own, so when the need arises, as it certainly will, just remember what my wife said: "Pry it off."

NEW OR CORROBORATING CLUES TO THE MYSTERY

- Paleo-Indian chert from several sites tested as radioactive, but only on the surface.
- Sediment from the Woods Site in Michigan tested as radioactive, but only in the Clovis era.
- Paleo-Indian chert from several sites tested positive for ^{40}K, a supernova marker isotope.
- The largest radiocarbon anomaly dates to around 41,000 calendar years ago.
- The Gainey sediment shows a major peak in magnetic grains in Clovis times.
- The Gainey sediment shows a major peak in magnetic spherules in Clovis times.

3

MAMMOTH UNDER
THE BLACK MAT

ON THE TRAIL OF A MAMMOTH

I flipped on my turn signal and turned into the entrance to the Murray Springs excavation, near Sierra Vista, Arizona. Upon hearing the name of that Clovis site, I pictured an oasis with flowing water, lush grass, and tall trees, even though I know the area to be arid. As we drove down the gravel road to the site, I saw around us the gently rolling San Pedro Valley, surrounded by the purple-and-beige peaks of the jagged Dragoon Mountains twenty miles away. With growing anticipation, I pulled into the unpaved parking area and stopped.

As we stepped out of the car, I got my first clear view of Murray Springs. Head-high desert-scrub mesquite trees and brittle brown grass surrounded me. Except along the San Pedro River in the far distance, I saw no big trees, no buildings, and no springs.

There was no one around except my companions, Dr. Vance Haynes, professor emeritus of the University of Arizona, and Jesse Ballenger, a doctoral candidate who was studying the Murray Springs site. Internationally known today as one of the most influential scientists in Paleo-Indian archaeology, Vance co-discovered the Murray Springs site in 1966, and it is one of the first and richest Clovis-era sites ever found. Vance and Jesse were here to show me the site, to help me take some sediment samples, to get samples for their own research, and to point out some Clovis-era puzzles.

One of those mysteries is what Vance calls the "black mat," which he describes as a dense, dark, organic layer, most likely formed from the explosive growth of algae during Clovis times. Archaeologists nearly always find the mat near the Clovis sediment horizon.

Vance explained, "We've identified the mat, or layers much like it, from about fifty sites over the United States from nearly the Atlantic to the Pacific, and there are more sites in Canada and Mexico. I've been puzzled by the mat ever since I came across it forty years ago, and I still don't know exactly what it is or what caused it." Vance published many papers on the black mat and the Clovis era, and he provided many of them to me prior to the trip (see the bibliography).

As I packed my backpack for the hike out to the site, I thought about the mysterious mat, which was completely new to me. Could it be connected to the Event? Vance continued, "The oddest thing is this: no skeleton of extinct megafauna has ever been found in or above the black mat, only below it, and no Clovis artifact has ever been found in or above the mat either. The American horse, dire wolf, saber-toothed tiger, American camel, mammoth, and mastodon, all of them disappeared in an instant before the black mat formed. When we dig up their bones today, the black mat covers them like a blanket."

I mulled over what he said and looked up to survey the horizon. A light haze shrouded the far mountains and foothills, smog probably, I thought. Other than that, this valley looks today much as it did 13,000 years ago, except it would have been colder and wetter, and the local plants would have been more like those now growing in the nearby mountains. At that time, a small band of Clovis people periodically camped at Murray Springs over a span of several years, hunting mammoth, bison, deer, and other now extinct Ice Age animals. Then suddenly all the megafauna disappeared, leaving behind skeletons but few clues as to what happened to them.

There have been many theories to explain their disappearance, yet so far no one theory explains all the facts. Vance favors a combination of climate and overhunting. Another theory, proposed in 1981 by one of Vance's students, C. R. Brackenridge, is that a supernova called Vela killed them all. However, it has since been determined that Vela probably occurred too far away to cause much trouble. As Vance talked, it was clear that he was open to the supernova theory but did not favor it. On the other hand, we thought Brackenridge might have been on the right track in suspecting a supernova to be the cause. He just blamed the wrong one.

I had already described some of our evidence to Vance and Jesse, mentioning that the black mat might be a consequence of our Event. Only a chemical analysis could tell. They were interested and polite, but I could tell they were skeptical and would need to see some solid evidence before they would accept our theory.

FINDING THE BLACK SAND

Impatient to get my hands on the black mat and take samples, I set off with the others down the trail to the excavation site. Vance had been there hundreds of times before, yet he seemed just as eager as I was to explore the site. In his late seventies, he appeared remarkably fit, and it was hard to keep up with his brisk pace down the winding trail.

Just above the mesquite across the valley ahead of me, I could faintly make out Tombstone, Arizona, made famous by Wyatt Earp and the shootout at the O.K. Corral. I thought of blazing guns and black-powder smoke and wondered how Earp would have responded that day if he had come face-to-face, not with the Clanton gang, but with a herd of trumpeting mammoths.

A few hundred feet along the site trail, we came to a dry streambed with seven-foot-high walls and a sinuous, flat, thirty-foot-wide bottom, along the banks of which I saw the jumbled remains of a footbridge. Vance sounded disappointed as he told me, "Heavy rains washed away the bridge some time ago. Because of budget cutbacks in Washington, the BLM (Bureau of Land Management) hasn't rebuilt it."

Murray Springs is special to Vance, and even though the government has protected the site, the lack of attention and funding from the BLM is painful for him. He has been lobbying the BLM to spend about a million dollars to create extensive, detailed exhibits for visitors to the Murray Springs site, but so far with no success.

As I wondered how we were going to get across, Vance moved to a sloping part of the bank and stepped over, slipping and sliding to the bottom. Jesse walked–slid down next and I followed him, raising a small cloud of dry, yellow dust that got into my mouth and nose. I grabbed for my water bottle, only to realize that I had left it in the car. I thought about going back, but Vance and Jesse were already striding down the wash. Reluctantly, I went after them, reminding myself that it is not a good idea to wander around in the desert without water.

Rounding the first bend of the wash, I was amazed to see long streaks of black sand or magnetite covering the bottom of the wash. "Wait a minute," I called after the others, as I hurriedly rummaged through my pack to pull out my rare-earth magnet. Bending down to place the magnet near the black sand, I was startled as the sand literally flew up from the wash to stick to the magnet, forming a large, bristly black lump of metal dust enclosing the magnet and my fingertips. I had never seen this much before, and I thought of Bill Topping at the Gainey site finding the magnetic grains. Were these the same type as he had found? I wouldn't know until we ran a chemical analysis, but they looked similar. As I shook and pried the black sand off the magnet and into a plastic sample bag, I imagined what the old gold miners would have done if they had seen this black sand, since it often is found with gold dust and nuggets. Turning it in the sunlight, I examined it more closely, looking for the sparkle of gold, but saw none.

FIRST SIGHT OF THE BLACK MAT

Vance and Jesse were waiting near a bend in the wash. I hurried to catch up with them, and as I did, Vance pointed to a long, thin black streak along the wash wall. "That's the black mat," he announced. The thin streak ran down both walls of the wash as far as I could see, up to the next bend (fig. 3.1). "We've found it at dozens of places around the San Pedro Valley," he said, "covering hundreds of square miles. Always, we find the Clovis artifacts and extinct animal bones right under it."

I glanced around, hoping to find a fluted spear point or a mammoth tusk sticking out of the wall, but I saw nothing but brown, twisted mesquite roots. As I pulled out my hammer to take a sample of the mat, Vance said, "Hold on. There's a better spot up ahead."

THE MAIN EXCAVATION SITE

We scrambled up another sloped bank onto a flat-topped point between two branches of the wash, overlooking the main excavation site. I started

Fig. 3.1. Murray Springs wash with the black mat showing at the arrow. Is this a "footprint" of the Event?

to read a small BLM display sign describing the site, but instead Vance provided a full narrative of what he had first found here in 1966.

Pointing to a twenty-by-twenty-foot area ahead of us, he said, "That is where we found 'Big Eloise.'" I was thoroughly puzzled by that until Vance explained, "Eloise was the name the crew gave to the young adult female mammoth whose skeleton we found there. Her bones were arranged in a lifelike pose with her head stretched out upright on the ground, just as she had fallen thirteen thousand years ago. A band of Clovis hunters ambushed her around a small muddy water hole over there." He pointed across the narrow wash.

I replied, "I assume you could tell she had been killed, rather than having died of old age."

"Oh, yes." He continued, "We found several broken spear points alongside her that almost certainly had killed her, and we found a few of the scrapers, knives, and other tools that the hunters had used to remove her hide and meat. And of course there were many cut marks on the bones, showing that human hands had been involved. After butchering the carcass, the hunters had hauled the meat off to the smoking pits about a hundred yards away." He pointed into the distance across the wash. "The hunters chopped off both of her back legs, one of which was up alongside her head. They carried the other one over by the cooking fires. That was the only thing missing from her—one of the legs."

Near Eloise, the crew had also found a rarely seen tool called a spear straightener, which the hunters used for making and repairing their spear shafts. Strangely enough, the researchers recovered it from the bottom of a perfectly preserved mammoth footprint, where it was broken into two pieces. As the Clovis hunters struggled to subdue Big Eloise, it most likely

Fig. 3.2. This BLM display at Murray Springs recreates the last moments of Big Eloise as she is attacked by Clovis hunters.

fell out of a hunter's pouch and she stepped on it. It probably was one of the last things Big Eloise ever did.

After years of meticulous excavations, Vance has pieced together many parts of the story of what happened with the Clovis hunters 13,000 years ago. He explained, "About a hundred yards away, we found the Clovis campsite and roasting pits." That was where Big Eloise was turned into mammoth jerky. "The people stayed until the roasting was finished and then packed up and moved south, killing several more mammoths at another water hole nearby. Altogether, we have found about half a dozen sites around the San Pedro Valley, showing that the Clovis hunters came into the valley periodically for several years. They made camp here while they hunted mammoths and bison."

Wondering where they came from, I asked Vance, who said, "We think they may have had a more permanent camp down in Mexico, but we really don't know. They seemed to travel a lot."

LUMPS UNDER THE MAT

Not long after killing Big Eloise, the Clovis hunters disappeared from the archaeological record and never showed up again. What happened? One would think that good hunting would have brought them back again, except that shortly after the death of Big Eloise, all the mammoths vanished from the entire planet—along with the American horse, American camel, saber-toothed tiger, short-faced bear, and dozens of other large animals. Some scientists believe the Paleo-Indian hunters killed them all (the overkill theory), but while the Clovis people certainly were skillful and deadly hunters, that scenario is not widely accepted as being the only reason, although it may be part of it. So what happened to them all?

One clue is the black mat. When Vance and crew first dug down and before they found bones, they came across the mat, which varied from a few inches to as much as a foot thick, and it was everywhere they dug. The most unusual aspect of the mat showed up when they dug completely through it into the Clovis level.

Vance recounted what they had found as they excavated, explaining that the mat covered everything, but it was uneven and bumpy. Some of the bumps were quite large, and one of them turned out to be Big Eloise (see figs. 3.3 and 3.4). The mat was draped over her bones like a thick blanket, staining her bones almost black and conforming to her skeleton as if it was shrink-wrapped.

Fig. 3.3. BLM excavators uncovering a possible relative of Big Eloise at Murray Springs. *Source: BLM Arizona*

Fig. 3.4. Big Eloise's skeleton as found by Vance and the University of Arizona team. The dark color of her skeleton is due to staining by the black mat that was draped over her. *Source: Dr. C. Vance Haynes Jr., University of Arizona; used by permission*

Vance told me, "A modern-day-elephant expert who analyzed the skeleton said that, based on the articulation of the skeleton, she could not have been dead for more than a few days or weeks before the mat was laid down. If it had been much longer, scavengers and even other mammoths would have pulled apart the bones, just as modern-day elephants do. After Big Eloise died, she was buried very soon under the black mat." The implications were stunning: within days of her death here along that ancient creek bed, almost no mammoths were left alive anywhere in the world. Big Eloise was one of the very last mammoths ever to walk the Earth. A chill came over me as I tried to comprehend the nearly incomprehensible force that could cause such a thing to happen within such a short time.

Vance continued, "We found around the water hole hundreds of mammoth footprints remarkably well preserved in the sandy soil" (see fig. 3.5). The black mat had conformed perfectly to each footprint, filling and preserving it for millennia. He explained, "When the crew carefully lifted off the black mat with dental picks, the footprints looked just as if the mammoths had walked by only a few days before." Instead, thirteen millennia had passed. "We knew the water came up very gently," he added, "otherwise it would have washed away the prints."

Fig. 3.5. The trail of mammoth footprints leads down to the water hole where Big Eloise was ambushed. *Source: Dr. C. Vance Haynes Jr., University of Arizona; used by permission*

ICE AGE CHILL

According to Vance, at the time of Big Eloise's death, the climate was poised to take a catastrophic turn. She had been alive during a time we call the Bolling-Allerod, a sudden warm period from about 14,400 to 13,000 years ago when the Ice Age appeared to be ending and the climate warmed up around the world. The ice was in retreat, the Earth was blossoming, and animals and plants were beginning to proliferate everywhere. Times were good.

Then all at once, 13,000 years ago, as if some giant freezer switch had been flipped, the bitter cold abruptly returned to nearly glacial levels. Scientists call this unusual and puzzling time the Younger Dryas, at the start of which all the giant animals disappeared. Did the bitter cold kill them off, as the chill theory suggests? While that may be possible, scientists now know that those animals had been through many similar severe climate changes in the past, and yet they had survived; this one should not have been any different. There must be more to the story than just the climate change.

In the San Pedro Valley, Vance found that the climate became considerably wetter as well, not with flash floods like those Arizona has today, but with very long, frequent, soaking rains that fell for perhaps days at a time, leaving lakes, ponds, and wet meadows all over the area. Within days after Big Eloise died, water levels rose, covering her skeleton completely. According to Vance, thick mats of algae began to choke these lakes and ponds, forming a floating blue-green scum on the surfaces, and as the algae died, they sank to the bottom, forming the thick black mat that buried Big Eloise and the Clovis campsites.

As I looked from the bank down to where they had found the skeleton, the extraordinary uniqueness of the Murray Springs site struck me. I was peering through a time window at the bones of Big Eloise. That scene spanned no more than a few weeks in ancient history, when a monumental catastrophe almost instantly wiped out tens of millions of animals, and most likely thousands of Paleo-Indians as well. As I stood there, it was frustrating not to know exactly what had happened. Like detectives arriving at a crime scene, we had few clues and few answers, and, like detectives, we just had to keep collecting evidence until the answers became obvious.

TESTING WITH THE MAGNET

Vance motioned for me to follow him across a makeshift fence, saying, "Let's go down into the arroyo. There's a great place to get samples down there."

Jesse and I followed him down and across the wet and soggy wash, pushing our way gingerly through thick thornbushes. Vance pointed to a band of tan quartz sand just above the base of the wash wall and explained, "During Clovis times, there was a small creek and a water hole right there that attracted the mammoths. Sometimes the stream got low on water, and we find shallow holes scraped out by the mammoths with their tusks and feet as they looked for water."

Underlying a dip in the black mat, I clearly saw coarse creek sand and

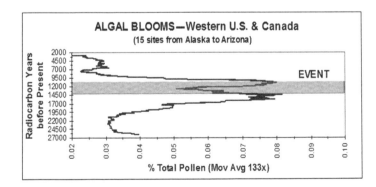

Fig. 3.6. Was the algal black mat widespread across North America? This chart is of pollen from fifteen Ice Age lakes from Alaska to Arizona. It shows a huge increase in Ice Age algal blooms at the exact time of the Event. Before and afterward, the levels were low. *Data from NOAA, World Data Center (2005)*

well-worn pebbles that a small stream had deposited in a U-shaped area that was only a few feet wide. Vance kneeled down against the wall and traced the tip of his trowel along the base of the thin line of black mat. "We ran many radiocarbon tests all over the valley on charcoal from just below the mat, and the dates average about thirteen thousand calendar years ago. That is when the extinction occurred, and that is when the mat began to be deposited," he said. The mat varied from about an inch to a few inches in that area, and was much thicker in nearby locations.

"How about the mat? Do you have dates above it, too?" I asked, curious to know how long it had lasted.

He said, "The average dates for charcoal just above the mat are about one thousand to twelve hundred years younger. We believe the mat continued to grow during all of the Younger Dryas."

One more thing was on my mind. I asked him, "Have you found any Folsom artifacts in the mat or above it?" Those artifacts, named after Folsom, New Mexico, where they were first discovered, were attributed to the Paleo-Indians who arrived on the scene about 600 years after the Clovis era.

"No, we have never found any," he answered. "The black mat here was culturally dead for more than one thousand years." Yet that does not prove no one was there; it only proves there were no traces of their existence at the already excavated sites. There may be some as yet undiscovered artifacts somewhere. However, our Event theory predicts that the reason no artifacts have been found in the mat is that there were almost no people left alive. Post-Clovis-era survivors slowly repopulated the area, and that took hundreds of years.

Looking at the thin layer of mat and the Clovis layer below, I knew from what Vance said that the dates of these layers matched those at Gainey. Therefore, if the catastrophic event that killed the mammoths produced the magnetic grains at Gainey, then the grains might be at Murray Springs too. Since Arizona is a long way from Michigan, however, I wasn't sure. This was the big test. I pulled out my magnet and took a deep breath. I hoped to find an increase in magnetic grains just below the black mat, when the last Clovis hunters were here. Vance and Jesse were openly skeptical, and Vance pointed out that since the surface wash was full of the magnetic black sand, it most likely was common throughout all layers in the area. I was worried about that too. Wondering what I would find, I touched the magnet about a dozen times against the cliff wall just along the bottom line of the black mat. When I finally pulled it away, a thin, unmistakable row of magnetic grains lined the edge of the magnet. I showed it to Vance, who slid his finger along the edge without saying anything. So far, so good. There should be far fewer grains above and below for the pattern to match Gainey.

After cleaning the magnet, I tested the light-colored sediment layer above the mat, and much to my surprise, as well as that of Vance and Jesse, there was nothing there at all: not a single magnetic grain. I tested the gray sediment layer below the mat and found nearly the same thing: almost no magnetic grains at all. Without a doubt, there was a major increase in the grains below the black mat in the Clovis layer—the same as at Gainey.

Vance was surprised, but still skeptical. He said, "I don't know about that. Every geologist knows that magnetite streaks are common in sediment like this. But if you can find the same thing at other Clovis sites, then maybe I'll be convinced." That seemed reasonable enough.

We still needed to do a chemical analysis, but so far I could not have hoped for better results. In later testing, we found about 100 times more particles in the Clovis layer than just a few inches above or below it. Clearly, an unusual occurrence had deposited them there, but we did not know what. Maybe there was some other earthly explanation.

A LAYER OF CHARCOAL

There was one more puzzling surprise to come, which did not become clear until much later. As I dug sediment samples out of the wall near the water hole, I came across a dime-sized piece of charcoal, valuable for dating sites using the radiocarbon method, and Vance eagerly collected the charcoal to take back with him. As I continued collecting sediment, I noticed small flecks of charcoal elsewhere in the Clovis layer, and none above it or below it (see fig. 3.7). This suggested increased wildfires during the Clovis period. Some scientists believe the Paleo-Indians purposely set wildfires to herd animals to kill sites and to cause fresh grass to grow to attract the animals they liked to hunt. Had something besides the hunters also caused the fires? Maybe the charcoal was another clue to the Event.

WHAT WAS IN THE GRAINS?

Later, we tested the Clovis grains by two procedures, neutron-activation analysis (NAA) and prompt gamma-ray activation analysis (PGAA), to determine the amounts of various elements. We also tested the black magnetic sand from the modern wash, which, although it looked identical to the Clovis magnetic grains under the microscope, gave us a major surprise. The Clovis grains and the black sand grains were very different in many ways,

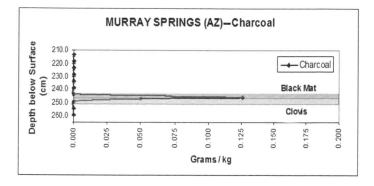

Fig. 3.7. Charcoal peaked just in the Clovis layer, suggesting that there were many more fires at that time.

and in fact did not seem even remotely related chemically. For example, we compared titanium, an extraordinarily hard metal that is used to make parts of satellites and the space shuttle. The surface black sand had 0.4 percent titanium, whereas the Clovis grains contained about 10 percent, twenty-two times more. Such high levels are rare anywhere in the world, let alone in Arizona. Zinc, cobalt, and chromium were also high in the Clovis grains, but not in the surface black sand.

Seeing such high levels, we decided to check for heavy-metal toxicity, and found that the levels of titanium, arsenic, and iron would be deadly if eaten in high quantities or if contained in creek water. Clearly, that Clovis-era material would have been unhealthy for animals and people.

Even more surprising, the Clovis grains were mildly radioactive, testing ten times higher in uranium and radioactive thorium than the surface particles. The results suggested that the two sets of grains had come from different places. In the oddest twist of all, when we compared the Clovis-era grains to known sources, they matched two other sites very well: Canada near the Great Lakes and the moon! Either source area was hard to accept at first, but if either was true, how did particles from Canada or the moon end up in Arizona? At that point, we were not sure. At one time, we had expected the composition of the grains to look like meteorites, and they did not. They were very different. This was unexpected, and we did not know what to make of it at all.

In addition, when inspecting the magnetic grains under the microscope, we discovered highly polished magnetic spherules just as were found at Gainey. They showed up only in the Clovis layer and not above or below it in substantial amounts. The polished spherules only form at temperatures of several thousand degrees, and they typically come from meteorites that crash into Earth or explode in the upper atmosphere. This did not fit our theory very well, since the magnetic grains from Murray Springs do not show a clear extraterrestrial composition, unless you count the moon. It was not clear to us where they came from and how they melted.

RADIOACTIVE TEETH AND BONES

Since we had found radioactivity at Gainey, I returned to Tucson later with a Geiger counter to run some tests. Sometimes buried bones attract radioactive minerals, so I arranged with Vance and Everett Lindsey, another professor emeritus of the University of Arizona, to test some of the university's extensive collection of bones from Murray Springs and other Clovis-era sites, especially the mammoth bones. When I saw the teeth and bones, one thing stood out: many showed signs of dark staining, usually on just one side (fig. 3.8).

When I asked Vance about the staining, he confirmed what I suspected. The dark side of the stained bones was the one in contact with the black mat, and the unstained ones were ones that they had found completely buried in the sand.

First, I tested the mammoth teeth and bones with the radiation detector, and as expected, many of those from Clovis-era sites were mildly radioactive, but only the ones that had been in contact with the black mat. The non-mat bones were not radioactive. Knowing that fact, I retested the teeth or bones that had both a mat side and a non-mat side. What I found next was so surprising that I did not believe it at first.

The mat side was more radioactive than the non-mat side. To be certain, I retested some teeth repeatedly, always with the same result. Clearly,

Fig. 3.8. Lower jaw of a San Pedro mammoth stained by the black mat (above the line). The tooth is both magnetic and radioactive. *Source: University of Arizona collection*

the radioactivity was associated with the mat and had not been in the environment before the mat formed 13,000 years ago. This fit perfectly with our tests of the magnetic grains, so the radioactivity mystery related directly to the magnetic-grains mystery.

That was only the first of three surprises about the mat. The next one came when I tested a mat-stained mammoth tooth with a tiny magnet on a string. If there are any magnetic materials in an object, the string magnet will swing and stick to it, and that is exactly what happened with the tooth. The magnet weakly swung over to stick to the tooth in a few places; this happened only on the mat side. The non-mat side was nonmagnetic. Surprised, I repeated the test many times on many teeth, always with the same result. The mat side was magnetic and radioactive and the non-mat side was neither. This suggested that the increased radioactivity and the peak in magnetic particles were somehow associated with the Event.

POISONED WATER HOLE?

There was one last surprise regarding the black mat. Vance had run several tests that indicated that algae might have produced it, and we knew that sometimes algae could explode into huge blooms, killing fish and animals. The red tide in Florida is a good example of an algal bloom that can be deadly. Sometimes the red tide will suddenly grow out of control near Gulf cities like Tampa and St. Petersburg. The rapidly multiplying tide releases algal toxins that kill fish and plug the shallow coastal bays with their floating bodies. Research into its causes has shown several key conditions that are necessary for the bloom to occur: warm water, for example, and a plentiful source of nutrients, especially iron. We reasoned that since the magnetic grains contained a lot of iron, they may have triggered a massive algal bloom that formed the black mat, much as happens with the red tide. It can also happen in freshwater, when blue-green algae produce powerful toxins that attack the liver and can kill large animals like cows within minutes after they drink it. These blooms are an occasional problem for livestock around the world today, including in the United States, Canada, China, and Australia.

Vance's analysis of the chemistry of the mat clearly indicates that there was an algal bloom in Clovis times. If so, then the water could have contained algal toxins then, just as local ponds do today. The toxins break down within months, so all traces would be long gone, but I wondered whether the algae-laced water poisoned the Clovis people and the large animals. To find an answer, we did research and uncovered a paper by Braun and Pfeiffer (2002), who studied a similar possibility at a lake basin in Neumark-Nord, Germany. There they found skeletons of many dozens of animals from different species in a mass die-off that dated to around 150,000 years ago. Puzzled as to what might have killed them, the researchers suspected toxic algae. As they predicted, when they analyzed the lake sediment, they found telltale chemical markers for toxins. They concluded that the poisonous waters had killed thousands of animals.

So what about the Ice Age animals? Could the toxic black mat be one answer to what had happened to them? Did the giant animals become extinct because they drank water containing high levels of algal poisons—or of toxic metals, like titanium and arsenic—or was their demise due to high levels of radioactivity? We know that the black mat appeared widely over North America almost exactly 13,000 years ago at the time when the megafauna went extinct, and it is rarely found before or after that time. If the mat did poison some of the animals, this would fit with the third commonly proposed theory for the extinction, the ill theory, which claims that some disease—or, in this case, poisons—killed them off. It was almost certainly not a major cause, but it could have been a contributing one.

In any event, even if the poison theory was true, we still had a major mystery to solve. Why had the poisonous black mat suddenly appeared all over North America at the same time, and what on Earth (or from elsewhere) had caused it? Most likely, the explosive growth of Ice Age algae somehow involved the Event.

WHAT DID IT?

Altogether, we know about the kill, chill, and ill theories for the extinction that occurred when Big Eloise died. Were any of those theories mainly responsible or was it something else entirely? We were not sure. What we knew was that we had found evidence for all of the theories at Murray Springs, and all seemed related to the black mat, which, in turn, seemed associated with the magnetic grains. More and more, the magnetic grains were emerging as the key to the mystery. We needed to discover where they came from and how those grains got to Murray Springs.

NEW OR CORROBORATING CLUES TO THE MYSTERY

- The mysterious black mat found at many sites in the United States, Mexico, and Canada is often draped directly over extinct megafauna bones.
- Right after the Event of 13,000 years ago, the climate turned wet and bitterly cold.
- The magnetic grains at Murray Springs are chemically similar to those from Gainey, 1,500 miles away.
- Murray Springs has an increased amount of charcoal in the Event layer, suggesting wildfires.
- Melted metallic microspherules found at Murray Springs suggest a high-temperature impact at the Event.
- There is some evidence suggesting that the black mat once contained dangerous heavy metals and toxins.
- Event-age mammoth teeth are magnetic and radioactive only on the side contacting the black mat, suggesting that the formation of the mat was somehow related to the Event.

4

HUNTING MAMMOTHS

FOLLOWING A HUNCH

As my wife and I entered the pavilion of the Gem and Mineral Show in Tucson, Arizona, I noticed two things immediately: a meteorite the size of a large TV and amethyst crystals the size of a bathtub, but I wasn't interested. I was there for something else.

It had all started a few days previously. My wife likes the huge Tucson show because she collects unusual rocks and gems, and every year she invites me to go and every year I decline—too many people and too much "stuff" to suit me. After declining her invitation again, I wavered as I recalled seeing at a past show a booth filled with boxes of mammoth tusks. Thinking about the iron grains embedded in flint from Gainey, I wondered if there might be similar evidence embedded in some tusks, but then I finally decided against going, since the odds of finding anything there would be astronomically small.

That's when something happened that changed everything: I felt a powerful hunch to go to the Tucson show. Whether you call it luck or intuition, I've learned to follow these hunches, because they often lead to good things. Thinking that it might be a dead end, I recalled one of the favorite adages of a former teacher: "In science, a dead end is just a place to change direction."

At the show, I headed for the fossil area, thinking that if the magnetic grains had survived, I might find them embedded in tusks, horns, or antlers, the only outer parts of an animal that could survive for thousands of years. Even though the survival of such items is rare, every year Canada Fossils Ltd. comes to Tucson from Calgary in Alberta, bringing tons of mammoth tusks, so I headed to its booth. Even if it did not work out, I knew that at least I would see some very unusual fossils, gems, and meteorites.

SIBERIAN MOLES

After I arrived at the tusk booth, the sales rep explained that Canada Fossils imports a few tusks from Alaska, although most of its tusks come from Russia. For hundreds of years, the native people of Siberia have made a lucrative business digging the tusks out of the permafrost and selling them throughout the world. After the ivory ban went into effect in 1986 to protect endangered elephants, the Russian mammoth-tusk trade increased considerably, because even though Siberian ivory is tens of thousands of years old, it is just as easy to carve as modern elephant ivory. With the shift to mammoth ivory for carving, African elephants are recovering. So, in an unusual twist of fate, the extinct mammoth, a relative of the modern elephant, is helping its living cousin avoid extinction too.

A legend involving mammoths comes from Russian explorers who entered remote areas of Siberia centuries ago. They made first contact with the Dolgan, a native tribe of reindeer herders, who warned the explorers to be cautious of the area's giant moles. The story goes that the huge moles hated sunlight and fresh air; whenever one accidentally broke through the surface of the frozen ground, it died instantly. Of course, the explorers were eager to see those giant moles, hoping to discover a spectacular new species. When they asked the Dolgan to show them a dead mole, the tribe hauled out piles of long rib and leg bones, and the explorers were duly impressed with the huge size of the mystifying moles.

The mystery took an unexpected twist when the Dolgan took the explorers into one of their tents to show them the long, curved front teeth that the moles used for digging. To their surprise, the explorers realized that the non-mole-like tusks were from some form of elephant, even though they knew nothing about mammoths at the time. However, try as they might, the men could not convince the Dolgan that those tusks came from anything except giant moles. The Dolgan just laughed as the explorers described huge, gray, chubby animals with giant floppy ears and long noses down to their toes. They liked their mole story better.

BOXES OF GIANT MOLE TUSKS

The sales assistant agreed to let me see all the "mole" tusks he had, most of which were in a spare motel room. Entering the room with a surge of excitement, I was overwhelmed. The floor and tables were stacked high with boxes of mammoth tusks and other fossils, making it hard to move around. There was at least a ton of tusks. I was in mammoth-mole heaven!

When I explained to the sales assistant that I was looking for tusks that were magnetic, he appeared doubtful and politely kept quiet. Frankly, I too had doubts that I would find any. The odds were clearly against me, since the number of mammoth tusks with particles stuck in them must be a minuscule fraction of the total tusks that ever existed. The one factor in my favor was that tusks decay steadily and disappear over time, so the fact that they had a room full of them suggested that all those tusks were relatively young. Even so, I wondered if even a single mammoth in that room had witnessed the Event.

Wrestling with the heavy boxes, I methodically went through each one, unloading and testing the tusks, each of which had been broken long ago into pieces between one and three feet long. Most tusks were labeled "Imported from Russia," although a small number came from Alaska, where gold miners exposed them while using huge water cannons to wash away the permafrost in the search for gold (fig. 4.1).

After an hour or so, I had gone through about three-quarters of the boxes without finding anything, and I began to think I was wasting my time. Then I lifted out a small, shattered tusk tip and put it on the table under the bright desk lamp (fig. 4.2). Something caught my eye, and I held my breath. I could see tiny, shiny black spots, each about one-eighth inch in

Fig. 4.1. Mammoth tusks uncovered by miners in Alaska around 1890. *Source: Library of Congress*

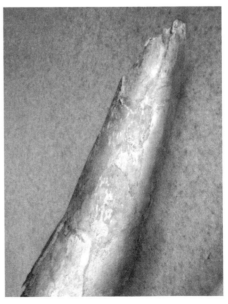

Fig. 4.2. The first tusk we found containing impact particles.

diameter, on the outside of the tusk on one side. Could this be it? Grabbing my magnifying glass, I peered down at one spot while swinging my string-mounted magnet close to it. Instantly, the magnet jumped to the black spot and caught there, sticking out almost at a right angle to the tusk. My heart began to beat faster.

Another spot was the same. Then I tested tusk areas away from the spots, with no response. Only the dark spots contained enough iron to attract the magnet. And I noticed one more oddity: each spot had a brown ring around it, which appeared to be charred (fig. 4.3). Had the particles been very hot when they came in? It would take more testing to find out, but it certainly looked like they had.

With renewed enthusiasm, I searched the other boxes, inspecting nearly seventy large pieces of tusk, but to my disappointment I found no others with the spots. Those tusks represented a maximum of seventy mammoths, so I had found particles in one mammoth out of seventy, at most. Undoubtedly, there are hundreds of thousands of tusks still buried in Alaska and Siberia that contain no magnetic particles at all, so I was satisfied that, against all odds, I had found one. I had been reluctant to go to Tucson at first, but the hunch had paid off.

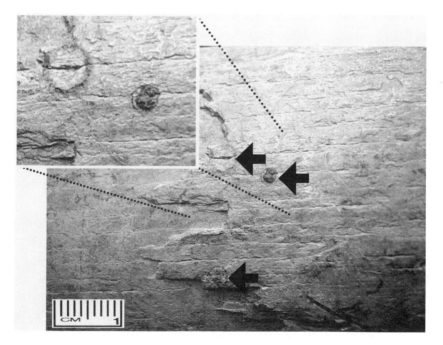

Fig. 4.3. Close-up with inset of particles in the tusk, one with a visible particle and one without. Both are magnetic, and both have a dark ring around them. The rest of the tusk is nonmagnetic.

THE GIANT SKULL

After finding the tusk, I was finished at the Tucson show, but my wife needed several more hours to shop for gemstones. She suggested that I look for more tusks, but I responded, "This is the only large-tusk dealer here, so it looks like my good luck is over for the day. I'll just sit and wait."

She countered, "Why don't you look around while I shop? Maybe you'll have another run of luck!" I was tired and ready to go, but I knew better than to expect her to leave the gem stores early, so to kill time I wandered around among the hundred or so fossil dealers. In several other showrooms, I found a few tusks, but none contained micrometeorites.

Then, as I rounded a corner, I saw a massive bull-like skull on the display table of a German fossil dealer. Excited, I recognized it instantly and quickened my pace to check it out. There, stretching from end to end of the table, I saw the skullcap of an extinct Ice Age bison with a four-foot horn span (fig. 4.4). When alive, the beast looked much like today's buffalo, except larger. It would have stood nearly seven feet tall at the shoulders, and its muscular body would have weighed more than 2,000 pounds—one bison would have provided many meals for a tribe of Clovis people. Millions of those bison once grazed the Ice Age steppes from England to Siberia and from Alaska to the American Great Plains until, after having been around for nearly a million years, they suddenly vanished 13,000 years ago at the time of the Event. I wondered if this Ice Age bison could have been one of the last to walk the planet.

Pulling out my magnifying glass, I leaned over nonchalantly to inspect the skull, trying not to attract attention. Sunlight glittered off the shiny bone so that even with unaided eyes, I could see them—large black grains embedded in the horn. Breathing faster, I fumbled for my string magnet, doubting the evidence before my eyes. Swinging the magnet near one large black spot, the string jerked sharply and the magnet stuck fast, even though it was on a vertical part of the horn. The black dot contained iron, all right. Lots of it!

Inspecting the metallic grain with the magnifying glass, I could see that unlike the iron particles in the mammoth tusks, it was less rounded and much more angular, like the surface of an actual meteorite. In the strong sunlight, I also saw reddish rust spots around it, another sign it contained a lot of iron, and it was embedded deeply into the bone, making it unlikely that it lodged there after the animal died. Quickly, I scanned the rest of the skull and found fourteen magnetic grains stuck in it, and they were only on the top side, not on the bottom or inside the skull. This is just as one would

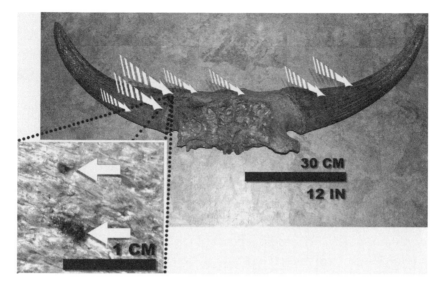

Fig. 4.4. Four-foot-wide bison horns and skullcap, with two visible embedded metallic particles, which are highly magnetic. The rest of the skull is nonmagnetic.

expect if the animal had been bombarded with high-velocity micrometeorites while standing.

Straightening up, I knew I had to have that skull, so I checked the price tag stuck on top. It was expensive for our research budget, but, hoping for a discount, I turned to the salesman nearby and asked, "How much would you take for the *Bison antiquus* skull?" I used the Latin name for the animal.

"It is sold," he said with a hint of a German accent. "Someone bought it just before you arrived."

I was shocked. Deep disappointment washed over me as I glanced over at that rare treasure on the table and realized that I could not have it. Anxiously, I blurted out, "Do you have any more? Or do you think the buyer would sell me that one?"

"No, it is the last one, and they are very hard to find," he said. "The buyer purchased it for a client." Dejected, I walked over to that incredible skull for one last look and then, sadly, walked away.

I went about twenty feet when a sudden impulse came over me. Perhaps I should not give up so quickly, I thought. Maybe I can talk the skull's new owner into selling it, if only the shop owner will give me his name. I walked back to the store again, just as a dark-haired man approached the shop owner and asked, "Do you have my bison skull ready?" What incredible luck, I thought. Here is the skull's new owner himself.

Before I could say anything, the shop owner pointed and said, "It is packaged up and leaning against the wall." He walked over, not to the skull on the table, but to a large package across the room. Bewildered, I watched as the buyer lifted the package and left. Irritated, I realized that the owner had two skulls after all!

Quickly, I pointed to the table and asked him, "Is *that* skull for sale?"

"The *Bison priscus* skull? Yes, it is," he replied. "But no more *Bison antiquus*. He took the last one."

Priscus! Antiquus! I laughed aloud, to the bafflement of the owner. The two extinct bison types look similar, and I had confused the Latin names. Greatly relieved, I agreed to a reduced price, and picked up my meteorite-laced extinct-bison skull. As I walked away with my prize, I reflected on the day's unusual events and how good "luck" had played a key role. I had been reluctant to go to Tucson at first, reluctant to look further after finding the mammoth tusk, and had almost given up on buying the skull. However, by following hunches, mine and my wife's, I had come across two terrific micrometeorite-filled discoveries. It had been a tremendous day.

TRAVELING TO CALGARY

Learning that Canada Fossils Ltd. had tons of tusks in Calgary, I flew there a few weeks later to look through what the company had, hoping to find more. A sales representative showed me a single, very large, complete Alaskan tusk that seemed promising. However, it had already been cleaned and polished, which for my purposes was not good. The cleaning process had removed a lot of the bark, or outer surface, and several tests with the magnet showed nothing. As I looked with wonder at that huge intact tusk, I began thinking of Big Eloise, and I decided this mammoth should have a name too. I began silently calling the mammoth Big Ed, although, frankly, I didn't have a clue as to its gender.

The odd thing was that Big Ed's tusk had about a half-dozen unusual dark elliptical marks on the surface of the tusk that faced upward when he was alive. John told me that the ivory originally had contained many shallow holes and that the polishing process had smoothed them out, leaving rounded marks. These could have been impact holes, yet when I tested with the magnet, I found nothing. Reluctantly and with disappointment, I moved on to test other pieces of ivory.

It was hard, cold, dirty work for two long days as I sorted and stared at a seemingly endless fifteen tons' worth of mammoth tusks. The worst part was finding nothing. Discouraged and dejected, I was about to pack up to leave when John suggested one last possibility, a small cache of high-quality tusks that they had set aside for a special ivory-carving project. So it was that on the last day, I found a new piece of tusk bearing about a dozen round magnetic particles. Within hours of finding that first tusk containing particles, I found two more. Then I had the biggest surprise of all.

Something kept pulling me back to Big Ed's tusk, so before leaving, I examined the tusk one last time very thoroughly. Hidden near the tip, I found an embedded metallic particle. The magnet swung in and held, sticking out sideways from the tusk. Elated, I stood up quickly and gazed over the huge twelve-foot-long tusk. The best tusk was the last one. After that, I went back to the circular marks and carefully tested them again. They were ever so faintly magnetic, and now I knew why I had missed that before. The grinding and polishing had removed nearly all the embedded metal grains from the tusk surface, including the ones in the shallow holes (fig. 4.5), so the effects were only faintly detectable.

When I looked at the alignment of the particles in the large tusk from Big Ed, I noticed something odd. We knew that no matter which direction a mammoth might have been facing at the time the particles hit, they would all have flown in parallel to each other from the same direction. We had always supposed that the particles must have emanated from some angle above the horizon, say, maybe 45 degrees or greater, since that is typical of meteorites, as we believed these to be. However, that was not the case here. To create the pattern in the large tusk, they must have come in at an angle just slightly above horizontal from the front of the animal, assuming the animal was alive and standing up, as shown in figure 4.6.

Fig. 4.5. Large dark spots on Big Ed's tusk caused by the apparent impact of incoming hot particles.

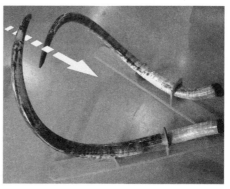

Fig. 4.6. Mammoth tusks showing direction of travel by particles.
© 2005 Canada Fossils, Ltd., at www. canadafossils.com; used with permission

In addition, to embed themselves so deeply in these tusks, the particles must have been traveling very fast, maybe at supersonic speed. If so, what had propelled them? An explosion? An impact? I was stumped. This new information did not fit our theory, nor would some of the chemical analysis that we ran later. We were getting startling and unexpected results.

Nevertheless, it was an incredible day. I had searched through tens of thousands of pieces of ivory from the tusks of at least 2,000 mammoths. After two frustrating days of finding nothing and being ready to go home in defeat, now I had four more pieces of tusk, each of which apparently came from a different animal, to go with the tusk from Tucson.

It is a curious pattern that, during our research, we often reached a dead end and were ready to give up when we made a major breakthrough. Scientific research may seem to be only about logic and facts, but so-called luck often plays a part. One well-known archaeologist remarked to me, "Every scientist hopes to be good at what he does. I also hope to be lucky."

Whether it is called luck, intuition, or coincidence, it has been a powerful factor in many of the most important scientific discoveries in history. Throughout our own research, as with finding the tusks, many of our most unusual discoveries or chance connections with key players happened through extraordinary scientific luck. As researchers, we grew to welcome those surprising events, since they often led to exciting new evidence.

THE SIBERIAN CONNECTION

As I packed up to leave with my prized pieces of tusk, John and Rene Vandervelde, then manager of Canada Fossils and a former geologist, came in to look at them. Explaining our theory to Rene, I saw that he was clearly intrigued.

His first comment was, "I think of all the tusks that have passed through our warehouse over the years, and I wonder how many others had these magnetic grains embedded in them."

I explained, "Our evidence indicates that the Event occurred over central Canada and the Great Lakes, so we are not surprised that it reached into Alaska, where . . . uh . . . that tusk was found." I almost slipped and mentioned Big Ed by name. I wasn't sure whether naming their mammoth would seem presumptuous.

Rene, who had worked for many years as a Canadian oil geologist, responded, "Well, your Event certainly went farther than Alaska."

Puzzled as to how he knew that, I said, "Pardon me?"

He picked up one of the other pieces of tusk on the table. "This par-

Fig. 4.7. The blue-tinted Siberian tusk. The entry craters from seven magnetic particles are shown. Arrows mark the direction of travel. The longest arrow shows the largest elongated crater formed by oblique impact. There are no craters on the other surfaces.

ticular tusk is not from Alaska," he said. "It is from central Siberia, so your particles must have made it that far."

Central Siberia! That was thousands of miles from Alaska. I was speechless at this revelation. Rene continued, "Do you see the bluish color on this tusk? That is from the mineral vivianite, and it means the tusk came from the Taimyr Peninsula or the Yakutia Province. That is where we get all our Russian ivory." As his finger traced the purplish blue mineral stains, I leaned over to get a better look (fig. 4.7). "The other tusks you have are not blue," he said, "and instead, they have this whitish mineral in many spots. That is calcite, and it is typical of Alaskan ivory. So you have four from Alaska and one from Siberia."

He knew his ivory, and this new evidence surprised me. If our particles had landed in Siberia too, then the Event must have been much larger than we thought. The evidence suggested that the Event had affected most of the northern part of the globe, maybe all the Northern Hemisphere, including Europe. If so, the Event was almost incomprehensibly massive. No wonder so many animals went extinct!

BIG ED'S LAST MOMENTS

As I packed my box of tusks, my mind kept returning to the strange puzzle of the nearly horizontal particles, and I tried to visualize the last moments of the mammoth I called Big Ed. I knew from reports on modern-day elephants that when threatened with danger, an elephant herd will gather around the younger ones and directly confront the danger. Swaying back and forth, they shake their fearsome tusks and swing their clublike trunks, presenting a formidable gray wall of defense against potential predators. Since mammoths are related to elephants, they had most likely done the same thing.

When the Event suddenly flared in the skies over the continent, did Big Ed turn to face the danger? If so, he turned to face millions of red-hot particles traveling at jet speed. They burned into his tusk, struck his eyes, and possibly blasted him backwards. He almost certainly perished that day, along with every other member of his herd.

Eventually, over the centuries, mud and permafrost buried his skeleton, and Big Ed vanished from view. His story became known again only after a modern-day gold miner washed Big Ed's tusk from the Alaskan muck.

RUNNING TESTS AND LOOKING FOR ANSWERS

Later, trying to solve some of the mysteries of the particles, we ran tests on the tusks, with mixed results. Long before I discovered the tusks in Tucson, one particle-laden tusk broke apart, splitting one magnetic particle roughly in half (fig. 4.8). After removing the broken half from the tusk, Dr. Ted Bunch, also of Northern Arizona University (NAU) and retired from NASA's Ames Research Center, and Dr. James Wittke from NAU performed a chemical analysis of the particle and found that it contained about 2 percent iron oxide, about 0.2 percent magnesium oxide, and a little manganese, with most of the rest just tusk material.

This new information was startling, since we had expected the particles to be high in iron, like meteorites. It was a serious setback for our theory that the tusks had been hit by particles. We were forced to consider that the

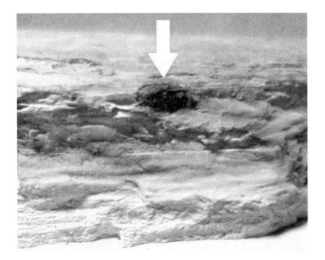

Fig. 4.8. Large ⅛-inch split particle embedded in the tusk.

black spots might be unusual yet natural occlusions in the tusk, rather than caused by impacts.

However, the answer became clearer when we received new NAA results for the magnetic particles from Gainey and Murray Springs. The ratio of iron to magnesium in the tusk was nearly identical to that of the magnetic particles from Murray Springs, and the ratio of iron to manganese was very close to that of Gainey. This was too close, we figured, to be merely coincidental. It suggested that all the particles had come from the same source region. In addition, further microscope work revealed that some of the magnetic spots in the tusks had small entry holes in their centers, surrounded by the brown charred rings. This led us to theorize that on impact, some of the particles vaporized into millions of tiny fragments, which dispersed throughout a tusk near the impact crater. If true, then this would explain all the evidence that we found in the tusks.

MORE LOW-TECH TESTING

We were curious about the likely velocity of the particles, so I asked Dr. Michael Whitt to run an unusual low-tech test to re-create the events of long ago. He agreed to shoot the mammoth tusk using a shotgun with a muzzle velocity of about 820 miles per hour and firing #12 birdshot, which is about the same size that we estimate the largest particles to have been before they vaporized. Michael and Frederick Kaiser, both skilled hunters, shot many small pieces of the same tusk material, and they got the same surprising results every time. Even from close range, the birdshot just dented the tusk before bouncing off. Not a single piece of birdshot stuck. Based on the comparative depth of penetration, we estimate that the original particles were traveling at an incredible 3,000 to 4,000 miles per hour, faster than all conventional jet planes and faster than most missiles. At that speed, the combined force of the particles was almost certainly enough to knock a mammoth trunk-over-heels backward.

When he was finished with the experiment, Dr. Whitt could truthfully claim, as few can, that he had shot a mammoth, and he kept a small piece of tusk as a trophy. However, after the experiment, we still could not explain the mystery of the circumstances that caused those particles to hit Big Ed.

ADVENTURES IN X-RAYS

To get a better look at the composition of the particles without destroying them, we decided to try X-rays. Thinking it would be easy to arrange,

I called several medical labs. I dialed the first one, and the receptionist answered, "X-ray."

"Hello," I responded. "I have an unusual request, I think. Can you do an X-ray of a mammoth tusk?"

Predictably, she answered, "A *what?*"

"A mammoth. You know; it's something like a big hairy elephant. I have an elephant tusk that I need to have X-rayed."

"Oh no!" she said. "We don't do animals."

"Well, it's not really an animal," I tried to explain. "It's just a tooth—sort of. The animal died a long time ago."

There was dead silence for a few moments, until she said, very emphatically, "We don't *do* animals . . . especially *dead* ones! We only do humans . . . and only live ones. Company policy."

Thanking her, I hung up and tried the next lab, realizing by then that X-raying a mammoth is an extraordinary request. The next receptionist seemed similarly flustered, so I asked to talk to the lab's medical director, Dr. Scott Fleischman, from whom I received a different response. He was clearly intrigued, and after admitting that this would be the first mammoth to visit his lab, he arranged to have the X-rays done by one of his experienced radiology technicians, Greta Hegeman. Before we hung up, he suggested that I discreetly bring the tusk in a covered container. He was unsure how his patients might respond to the sight of a mammoth tusk coming in through the front door.

The next day, I arrived at the lab and went to check in. Dr. Fleischman was away but had alerted the receptionist that I was coming. She greeted me warmly and admitted that when she had first heard my request over the phone, she thought it was a crank call. However, she still did not quite grasp the unusual nature of my visit, since she handed me a patient questionnaire and explained, "Please fill out all the yellow highlighted sections, including the one for health insurance."

After glancing at the yellow-marked form, I explained to her politely, "I think there's a little misunderstanding. The X-ray is not for me." Then I pointed to my bag containing the tusk and spoke softly out of consideration for Dr. Fleischman's nearby patients. "The actual 'patient' is a mammoth. And I don't think it can qualify for health insurance, since it's . . . you know . . . extinct."

"Ah-h-h-h," she said, as the full picture dawned on her, but she was stymied by how to enter my mammoth into the computer system. Several times she started to type but then paused in uncertainty. Finally, she got up and hurried off to consult with someone in the back, returning a few min-

Fig. 4.9. Photo and matching X-ray of two impact particles, one that was visible and one that was not visible, but pitted. This suggests that both spots formed the same way.

utes later. After entering something rapidly into the database, she turned me over to Greta.

It took several tries to get the X-ray machine calibrated for the tusk, since it is many times thicker and denser than human bones. Finally, Greta got the machine set up perfectly, and she took a number of good X-rays of four of the tusk pieces, two of which showed up well (fig. 4.9). Each visible round black micrometeorite in the tusk produced a corresponding white mark on the X-ray film. On one X-ray, we were able to get a good sideways view of an impact spot, and as expected, the meteorite was roughly spherical, confirming that the spots extended deep into the tusk.

Satisfied with the results, I checked out at the front desk, and the receptionist handed me a receipt for the X-rays. As I turned to leave, I saw for the first time how she and her supervisor had entered my unusual "patient" into the computer database. The receipt simply read, "Patient: Mr. Hairy Mammoth. Service: X-rays." Smiling as I walked out, I wondered what she had put down as the patient's date of birth—the Ice Age?

NEW OR CORROBORATING CLUES TO THE MYSTERY

- Five mammoth tusks display embedded magnetic particles with raised charred rims.
- One tusk was from Yakutia Province, Siberia, and the others came from the Yukon in Alaska.

- If the Event hit both areas, then it had a radius of 2,000-plus miles and covered 10 percent of the planet.
- Evidence suggests that magnetic particles affected most of the Northern Hemisphere.
- The entry wounds on one tusk suggest the particles traveled just above horizontal.
- Based on depth of impact, the particles most likely traveled at 3,000 to 4,000 miles per hour.
- X-rays show that the embedded particles are roughly spherical.
- Some X-rays reveal hidden iron buried inside lighter-colored craters.

5

BRIGHT YELLOW BONES

CROSSING THE HIGH PLAINS

Making an early start from the motel, by 6 A.M. I was on the road to Black-water Draw, a well-known Paleo-Indian site near Clovis, New Mexico, where the first Clovis point was found. The sun rose slowly, painting golden streaks on the eastern horizon, while overhead, a shimmer of stars still lingered in an ink-blue night sky. I passed just a handful of other cars on the ruler-straight back road, and soon the rhythmic hum of the tires began to make me drowsy, so I rolled down the window. Shivering in the frigid blast of March air, I looked out over the rolling High Plains of eastern New Mexico as they slipped past beyond my headlights. It was too dark to see much, but there isn't much to see in these nearly treeless, high desert flatlands.

Stretches of grass, greened by the recent rains, showed up in the low light. An occasional cow appeared on the side of the road, and beyond the two pools of light, I faintly made out herds of cattle. I began to imagine being there 13,000 years ago, during Clovis times. Although the climate was much colder then, more like Canada today, the landscape would have been similar, with verdant savannas filled with tall grasses interspersed with hardy bushes. Then, as today, it would have been home to herds of grazers rather like our cows and horses. The now extinct American horses that roamed those plains were about four to five feet high at the shoulder, much smaller than our modern horse. In addition, instead of the cattle that we know, there were millions of shaggy bison, also known as buffalo, though they were mostly larger than the few that remain in the West today.

Clumsy, one-humped camels ranged over the plains too, along with elephant-like mammoths with curved tusks, pulling up thick clumps of grass with their trunks.

With a start, I realized that if I had been there when the now extinct horses and mammoths roamed these grasslands, I would have been in mortal danger. Hidden in the dense grasses surrounding me, aware of my presence yet invisible to me, there might well have been a 600-pound, four-foot-tall saber-toothed tiger, or a 200-pound hungry dire wolf with a head as large as a bear's. Although both powerful predators would have outweighed me and been far stronger, they might have passed me by as being too small for a snack; they preferred buffalo and mammoth meat. Visualizing the snarling face of a saber-tooth, I snapped fully awake and instinctively rolled up the window, as if to put something between that ghostly visage and me.

The Clovis hunters would have been nearby too. Like the tigers, they preferred mammoth and buffalo for lunch, and it was not the meat alone that was important. To them, hunting those animals was like a trip to Safeway and Home Depot rolled into one. They cut up the hides for clothing, bags, and shelter coverings, or sliced into it strips for making ropes and straps. Bones and tusks turned into shelter support beams, and the cracked bones became tools and weapons, while at the same time yielding nutritious marrow. The Clovis people processed sinew and ligaments into belts, twine, straps, shoestrings, and bindings for spear points. The list of useful buffalo and mammoth products would have been long, limited only by the remarkable ingenuity of those people, who were highly skilled, resourceful, and inventive. They were as intelligent as we are, although they were quite likely far stronger and hardier than most of us. If any were still around today, they could compete respectably well in many modern athletic events, especially for endurance and strength events like weight-lifting, wrestling, and football. After years of tossing spears at charging mammoths and buffalo, they would be good at throwing the javelin, too. Because the Clovis people ranged over large distances, they trotted a lot, as some modern-day tribes still do, so they would be great marathon runners. Now, that would be a sight, I thought, to see a band of Clovis people jogging in the Boston Marathon. Unlike me, they would surely not be in the back of the pack.

Conversely, we probably would not do well in their world, living according to their rules. For example, imagine how willing you would be to grab a spear and stalk a semi-truck-sized, 12,000-pound, twelve-foot-tall angry mammoth bull, kill it at close range, and then cut it up into food and shelter. Even if our lives depended on it, we would find it extraordinarily difficult to adapt to life in Ice Age New Mexico without all the props of civilization. Most of us would quickly run out of food, or become food for something else.

ARRIVING AT BLACKWATER DRAW

The sun had come up, and, glancing at my map, I realized I was close to Blackwater Draw, a long valley that runs from New Mexico far down into Texas. In Clovis times, the draw held a flowing river most of the time, along with a string of springs and shallow pools. The fresh water, lush grass, and shady trees drew herds of animals, which in turn attracted the Clovis hunters. For hundreds of years, they came and went from this area, leaving behind spears and tools as a record of their presence. When scientists found some of these highly distinctive tools in 1929 and the early 1930s, they provided the first conclusive evidence of the presence of early humans along with the animals that later became extinct, a major step forward in American archaeology. Those people and their unique, deeply fluted spear points received the name Clovis in honor of this site.

Seeing the Blackwater sign, I entered the open gate and drove to the office to meet Joanne Dickenson, the site's curator and resident archaeologist, from Eastern New Mexico University. She gave me a tour of the excavation right away.

The sprawling site covers about forty acres spanning the old riverbed, and not much is visible from the office, since most of the site is in an excavated pit below the level of the plains. As Joanne drove me over the edge of the bank and down the steep Jeep trail into the site, I saw the main diggings, which included steep, twenty-foot-high embankments and deep pits. Well-marked visitor's trails snake around the site; about twenty large photo display signs spaced at intervals explain the history and significance of important parts of the site.

For decades, the site had been an operating quarry, and during the early years of digging out the sand and gravel for making roads in New Mexico, the operator began uncovering bones. He considered them a major nuisance because he had to sift them out of the sand before he could sell it. After he built up a sizable pile of sifted bones, spear points, and stone axes, people began offering him a few quarters or dollars for the best ones. Catching on to a new business opportunity, he set up a roadside stand to sell "Blackwater Bones and Fossils."

Joanne explained, "Several scientists who stopped at his roadside stand noticed the rare fossils and realized the importance of the site. That's when the controversy began, along with the race against the bulldozers. The scientists made a deal with the owner to excavate ahead of the mining operation, and while the owner was helpful and sympathetic at the beginning, with increasingly difficult deadlines to meet, he became less tolerant of the 'bone collectors.' "

I said, "That must have been tough, working in the midst of bulldozers and draglines."

"It was," Joanne said, as the truck hit a bump and bounced us both around sharply. I grabbed the armrest, but Joanne was used to it and continued without noticing. "You know, archaeologists always like to take their time with dental picks and paintbrushes scraping away a little soil at a time. But they couldn't do that in those early days on the mining site. It was crisis archaeology, working at a breakneck pace" (fig. 5.1).

As time went on, the owner was increasingly pressed by production deadlines, and the archaeologists sometimes had to get out of the way at the last second as a giant bulldozer scraped away their partially excavated bones and artifacts to reach the valuable gravel below. Joanne continued ruefully, "It will never be known exactly how many beautiful, priceless Clovis spear points lie buried inside New Mexico's highways. There must be thousands of them, enough to fill entire museums."

In spite of the difficulties, the archaeologists made startling discoveries, and pressure grew to preserve the site, especially after many more remarkable Clovis spear points appeared. However, the owner had a lucrative sand-and-gravel business, and he rejected offers to purchase the site, always demanding a steep price for the land.

The solution came in an unusual way. Even after screening, the sand contained small pieces of crumbling bone, making it less than ideal for road building. Eventually, road builders found better sources of sand that contained no bone. As the owner's market began to shrink, he gave up and sold

Fig. 5.1.
Excavators
struggled to
stay ahead of
the bulldozers.
*Source:
Blackwater
Draw Archives,
Eastern New
Mexico University,
Portales*

the land to the state of New Mexico. In the end, ironically, it was not the "bone collectors," but rather the bones themselves that played the determining role in preserving Blackwater Draw for the future.

THE MYSTERY BONES

Near the end of the tour, Joanne told me about several interesting finds that she had back at the office. As she parked the truck and we reentered the office, I was eager to see them, since from what she had told me, I suspected that they might tell us something about the Event. First, she brought out several boxes of bones, and at a glance I could see one highly unusual fact—they were bright yellow. There was one long bison bone, along with the upper leg bone, part of a skull, and a hump vertebra from a mammoth. All appeared to have been dusted with vivid yellow talcum powder, which stuck to the surface and was hard to remove.

Joanne explained, "These bones were pulled out of the sands of an old dry pond along with Clovis spears and scrapers, so we know they are about thirteen thousand years old." At first, the excavators did not know what they had. Their later testing suggested that the yellow mineral was a form of radioactive potassium-uranium ore called carnotite.

I thought of what we had found at Gainey and Murray Springs, and here it was again—radioactive minerals in the Clovis layer.

I knew, however, that New Mexico is famous as a source of uranium, and since uranium minerals are typically water soluble, it is normal to find them where water flows. I asked Joanne, "Is there a lot of uranium around here? Are any mines nearby?"

"No mines," she replied, "but there is a big deposit upstream from here."

That would certainly explain the radioactive bones, yet if it is common in the environment, it should be in other layers above and below the Clovis layer. When I asked her about that, Joanne found two reports on Blackwater radioactivity, one by Sarah Kruse (2000) and one by James Fitting (1963), a scientist who studied the site. She pointed to one of Fitting's charts. "He analyzed dozens of bone samples from above, below, and within the Clovis level," she said, "and he found that the radioactivity was low at the deepest layers, then increased in the Clovis era" (fig. 5.2).

I was amazed at his figures; the radioactivity shot up an incredible 800 percent in the Clovis layer and then rapidly declined to its original levels in the sediment above. Far from being evenly distributed in the Blackwater

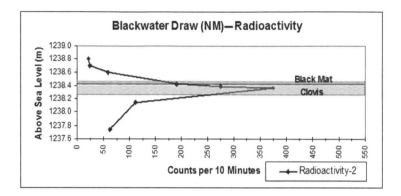

Fig. 5.2. Radioactivity in bones by depth. *Source: Fitting's study (1963)*

riverbed, it was highly concentrated in the layers that dated to around 13,000 years ago. This implied that the radioactive minerals are not commonly found in the area. Something unusual had happened at the site, something like the Event.

Eager to test for myself, I got my Geiger counter from the car. After putting on the headphones and checking the meter for calibration, I set the sensitivity to HIGH and placed the detector near the first bone. The meter needle pegged sharply off the scale, and I jerked the probe back as the staccato rattle of the meter pounded loudly in my ears. I could not believe what I was hearing—those bones were extremely radioactive! Turning the sensitivity down, I gingerly moved the sensor back over the bones. The noise still rattled as I watched the needle rapidly climb to nearly 400 counts per minute. Testing the others, I found the same thing.

Joanne said, "Would you like to take some samples to test?" I nodded eagerly, and she disappeared into the back room. Returning with a handful of plastic bags, a dental pick, and other small tools, she carefully scraped away small flecks of the yellow dust, bagged them, and labeled the bags. Gratefully, I took them and told her that I would do NAA and PGAA testing, just as we had done at Gainey and Murray Springs. I suspected that the tests would find uranium, and I wondered if there might be other radioactive isotopes in there as well. When the tests came back, they confirmed high levels of uranium in the yellow dust, which contained nearly 2,900 parts per million, an incredible 1,600 times more than normal soil, making it rich enough for a world-class uranium mine. Something had made those bones "hot."

THE BLACK MAT REAPPEARS

Vance Haynes had told me at Murray Springs that the black mat also appears at Blackwater. Confirming this, Joanne took me over to the Interpretive Center, a large building covering one of the key excavation sites to protect it from the weather and from looters. When we went in to get sediment samples for NAA and PGAA analysis, I stepped over for a closer look at the site's black mat. Not as dark as at Murray Springs, it was otherwise nearly identical in texture and appearance, and it was only a few inches thick.

I slid my finger along the joint between the dark, smooth-textured mat and the lighter Clovis layer just below it, knowing that my finger rested on the extinction layer where the Event happened. Below it, lots of mammoths and Clovis people; above it, they were gone.

"So that's where it happened," I said to Joanne. "That's the extinction event." It was such a thin line for such a dramatic effect. Joanne nodded and continued looking over my shoulder silently. She had long been curious about what had happened at Blackwater 13,000 years ago.

Still wondering what effect the Event had on the people back then, I asked her, "This line is the last time you see the Clovis evidence, but where does the next group of people show up?"

Joanne pointed. "See how the formation turns light gray just above the mat? Then see the next color change above it?"

I nodded and slid my hand up the wall about eighteen inches to where she was pointing.

She continued, "That's the first sign of people after Clovis. We find a lot of bison bones and Folsom points there."

After standing up, I saw that the excavated ledge was jammed with spears, tools, and bones. The new people after Clovis had been very busy. All the space in between, however, meant that there was no evidence of any people at Blackwater Draw for a very long time. Eight radiocarbon dates indicated that no humans had visited the Blackwater Draw for more than 1,000 years.

That was astounding to me. Back then, the site had water and thick plant growth. It certainly attracted animals, and *they* should have attracted people, but there were none. Did the Event eliminate every living person in the area? There was not enough evidence to answer those questions—not yet—although the culturally barren earth beneath my hand hinted that it had.

MAGNETIC GRAINS AND SPHERULES

Reminding myself that there was still a lot to do, I moved to test the wall with the Geiger counter. As with the bones, the Clovis sand layer was highest in radiation, although much less so than the bones. Undoubtedly, the bones had attracted and concentrated the radioactive minerals, but still the bulk sediment was noticeably more radioactive than that above or below it.

Next, I pulled out my magnet and placed it against the exposed wall, sliding it over the Clovis layer and the spring sand layer below it. When I pulled the magnet away and checked it, Joanne leaned over to look too, curious about the process. We saw that the magnet's edge held tiny flecks of what looked like black pepper or gunpowder grains—thousands of them, more than I had seen at Murray Springs. Here was proof that the Clovis layer matched the Arizona site, but what of the layers above and below? Testing them next, I found them to be much lower in grains as well. The pattern was the same as at the other sites.

Last, I collected sediment samples for chemical analysis, being sure now that this site matched those in Michigan and Arizona. Nevertheless, when we got the analysis back later, it showed some surprising differences, especially regarding radioactive elements. In addition to the high uranium, the Clovis magnetic grains contained nearly three times more radioactive thorium than normal.

So far, every site had been similar to the others in major ways, yet each provided at least one odd and intriguing new clue that kept us guessing and propelled us forward. Every time we thought we had a solid answer, the clues shifted again to throw us off balance.

KNEELING BY A CLOVIS WELL

After packing the samples, Joanne led me next door to the small A-frame building covering one of the famous Clovis wells, the oldest hand-dug wells ever found in the Western Hemisphere. In the center of the small, low-ceilinged, dimly lit building, I saw a shallow hole that was about 2.5 feet wide and 5.5 feet deep with a slightly raised rim. The band of Clovis people had dug it about 13,000 years ago to reach cleaner water below. It probably still looked much the same as it did back then.

Katz (1997) tells about the discovery by Earl Green, one of the Blackwater archaeologists, of several of these wells, which made headlines in the area in the 1950s. Earl later said that a local woman came over one day with several young kids, saying that she "wanted to show her grandkids the whales." Earl, assuming she must be referring to the large tusks and other curved bones, explained to her that no whales had been found there, only mammoths and other land animals. She was adamant that she was right, because, she said, "I was there when they dug 'em up." Earl insisted that there were no whales, and the woman left in a huff. He realized only afterward that he had misunderstood her local accent, and that she had been referring to the "wells," rather than to "whales."

As I stood there looking at that ancient steep-sided "whale," I was puzzled. The tribe could easily have gotten water from the pond and stream nearby, so why did they need a well? Were they getting sick from water laced with radioactivity and algal toxins? After realizing that the pond contained poison, did they try to find a cleaner source? If so, it would not have worked, since the toxins and radioactivity would have traveled easily from the pond through the sand and would have contaminated the well too. They had fire and could have boiled the water, but boiling does not remove those poisons.

While there are not enough clues to explain the well, it may be that they dug the well simply to avoid having to drink the dirty water caused by herds of mammoths and bison stomping through the water hole. No matter what the reason, the presence of the well up on the bank suggests that they clearly understood a great deal about water quality, just as modern people do.

I knelt around the rim of the well, being careful not to collapse the wall or to fall in. Later, I ran a test with the Geiger counter, which showed that the soil was radioactive, although much less so than the bones. Nevertheless, it gave the highest reading of any sediment that I tested at Blackwater. Odd, I thought, that the well bank itself was the most radioactive. Was it because the Clovis people had splashed the radioactive water repeatedly onto the bank? Or had the radioactive particles fallen directly on this undisputed Clovis living surface?

DIGGING DIRT FROM THE WELL

After that, Joanne helped me take sediment samples for elemental analysis by carefully scraping away the top layer in several places around the well with small dental picks. Later, when I analyzed the well sample, I was surprised to find that every pound of sediment contained hundreds of tiny magnetic microspherules, along with thousands of small magnetic grains that were similar, except that they were not spherical although still somewhat rounded. Such things typically come from meteorite impacts, but the elemental ratios did not match. The Blackwater chemistry was more like areas of Canada around the Great Lakes, or, strangely, like the moon. Neither source made much sense for these grains found on the bank of a Clovisera well in New Mexico.

As I knelt at the edge of the well and considered all the puzzling information, an eerie feeling came over me. I was kneeling in the same "kneeprints" as the Clovis people who had used that identical spot to draw water 13,000 years ago. I felt a strong connection to those vanished people, who lived at a difficult time. But the strangest feeling came from realizing that in all the sand deposits higher than the lip of that well on which I knelt, no one had ever found Clovis spears or tools. I was standing on the original surface, the very last bit of earth from which they vanished. Had a Clovis woman knelt down to draw water at exactly that spot on a day long ago when the Event happened? Perhaps she looked up just in time to see trouble coming from the northeastern skies.

PITS IN THE FLINT

After finishing at the well, Joanne and I returned to the office, where we made more discoveries. Knowing that there were microspherules, I wanted to look at some of the flint points and flakes from both the Clovis levels and the Folsom levels, which came after the Event. Using my microscope, I found that some Clovis flakes showed evidence of tiny impacts containing what looked like once molten, shiny metallic splatter strewn around them, similar to what Bill Topping had found at Gainey and many other midwestern sites. These craters seemed much shallower, however, as if Blackwater Draw was farther away from whatever had happened. Some of the craters even appeared to have tiny, raised, shattered rims around them, just like a typical large-impact crater. The craters were only in the 13,000-year-old Clovis flakes; I could find no such impacts in the Folsom flint, which dated nearly 1,000 years after Clovis.

THE HAND-CHOPPED TUSK

After finishing with the flint, Joanne lifted the last unusual item from the case and handed it to me, point first. Initially it seemed to be just a typical small mammoth tusk, but when I turned it around, I saw how remarkable it was. The base of the three-foot-long tusk showed very distinct cut marks all around the circumference, passing completely through the tusk (see fig. 5.3). Clearly, a Paleo-Indian had used a flint ax to hack the tusk from a dead mammoth, just as if he were lopping off a tree branch.

Joanne explained that they had found it in one of several areas of the pond that concealed seemingly harmless pockets of quicksand, which had formed when the springwater bubbled up from below. These spots were deadly hidden traps; whenever unsuspecting mammoths wandered into them, they sank and drowned. The excavators found some of the mammoth skeletons buried in a standing position, with their feet curled under and their heads arched back in poses typical of drowning victims. Some showed signs of having been butchered in place, as the Paleo-Indians removed what they could reach on the top but not on the sunken part of the carcass. That is just how they found the skeleton of the young juvenile mammoth, which was upright in the quicksand with the tusk missing. The Paleo-Indians butchered the animal in the quicksand, and then one of them used a flint ax to chop off the tusk before the animal sank out of sight into the sand.

Later, the excavators found the severed tusk lying on what had been the bank of the pond just above the waterline. Evidently the Paleo-Indian had turned away from the others, who were busy butchering the stranded mammoth, and waded out of the pool to place the tusk on the bank.

Here is where the story takes a puzzling turn. Why did he abandon the tusk? Tusks like that were difficult to obtain, and they were highly prized for the valuable ivory. Just to abandon it would be like a gold miner striking it rich with a handful of shiny gold nuggets, piling them on the stream bank, and walking away. It is unbelievable that a Paleo-Indian would leave behind that tusk except under extreme circumstances. Did tigers or wolves attack the tribe, or did the rest of the mammoth herd gang up to chase them off? None of those scenarios seems likely, since the people would have returned later to reclaim the tusk.

Fig. 5.3. Mammoth tusk with ax marks all around it. After chopping it partway through, the person snapped off the tusk like a tree branch.

There is another chilling possibility. Did the Event occur just at that moment? What if there were no Clovis hunters left alive to carry it away? There is one slim, circumstantial clue. When the cut tusk was found lying on the ancient riverbank, it was directly under the black mat, which draped over it in full contact like a dark blanket. Before that layer formed, mammoths and Clovis people freely roamed Blackwater Draw. When the black mat appeared, they disappeared. The ancient hunter could not have left that tusk on the bank very long before it was covered by the mat.

So did the Event strike exactly at that moment when the Clovis hunter stepped onto the bank, holding his prized mammoth tusk? Seeing that monstrous Event unfold in the northeastern sky, did he drop the tusk and run, or did he fall, only to have his own fragile skeleton turn to dust? There is no way to know for certain, yet this scenario explains the facts exactly as they were uncovered at Blackwater Draw. Dropping his valuable trophy tusk was likely the last act of that Clovis hunter, who disappeared in an eyeblink along with millions of mammoths and thousands of other Paleo-Indians.

After the Event, Blackwater Draw was quiet and empty for 1,000 years before any other people showed up in the area. By the time the Folsom people arrived, the Clovis hunters were long gone, along with their way of life.

NEW OR CORROBORATING CLUES TO THE MYSTERY

- Bones from the Blackwater site are highly radioactive only around the time of the Event.
- It appears that there were no people at the Blackwater site for 1,000 years after the Event.
- The site's Clovis-era sediment is highly radioactive.
- As at other sites, Blackwater's mysterious black mat drapes directly over extinct megafauna bones.
- As at Murray Springs, there is some evidence suggesting that Blackwater's black mat once contained toxic chemicals.
- Blackwater's magnetic grains are chemically similar to those at Gainey, 1,500 miles away.
- As elsewhere, Blackwater's magnetic microspherules peak sharply in Clovis-era sediments.
- Blackwater Clovis-era flint shows pits similar to the Midwestern Clovis sites.
- A Clovis hunter abruptly abandoned his hand-chopped tusk 13,000 years ago.

6

TO CLOVIS FROM CZECHOSLOVAKIA

SUNRISE AT BUCK LAKE

In quest of another Clovis site in Canada, I drove north from Calgary toward Edmonton, Alberta, and reached a cluster of homes overlooking Buck Lake. After checking into the lakeside motel, I called Anton and Maria Chobot, a couple formerly from Czechoslovakia. They had found an ancient Clovis-era campsite alongside their lakeside home, and Anton had agreed to meet me that morning by the lake and show me what they had found.

Up before the sun on a frigid day, I waited for Anton in my rental Pontiac with the heater going full blast, nursing a very bad cold while gazing out over the stunningly beautiful, very cold Buck Lake. Wisps of steam rose from the long, narrow lake, nearly glass-smooth except for the wakes of a few mallard ducks paddling around. Just beyond them was a rough jumble of branches forming a beaver lodge, and to the west, the Rockies were about seventy miles away, although I could not see them over a row of trees.

Just then, Anton drove up and got out of his Toyota SUV. Of medium height and dignified, as seems somehow appropriate for someone from Old World Czechoslovakia, Anton shook my hand firmly, welcomed me in fine English, and pointed in the direction of his home across the lake, to which we then went together.

CABIN BY THE LAKE

Once inside the high-ceilinged cabin, Anton introduced me to his wife, Maria, and I was treated to Old World hospitality as they offered me cookies and tea. While they told me their story, I could see through the open door into the spare bedroom where knotty-pine-paneled walls held dozens of display plaques with hundreds of spear points, axes, and arrowheads.

Astounded to see so many, I was eager for a closer look, yet managed to summon a little New World patience.

As we sat sipping tea, Anton told me some of their unusual story. "Originally, Maria and I lived in Czechoslovakia, where I was trained as an electrical engineer. The Soviet Bloc was keenly interested in the Middle East, so I traveled there a lot, especially to Syria. While there, I became interested in archaeology, taught myself how to excavate correctly, and did some excavations at the ancient site of Ugarit, north of Beirut on the Mediterranean, which was good training for what I would unexpectedly come across here in Canada."

"More cookies?" Maria asked. I shook my head as she added, "We returned home in 1962 to find that times were not so good in Czechoslovakia. There were tanks everywhere."

Anton explained, "The Hungarian revolt had led to a Russian crackdown, and there was trouble in the streets. Life became much more difficult." Maria nodded in agreement, as Anton's voice trailed off. Clearly, the memories were not pleasant.

"One day," he said, "we went on vacation to Yugoslavia, and we never returned home. We slipped away and came to Canada, to Edmonton." He got quiet again, and I thought of all the friends and life memories they must have left behind.

Changing the subject suddenly, he said, "Here, let me show you something." Thinking he was going to show me the artifact-lined back room, I jumped up. Instead, he took me to the twelve-foot-tall picture window and pointed out toward the lake. "Our cabin is on the higher upper terrace, but do you see the low terrace beside the lake?" I nodded, looking past a five-foot-high, clearly defined shoulder along the lake, which separated the two levels. "That is where the flint cobbles are. The lake was here thirteen thousand years ago, just like today, and the waves would expose the cobbles after a storm. The Clovis people would collect them along the lakeshore." As Anton talked, his voice had become softer. "Then they carried them up here to chip them in their workshop."

He continued, "It was a flint factory. I stopped counting at around seventeen thousand spear points, arrowheads, axes, knives, and flakes." The amazement must have shown on my face. "Would you like to see some of them?" he asked. Would I ever!

THE CHOBOT "MUSEUM"

We spent the next two hours going through the Chobots' entire museum-quality collection of artifacts, which they had mounted on boards in several bedrooms. They had recognizable Clovis points, plus a full succession of other points dating almost up to modern times. There also were points that Anton believed were pre-Clovis. Some Paleo-Indian experts, however, do not accept such older artifacts. They still subscribe to the Clovis First theory, which states that the Clovis people were the first humans in the New World around 13,000 years ago, and that there were no other people here when they arrived.

But there is a growing body of archaeological evidence suggesting that there *were* other people here, although most likely not very many of them. It appears that the Clovis people were not *the first people*, but rather *the first big wave of people* to arrive.

Anton had found an impressive array of very rough and simple axes, knives, and other tools associated with those earlier, pre-Clovis people. All the tools were from levels several feet below the oldest Clovis level, and being rough and crude, they bore no resemblance to the beautifully flaked Clovis tools that are recognizable around the world. As I viewed the Chobots' extensive collection, the pre-Clovis tools certainly looked primitive in side-by-side comparison with those from Clovis times.

One small doubt crossed my mind: The Clovis points from Buck Lake looked Clovis, yet they had less-defined fluting and were rougher than most Clovis points I had seen. I asked Anton if anyone had verified them.

"Oh, yes," he said. "Dr. Bruce Ball, a professor from the university in Edmonton, was one of the first. He confirmed that those with the fluted base are Clovis, and he helped us catalog them. He and other university people have done some excavations of their own here and identified the exact level of the old Clovis surface. One of their excavations is still exposed, and I will show it to you later. Then, there was Dr. Alan Bryan from the University of Alberta. He confirmed the Clovis points too.

"I registered the site with the government in 1981, and they recognized it as a provincial archaeological site, calling it Chobot Site #FfPq-3. They even gave me a certificate, which is hanging on the wall in the basement. When we go downstairs, I will show it to you.

"Also, I told the Provincial Museum in Edmonton about the site, and they were interested in excavating it, but because of budget problems, they could do nothing. I told another museum too, I won't say which, and they wanted to display our artifacts in an exhibit. I told them, 'All right, as long as you display the pre-Clovis ones too.' They declined, as it was too controversial for them.

"Would you like to see more of what they turned down?" he asked.

MAGNETIC DEER ANTLERS

In the Chobots' basement were a great many boxes, maybe more than a hundred, filled with flint tools and flakes. In one box, I found dozens of bones and antlers of various animals, some from Clovis times and some after. Remembering the magnetic bones in Murray Springs, I tested about a dozen with the magnet and found none of them to be magnetic. As I continued to move the magnet slowly over the tray, however, one Clovis-era antler tip rattled slightly in response. Lowering the magnet made the tip jump up to stick to its bottom.

With the antler stuck fast to the magnet, I asked Maria to hold it up for some photographs. As a backdrop, we decided to use the white side of the refrigerator, and remembering the file cabinet, I first cautioned Maria to hold the powerful magnet very firmly while it was near the refrigerator.

As she steadied her hand against the metal side, the magnet flew out of her hand and slammed into the fridge with a loud bang. Startled, she stood there speechless. I apologized and showed her how to get a firmer grip and to hold the magnet farther away next time. I also cautioned her to keep her fingers out of the way, since the magnet is strong enough to pinch skin. As she held it up one more time—bang!—it happened again. The magnet was so powerful that it jumped out of her hand each time. Finally, we found a safe distance and managed to snap some photos of the magnet with a deer antler dangling from the bottom, as you can see in figure 6.1.

I wondered whether, like the tusks, enough high-speed magnetic grains had slammed into the antler to make it magnetic. Looking closely, I could not see grains. It was more likely that the magnetic iron had soaked into the antler from the sediment. Whatever the source, I now knew that there were many magnetic particles in the Clovis layer, and I was ready to find some.

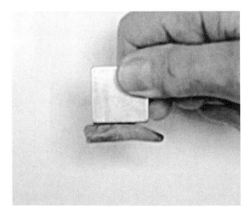

Fig. 6.1. Magnetic antler.

UP TO MY EARS IN MUD

After the tour, Anton and Maria led me out to the plywood-covered shallow excavation pits near their house, many of which were flooded with water from the recent heavy rains. We soon found an empty one about four feet square and two feet deep with a flat shelf at one end, about halfway down. Walking up with an armload of tools and buckets, Anton looked into the pit and said, "That ledge is the Clovis layer. The university excavated down to it when they were out here last time." At that moment, Maria walked up with a handful of plastic bags for the samples.

Hopping down into the muddy pit to get a closer look, I sank up to my boot tops in the sticky gray mud. As I struggled to pull my feet out, Anton handed me a small piece of plywood. Dropping it on top of the mud, I knelt down for a closer look at the wall, which was well hidden by mud and roots. After chopping them away and carefully scraping the wall, I got my first clear look at the Clovis layer.

What I saw made my pulse race. There was the black mat! A thin black streak ran across the pit just above the level of the Clovis ledge. I couldn't believe it. There I was, more than 1,500 miles away from Murray Springs, and yet it looked identical to the black mat there. Could this be the same stuff as in Arizona? I would not know until we analyzed it chemically.

Carefully, I dug out muddy sediment samples and bagged them, covering the entire sequence from the surface down into the Clovis layer. Then, as I shoved the trowel gently into the wall at the bottom of the mat, I heard a dull click. Instantly alert, I knew what it was—the sound of metal hitting flint. Carefully lifting out the muddy sample, I worked my fingers through it until I found something hard. Picking it out of the lump of soil, I rubbed off the dirt. As sunlight reflected off the wet surface for the first time in 13,000 years, I saw a triangular flaked tool, probably a scraper, or maybe a knife blade. It seemed to be from the Clovis layer, so I held it up to Anton. He took it in his hand, flipping it over several times before responding.

"It is Clovis," he said with quiet certainty. "I have dozens from the same level that look much like it." He paused, looking at the tool with the glow of discovery. Clearly, he loved doing this work.

He handed it back. "It is yours," he announced. "You found it."

Thanking him, I flipped the tool over several times too, observing how it had been flaked and thinking of the Clovis man who made this tool on this lakeshore so long ago. How had he used it and how long did he keep it? Did he lose it here? So many small questions for which there would never be answers.

Bagging the tool, I returned to collecting the last samples. Near the Clovis tool level, I noticed several small pieces of charcoal stuck in the wall. Carefully, I dug them out and put them in a separate bag. That was the last of the samples, so I climbed out of the pit and tried kicking the clinging mud off my shoes, a seemingly impossible job.

Finishing up, I held up the bag with the two small pieces of charcoal. I told Maria and Anton about finding the same thing at Murray Springs and mentioned that there may have been widespread forest fires at that time 13,000 years ago. I explained that we did not know what had caused the fires, although we knew there had been some kind of catastrophe in Clovis times.

Anton responded, "It could be from forest fires, or it may be wood from a fire pit. We have uncovered three Clovis campfire sites with charcoal; later I will show you where we found one, and we also have a bag of charred wood in the freezer." A bag of charcoal—in the freezer! A freezer made me think of food, and soon we broke for lunch.

THE CLOVIS CAMPFIRES

After some of Maria's fine Czech cooking, Anton told me about the highly unusual campfire sites in their yard. We left the table and stood at the picture window again. He pointed to a clump of maple trees about halfway down the slope toward the lake. "One of them was right there," he said. "One night, we had a very bad storm with lots of wind, so bad that the whole cabin shook. The next morning, we looked out to see that a full-grown maple tree had blown over, luckily missing the house. When I went out to cut it up with the chain saw, I noticed some black soil stuck in the upturned roots. I pulled some out and crushed it between my fingers, realizing it was charcoal, and in the bottom of the hole, there was more of it, a lot more. I knew it had to be from a fire, since the black area was only a few feet across inside a circular ring of stones. In addition, from our excavations nearby, I knew that this was exactly the right depth for Clovis. I believe this is a perfectly preserved Clovis campfire."

After cleaning off the table, Maria walked up carrying a plastic sandwich bag frosted over with a thin layer of ice. Offering it to me, she said, "Here is some of the charcoal from the tree roots. I have been keeping this in my freezer for far too long. Now, it is yours."

When I glanced at Anton to make sure he was willing to part with their charcoal, he said, "Study it, and do what you can with it. Please take it." So I did, holding the ice-cold bag as if it contained black diamonds. As I thanked them repeatedly, I already had several tests in mind.

Anton continued, "Later, we found two more campfires while digging closer to the house. Both were at the correct depth for Clovis. One of them gave us quite a surprise, when, next to the fire, we found several small piles of flint flakes. You could almost make out the outline of the man's legs as he sat next to the fire while chipping a new spear point. The flakes fell in small piles around and between his crossed legs. When we found it, the place looked as if he had just gotten up and walked away."

A chill went up my back as I listened to Anton's story. Holding that bag of frozen charcoal, I felt much closer to those Clovis people. Huddled around that small campfire to keep warm, they could not have conceived that a man from the far distant future would hold charcoal from their long-cold campfire in his hands, look back through time, and try to understand a little something of their world.

Before long, the daylight began to fade. Taking my leave, I felt a strong kinship with that remarkable couple. For years, they have carried on a major amateur excavation of an important archaeological site, without professional help and without outside funding. While the outside world paid little attention, they have amassed a museum-quality collection of prehistoric artifacts. They are still at it today.

ANALYZING THE SAMPLES

Back at home, I ran several tests on the sediment samples from Buck Lake. As at other Clovis sites, they were loaded with thousands of magnetic grains, and there were considerably more in the Clovis layer than above or below it. The same was true with radioactivity vis-à-vis these samples.

The widespread pattern was holding, and it had just become much wider. Murray Springs, Arizona, is about 1,500 miles from Gainey, Michigan, which is the same distance from Buck Lake. Now we had a triangle-shaped area 1,500 miles to a side, covering an incredible million square miles of North America, from the Great Lakes to the American Southwest to the Canadian prairies. At every site, we had found a major spike in magnetic grains in the Clovis-era layer.

There was only one conclusion: the Event had been huge, most likely affecting all of North America from the Atlantic to the Pacific, and possibly extending from the Arctic Ocean across the Caribbean into South America. We were in awe of the immense forces that spread these tiny magnetic grains across an entire continent.

After the grains, I began to look for charcoal. The Buck Lake sediment samples had partially dried into hard, moist lumps, so I soaked them in water to break them up. Charcoal, sticks, seeds, and other junk floated to the surface, and I skimmed that off with a sieve. After the floating fraction had dried, I put it on a flat lab plate for picking out the bits of charcoal, and that's when I got a big surprise.

As I dumped the debris onto the plate, to my surprise, tiny spherules rolled across the plate—about a dozen of them. Seeing them meant only one thing: they had been floating on the water. *That's impossible*, I thought. *Spherules don't float.* To check the result, I picked up a few with lab tweezers and dropped them into water. I watched in astonishment as they floated like tiny fishing bobbers.

Under the microscope, they all looked nearly spherical, although their surfaces were rougher than the magnetic spherules from Gainey and Murray Springs (fig. 6.2 on page 87). They all looked similar, except for a variation in size; the largest ones were about the size of the small letter O but most were about the size of the period at the end of this sentence. Methodically, I checked all the layers from the Buck Lake site, and, as I expected, the spherule level peaked sharply in the Clovis layer, with few of them above or below (fig. 6.3). The abundance of spherules at the Buck Lake site was clearly a Clovis-era phenomenon, and I began to wonder if there were similar hollow spherules at the previous Clovis-era sites we had investigated.

RECHECKING OTHER SITES

Rechecking our sediment from other sites, I found the spherules at Gainey in large quantities, more than 1,000 in every 2.2 pounds (1 kg) of sediment, the most of any site. We had missed that. After that, as you will see, I found floating spherules at nearly every site from Canada to the Atlantic Ocean to the Great Lakes, spanning the entire continent.

With floating spherules from several locales, I began to look for clues

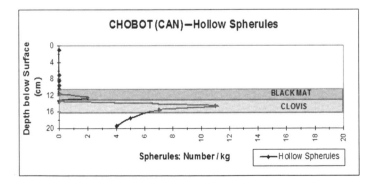

Fig. 6.3. Large peak in hollow spherules in the Clovis level.

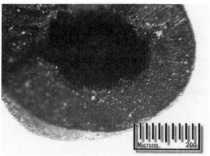

Fig. 6.2. Rough, lightweight spherules. Some are hollow and some are not, yet all of them float on water.

Fig. 6.4. Broken thick-walled "cantaloupe" spherule, one of the largest we found. The rim, composed of carbon, contains tiny bubbles, making it light enough to float.

to their origin. In the Buck Lake samples, I found several spherules cracked open, so I could look inside. One had a thick wall and was hollow like a cantaloupe (fig. 6.4). Another one was completely empty with a very thin shell and a smooth inside surface. It looked somewhat like an empty eggshell, and I wondered if they could be insect eggs of some kind.

I looked for other broken ones and found several hollow ones with very thick shells. They looked like all the hollow ones on the outside, but inside, the thick rims were made of some material with thousands of tiny voids. They definitely were not eggs.

Breaking one with tweezers, I found that it crunched like soft rock, suggesting meteorites, and they certainly looked like something from another planet. I scanned the literature, but while scientists mention hollow cosmic spherules occasionally, none looked like these, and nearly all meteoritic spherules were metallic, unlike these.

Our chemical analysis indicated that they were a mixture of nearly pure elemental carbon along with a little rock, a very unusual mixture. Where had the carbon come from? Some unknown type of organism? A firestorm? We did not know. We had only a few facts: first, the floating spheres were at nearly every site that we checked; second, they were only in the Clovis layer; and third, they contained almost nothing but carbon. With those facts, we were almost certain that they were related to the Event, but we were unable to solve the mystery of the spherules, so we set them aside. Later, analysis with a scanning electron microscope would provide key information to understanding the spherules and revealing the devastating power of the Event. More about that later.

GUNMETAL-BLUE SPHERULES

After that, I looked at the charcoal sample from the Clovis campfire. First, I used the magnet to pull out magnetic grains and found thousands of them. When I looked at the grains under the microscope, I saw more spherules, lots of them, about 300 per pound of sediment. These metallic ones were very different from the hollow ones, since they were attracted to a magnet, and they did not float. In addition, some looked blue-black like a gun barrel, and others appeared to have an odd dull black crust on them. Turning up the microscope to higher magnification, I saw that the crust appeared to be charcoal, except that it was not loose but rather had fused onto the surface.

I wondered if the charcoal crust was there simply because the spherules had mixed with the charcoal, and it had stuck to them. Looking at the gunmetal-blue color, which is often a sign that fire has heated the metal, the true answer dawned on me. That campfire had *cooked* those spherules! Either they were on the ground before the campfire was lit, or they fell before the hot charcoal embers from that Clovis-era campfire had cooled.

Was I looking at spherules and charcoal that recorded the exact instant of the Event? With my mind spinning, I reviewed what I already knew. The spherules, the magnetic grains, and the mat showed up at every site at exactly the same time. When they appeared, the megafauna and the people disappeared. If a blazing Clovis campfire had indeed cooked these spherules, then it might have happened while the unsuspecting Clovis people sat huddled there. Thousands of red-hot iron grains could have showered down upon them . . . at 3,000 miles per hour.

NEW OR CORROBORATING CLUES TO THE MYSTERY

- A Clovis-era antler tests as magnetic, just like the Murray Springs mammoth teeth.
- The Clovis black mat shows up in Canada, 1,500 miles from Arizona.
- Hollow spherules show up for the first time. They may be of cosmic origin.
- Many magnetic microspherules appeared to have fallen into a burning Clovis campfire.

7

THE HORSE AND THE CLOVIS SPEAR

SOUTH TO CALGARY

After leaving Buck Lake, I drove south to collect something unusual—brain-case dust. To do so, I headed to see Dr. Brian Kooyman, from the Department of Archaeology at the University of Calgary in Alberta, Canada (fig. 7.1). Unsure whether he would be available, he had left something for me with his department secretary: a small bag of sediment. But this was no ordinary sediment. It came from the brain cavity of an extinct horse, and I was eager to see the stuff. I was sure it was some of the most unusual sediment on the planet. Few others besides Brian had ever seen extinct horse "brain dust."

Fig. 7.1. Brian Kooyman holding a Clovis point from the horse site at the St. Mary's Reservoir. *Source: University of Calgary*

I was curious about whether the dust contained magnetic grains. Impatient to know, I opened my briefcase in the rental car and took out my supermagnet so I could run a quick test as soon as I got it. I carefully slid the magnet into my pants pocket, making certain it was safely far away from my credit cards and driver's license—it would have wiped clean their magnetic strips. After shutting the front car door, I opened the back to take some paperwork with me, and that's when I heard a solid *thump*. Startled, I thought something had hit the car, and straightened up, only to have my pants twist sideways. *Not again!* That darned magnet was stuck to the car door, although I managed to work it free without ripping my pants. Reminding myself to stay away from large metal doors inside the university, I set off to find Brian.

ST. MARY'S RESERVOIR

The story of Brian and the extinct horse had begun a few years ago at St. Mary's Reservoir near Calgary. The western part of North America had long been in drought conditions, and finally the falling water level had exposed areas of the reservoir that had been hidden beneath its surface for years. High winds created occasional dust storms that carried away part of the loose cover of dust, and some odd things appeared—long trails of huge footprints (fig. 7.2).

The giant prints and other, smaller ones came to the attention of scientists from the University of Calgary. Inspecting them, they were astonished to find that the prints were from extinct mammoths, camels, oxen, and horses. Such prints are exceptionally rare; in fact, the camel prints are

Fig. 7.2. Dr. Paul McNeil points to a line of 13,000-year-old mammoth footprints that look as if they were made yesterday. The prints are twice as wide as Paul's hat. *Source: University of Calgary*

the only ones known in North America. But what were they doing in the reservoir?

St. Mary's Reservoir covers an area of about 1.5 square miles (4 sq km). The area includes part of an old river plain that had existed since long before and throughout Clovis times, so it was a favorite site for Clovis-era grazers, with ample water and lush grass. Because of high usage of the site, the Ice Age animals left many prints, but that was not all. They left bones.

Not far away from the prints, the Calgary excavators found part of the skeleton of an extinct North American horse, including the skull and several crushed vertebrae. When they looked closely at the horse bones, the scientists were excited to find that the vertebrae showed distinct butchering marks, indicating that Paleo-Indians had hunted the horses. If that surmise held up, it would be an important find, the first direct evidence of such hunting. However, the butchering marks were not fully compelling on their own. The scientists needed more evidence to support the idea.

Looking further, they found a few Clovis points several hundred yards away, but they were disappointed to find no horse bones with them. In a moment of inspiration, several graduate students suggested testing the points for animal protein, since sometimes, after ancient hunters used a tool to kill or butcher animals, telltale traces of blood and dried tissue remained on the flint even after 13,000 years. It was a long shot, but worth doing.

The result of the test was conclusive: there was horse protein on the points. Evidently, the Clovis hunters had tossed spears tipped with those flint points at a herd of the now extinct horses, and at least one spear found its mark. Those Clovis people could never have imagined that 13,000 years into the future, some of their very distant relatives would use unimaginable technology to find out about that horse hunt so long ago.

GETTING THE DIRT

At the reception desk, Brian's secretary handed me a small package. Looking with fascination at the small bag of pale gray dust, I read the label indicating that it came from inside the St. Mary's horse skull. Earlier, Brian had mentioned by phone that it was the only remaining sediment from the excavation.

Eager to test it with the magnet, I turned to leave when someone called out. With a friendly smile, Brian hurried up to introduce himself. He had just a few minutes before conducting a student exam and wanted to say hello. I thanked him and briefly told him that his horse head might contain clues of a distant supernova or a celestial impact.

Tapping the bag, I said, "The importance of this is that you have very good dates that tie it firmly to the Clovis era, but even more important, it came from inside the skull. We can be very certain that this dust is from the same surface on which the horse walked. If a cosmic event happened then, there should be evidence of it in this stuff."

Brian was intrigued and seemed eager to know more, but he had to attend to his class, so we shook hands and he rushed off.

TESTING THE DUST

As I walked away, I looked around furtively for a bathroom where I could test the dust. This could be one of our closest connections to the actual Event itself, since the horse skull must have been on the surface when it happened.

I considered testing it by the bathroom sink, but then realized how that might look to a student or a professor who came in. There I would be with my hand stuffed into an open plastic bag filled with suspicious-looking powder. I imagined protesting to the Canadian Mounties that it was "horse dust" rather than illegal "angel dust." For privacy, I locked a bathroom stall, opened the bag of dust, and pulled out my magnet, while cautiously keeping the magnet well away from the metal stall.

Dipping the plastic-bagged magnet into the dust, I shook it around lightly and pulled it back out. Even in the dim light, I could see particles gleaming. It was loaded with them! Thousands of magnetic grains stuck firmly to the magnet's edge. I had never seen so many after one pass with the magnet. Ecstatic, I packed up everything and headed back out to the car. The horse hunt had been a great success.

FINAL TESTING

After I returned home and was able to test the sediment thoroughly, the initial results from Calgary held up. The St. Mary's Reservoir horse had the highest concentration of magnetic particles of any site that we had visited; it contained about thirteen ounces per 1,000 pounds (8 g/kg), meaning that nearly 1 percent of it was magnetic grains. In addition, we found microspherules inside the horse head—not as many as at the other sites, but still a significant number.

The most surprising evidence came from NAA tests on the sediment. We found that it was significantly higher than normal in potassium-40, or ^{40}K, and its presence is a direct link to a massive ancient supernova. Most likely, the exploding star created and ejected the ^{40}K out into space, and, after a long journey, it rained down around the dead horse in Calgary 13,000 years ago. Or there is another possibility. The supernova may have bathed a meteorite or comet with powerful radiation that altered its chemistry to form the ^{40}K that was carried to Earth in an impact. Either way, eventually the radioisotope found its way into the horse's braincase and then into the bag of dust that Brian handed me in Calgary. It was a long journey from the supernova to the plains of Calgary, but the presence of magnetic grains, magnetic spherules, and potassium-40 gave us invaluable clues to the mystery of the Event.

After we tested the Calgary horse dust, other sediment samples came up that gave us some surprising new clues to the Event.

MORE CANADIAN SEDIMENT

Around the time I went to Canada, we came across a scientific paper by Matthew Boyd and coworkers (2003) regarding sediment sampling in Lake Hind, a glacial lake that bordered the ice sheet. Located in southwestern Manitoba, Canada, the old lake covered about 1,500 square miles (4,000 sq km). The research of Boyd and his colleagues suggested that prior to Folsom times, the ice dams had burst, allowing catastrophic downstream flooding from this meltwater lake and others like it. Suddenly, the lake became shallower, allowing algae and other plants to flourish. Then, as the lake waters finally disappeared completely, both grazing animals and Folsom hunters moved out onto the rich former lake bed.

As we read the report on Lake Hind, we wondered whether the sediment might have preserved a snapshot of the Event, since a number of key points of interest caught our attention.

First, the authors had several radiocarbon dates, the oldest a maximum of 12,700 years ago, only shortly after the Event occurred, around 13,000 years ago. The sediment record showed that only a few inches (5 cm) below that date the ice dam had failed, and the narrow interval put the probable date of the dam failure into Clovis times. This scenario in the Boyd paper fit with our theory that a massive aerial explosion during the Event collapsed these dams, causing catastrophic flooding across what are now Canada and the United States.

Second, the paper described the shallow lake sediment in Folsom times as containing much more organic material than that in Clovis times. This sounded very much like the black mat from Murray Springs, Blackwater Draw, and the Buck Lake site, so I was curious to see how it looked.

It sounded promising, and we wanted to check it out, so I called Matthew and asked if he had more sediment available. He did, and I arranged for shipment. I found the package waiting for me when I returned from Alberta.

TESTING LAKE HIND

We found significant amounts of all our key Event ingredients: grains, spherules, radioactivity, and the black mat. In addition, we found a strong link to a supernova: NAA analysis of the Lake Hind sediment showed nearly 300 percent of the potassium-40 that we normally expect. It is the highest level of ^{40}K that we have found so far, and it occurs exactly in the Event layer at the end of Clovis times.

We analyzed the Lake Hind samples, and we found that the magnetic grains were up to twenty times more abundant during the Clovis era than just a few hundred years afterward. The hollow, floating spherules show an even more dramatic rise and decline; they go from nearly 200 per 2.2 pounds (1 kg) immediately after Clovis times to zero shortly after that, where they stay at zero for the next 2,000 years. Both results are consistent with some major and unusual situation 13,000 years ago.

CLOVIS-ERA FIRES

Boyd and his colleagues sampled the Clovis-Folsom section of the core for both charcoal and wood fragments and fires during Clovis times. Both showed up clearly in the Lake Hind record. The combined total of woody debris and charcoal peaked dramatically just after Clovis and declined steadily over the next several thousand years, which is the span of the core sample. At its maximum, the total was twenty times higher than several thousand years later. There were lots of dead plants and charcoal around just after Clovis times, supporting our theory that the Event decimated millions of trees and plants and touched off widespread wildfires.

ALGAE AND THE BLACK MAT

The Lake Hind samples look just like the black mat at Murray Springs and Blackwater, and they cover the same period. Could they be the same thing? Vance Haynes indicated that he had found the black mat from Canada to Mexico, so it seems likely that they are the same. In support of this, Boyd tested the core for algae, finding that algal spores compose nearly 80 percent of the total pollen and spores just after Clovis times, meaning that for many hundreds of years, almost nothing but algae grew in or around Lake Hind. Similarly, according to Haynes, the algae that formed the black mat at Murray Springs grew abundantly for more than 1,000 years.

ANALYSIS OF THE CANADIAN SITES

At the Chobots' site, St. Mary's Reservoir, and Lake Hind, we found vital clues that extended the reach of the Event into Canada, right to the Clovis-times edge of the ice sheet. Clearly, the Clovis catastrophe was shaping up as a continent-wide affair.

In addition, in Canada, we have another link between the Event and the mat, a connection that spans the continent. First, at the Buck Lake site and Lake Hind, all the Event markers rise substantially in Clovis times to their highest levels, followed by the explosive growth of algae and the mat. We suspected that all those signs were connected somehow, but at that point, we were not sure how. Later, we would find a link that was very unexpected. We found the answer in research on the dinosaur extinction, which was just a larger version of the Clovis megafauna extinction. At that time, 65 million years ago, researchers found an unlikely connection between the demise of the gargantuan dinosaurs and the rise of diminutive algae. More about that later, but for now, here's one hint: it involves the floating spherules from the Buck Lake site. Before we look at that, however, we have to explore another Ice Age mystery in Canada.

NEW OR CORROBORATING CLUES TO THE MYSTERY
- From St. Mary's Reservoir near Calgary, a Clovis-era horse skull contains microspherules.
- It also contains the highest levels of magnetic grains we have found at about twelve sites.
- The St. Mary's specimen also shows elevated supernova potassium-40.
- Glacial Lake Hind shows the highest levels of supernova potassium-40 detected thus far.
- Glacial Lake Hind contains a likely portion of the continent-wide black mat.
- Widely distributed evidence suggests the Event affected all of North America.

8

THE MYSTERY OF THE DRUMLINS

THE DRUMLIN PUZZLE

While in Alberta, I was eager to pursue another possible clue to the Event. West of Calgary, near the edge of the Rockies, there are mysterious features called drumlins that were left behind by the melting ice sheet (fig. 8.1). Appearing as boat-shaped piles of rock, sand, and gravel, drumlins range from a few hundred feet to about five miles in length and up to about a mile wide. Most are less than 100 feet tall.

As I drove the freeway toward Morley, I thought about these puzzling drumlins, a name that sounds more like a band's musical instrument than a pile of gravelly sand. Although scientists disagree on the exact creation process for drumlins, most attribute their formation to one of two methods, both of which relate to the glaciers, which caused all of them in a given area to line up in the same direction (see fig. 8.2). The first theory proposes that drumlins were formed from the natural movement of an ice sheet, and the second theory suggests that the drumlins resulted from meltwater flooding.

Fig. 8.1. This half-mile-long drumlin is near Morley, Alberta. Its keel-like shape is similar to the inverted hull of a sailboat.

We are more interested in the second theory, according to which water pressure periodically builds up to a critical level underneath a heavy ice sheet, then bursts out in a catastrophic flood. When that happens, the muddy, abrasive meltwater carves the drumlins.

The second theory fits well with our evidence for a cataclysmic Event at the end of the Ice Age. If there had been an airburst or a cosmic impact at that time, the high pressures would have exerted an immense downward force on the ice sheet, causing the water beneath it to surge out, creating the drumlins.

There is one interesting fact that fits with the catastrophe link: no one can find undeniable cases of drumlins forming today. Partly, this is because most of the continental ice sheets are gone, but that cannot explain the mystery completely. Drumlins were created at one time in the past and never again, making them one of the enduring mysteries in the study of glaciers.

BIRTH OF THE DRUMLINS

Dr. John Shaw of the University of Alberta thinks that the high-velocity surge of subglacial muddy water did two things: first, it carved depressions in the bottom of the ice sheet in the shape of drumlins. Then, when the flow stopped, the sand and gravel filled the cavities, creating the familiar drumlin shape, as shown in figure 8.3.

Fig. 8.2. A U.S. Geological Survey (USGS) aerial photo of a drumlin field near Powers, Michigan. The long spindle shape is typical of these glacial landforms.

Fig. 8.3. We tilt up the ice sheet to show how drumlins form. Glacial meltwater, surging between the ice sheet and the ground, creates the drumlins. *After Shaw and Gilbert (1990)*

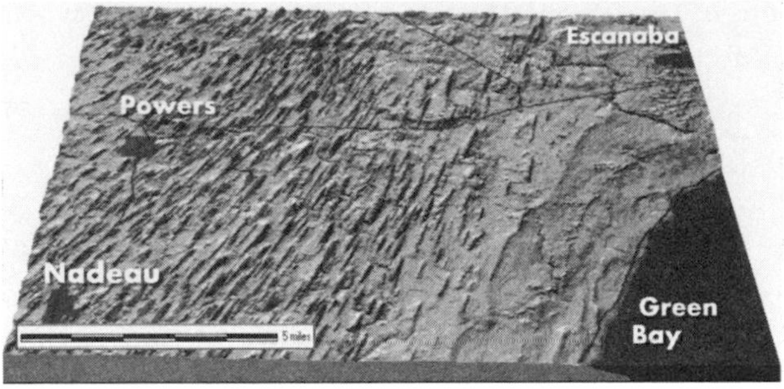

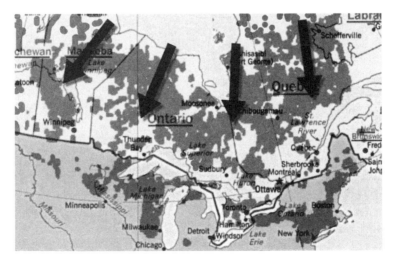

Fig. 8.4. In this USGS digital elevation mode (DEM) image, a dense field of miles-long drumlins lies west of Lake Michigan near Escanaba, Michigan.

Furthermore, Shaw found unmistakable evidence for two massive surges that led to two episodes of drumlin formation. This is important for our theory, as you will see in a moment.

Shaw's work and that of many others have shown that drumlins blanket huge expanses of North America, including large parts of many provinces in Canada and in most states in the north-central United States (figs. 8.3 and 8.4).

IS THERE A LINK TO THE EVENT?

You may recall that the chemistry of the magnetic grains suggests a link to Canada. As it turns out, the largest drumlin fields in the world are around Hudson Bay and the Great Lakes (fig. 8.5). Because of the location of these

Fig. 8.5. This map shows the major drumlin fields in Canada and the northern United States. Note that the fields are thickest around Hudson Bay and the Great Lakes. Most drumlins in the fields are oriented with the arrows. *After Colgan (2000)*

drumlin fields, we believe that many drumlins, but not all of them, may have formed when the Event happened. We already know that the Event produced enormous pressures that drove particles across the continent and embedded them in flint and tusks. We believe that same pressure caused the ice sheet to surge forward, forcing high-velocity, high-pressure meltwater to pour out in massive sheet floods, which created the drumlins.

IS THERE A EUROPEAN CONNECTION TO THE EVENT?

In addition to those in North America, drumlin fields exist in Ireland, Lithuania, Latvia, Estonia, Poland, Finland, Denmark, and other parts of Europe. There are no reported drumlin fields in Africa, Australia, or South America.

Since we have linked the Event to North America, you might be surprised to hear that it could have affected other parts of the world. Reading about drumlins in Ireland and Lithuania, you might wonder how the Event had anything to do with those countries. However, the drumlins there formed at exactly the same time as the drumlins in North America. Is this a coincidence, or is there a connection?

We know from the tusks that the Event affected North America and Siberia, and now we are suggesting a link to Europe. Is it possible that the Event sent particles flying and meltwater surging across all the continents in the Northern Hemisphere? If the drumlins formed because of meltwater flooding, then that is possible.

DATING THE DRUMLINS

Although the exact date of drumlin formation is unclear, nearly all scientists agree that they formed about 16,000 to 13,000 years ago at the end of the Ice Age. They know this because of radiocarbon dating and because most drumlins are made of loose sand and gravel. If the ice had moved even several hundred yards over those soft, sandy features after they formed, it would have flattened them beyond recognition. That they exist at all today proves that the ice sheet did not advance any farther; instead, it melted away from above them, leaving all the huge drumlin fields we see today. That happened only at the end of the Ice Age.

At that point, we could not be certain that the Event created all the drumlins; the timing might have been coincidental. We were certain that they formed at the same time as the Event, and that none has formed on the planet since then, but we needed more evidence to confirm a connection to the Event.

A NEW THEORY OF DRUMLIN FORMATION

I was eager to find Event evidence in Morley, Alberta, and to test our theory about the formation of drumlins:

- Sixteen thousand years before the present, the ice sheet still covered the area around Morley, and there were no drumlins yet.
- At that time, the last phase of the cosmic Event began, causing huge surges in the ice sheet that created large fields of drumlins. This was the first event for which Shaw found evidence.
- Then, about 13,000 years ago, the cosmic Event peaked in intensity, sending high-speed particles crashing into and across the top of the ice sheets.
- At the same time, the Event created a nearly instantaneous surge in the ice, causing the glaciers to flow out of the mountains and to slide across the plains around Calgary.
- Simultaneously, meltwater floods roared beneath the moving ice, sculpting the drumlins in the area. This was Shaw's second wave of drumlins.
- Not long after the ice came to a halt, the cumulative effects of the Event, including hot particles and climate change, caused the glaciers to melt away gradually. They never moved forward again, thus preserving the drumlins.
- As the ice above the drumlins melted, the magnetic particles and spherules contained in the ice came to rest on top of the drumlins.

TESTING THE THEORY

If our scenario was right, I expected to find many particles and spherules on top of the drumlins. The top was important to test, because magnetic grains are heavy and tend to move from a high point to the lowest point in an area. If we found them on top of a drumlin, it was unlikely that wind or water had carried them there and more likely that they had melted out of the glacier.

The hardest job was finding a drumlin exposed by a road cut. Most roads in the area curve around them, because they are steep. But finally, I did find one that had been exposed. Digging out my tools, along with the supermagnet, I found a good spot along the cut face and prepared to test with the magnet at the exact spot where the modern soil lay in a sharply defined layer on top of the glacial gravel. If our theory was correct, that spot was the top of the drumlin at the end of the Ice Age just after the Event.

Patting the magnet along the wall for about a minute, I checked the edge for grains. Thousands of them! From end to end, they covered the magnet. So far, the test was good. The final key tests would determine how many grains were above and below the drumlin top.

I tested the dark organic soil on top—almost no grains. Next, I sampled the level below the first one, which would have been down inside the original drumlin. There I saw grains, but far fewer than at first. The predictions were holding up, although I would not have a clear answer until I finished the complete grain separation back home.

THE FINAL ANALYSIS

The final tests held to the same pattern. Both the grains and spherules peaked sharply in the original top of the drumlin, just as at other Clovis sites. In fact, the level of magnetic grains was the third highest we have recorded to date, equaling four ounces of grains for every 100 pounds of sediment. The magnetic spherules occurred at the second-highest level that we have ever found. There were more than 1,000 of them for every 2.2 pounds (1 kg). Not only did the grains and spherules occur as predicted, they also occurred at a high concentration. Of the nearly two dozen sites that we tested, the overall results from the Morley drumlins were second only to the totals from Gainey, perhaps because both sites were closer to the ice sheet than were other sites. In addition, they are closest to central Canada and the Great Lakes, which the evidence suggests were a focal point for the Event.

CANADIAN OVERVIEW

At Lake Hind, at St. Mary's Reservoir, and at Morley, we found excellent sites at which all our key parameters showed predicted peaks around Clovis times. Those three sites are about 600 miles (1,000 km) apart, showing that similar effects occurred at the same moment across all of southern Canada south of the ice sheet. When we include Gainey, which is near Toronto, Ontario, the distance expands to 1,500 miles (2,400 km). The reach of the Event most likely covered the entire North American continent, extending out into the Atlantic and Pacific Oceans. The drumlins in Scandinavia and Ireland suggest that the Event's reach was far greater than that.

NEW OR CORROBORATING CLUES TO THE MYSTERY

- A Morley, Alberta, drumlin has the second-highest combined level of spherules and grains thus far detected.
- The peaks in grains and spherules suggest that drumlins formed near Clovis times.
- The pattern of drumlins matches our theory for a central Canadian focal point for the Event.
- The drumlins support our theory of surges of glacial ice during Clovis times.
- The drumlins support our theory of Event-related massive meltwater floods.
- The presence of drumlins in Scandinavia and Ireland suggests that the Event reached Europe.

9
OUT ON A LIMB IN CAROLINA

STORM WARNING

Preparing to leave my motel in Alberta for an archaeological location called the Topper site near Allendale, South Carolina, I watched the Weather Channel with concern as fluorescent green radar blips flickered across the southeastern United States, indicating extremely heavy rain clouds. A stalled weather system was drenching the already soggy region with torrents of rain. During the past week, parts of the area had received a year's worth of rain and it was only June. With disappointment, I thought briefly about canceling the trip, even though I knew that the Clovis excavation I was to visit was open only a few weeks each summer, and if I missed it, I would have to wait another year. I was forced to take a chance.

As I drove from Barnwell, South Carolina, to the site, I kept my windshield wipers on high, but they were barely able to keep up with the relentless downpour. I wondered if the excavators had even come out today; unlike me, they were probably still in their warm beds. My spirits were as dismal as the weather. But to my surprise and relief, as I approached the entrance to the Topper site, the rain suddenly lessened to a slow drizzle.

The site is an ongoing Clovis and pre-Clovis excavation located on the grounds of Clariant Corporation's chemical plant near Allendale, so I had to check in with plant security before going to the dig. As I signed the guest register, I saw the recently written name of Dr. Al Goodyear, the site excavator from the University of South Carolina, so I knew the crew was already there in spite of the weather.

When I had first heard that the site was a chemical plant, I imagined billowing smokestacks and massive factory buildings, but the opposite was the case. There were no smokestacks at all. Leaving behind the one huge plant building, I drove along a one-lane road overarched with a thick fringe of trees, creating a tunnel of leaves and tinting the dim daylight green. Ahead of me lay an 8,000-acre Savannah River forest preserve, dense with elm, sycamore, oak, and pine. Somewhere in the midst of all that, I was to find Al Goodyear and the Topper site.

The pavement soon changed into a dirt road that snaked through the forest, and as the pavement ended, so did the rain. Before long, I came across about a dozen trucks and cars pulled off the road among the trees, with no people in sight. After parking and gathering my gear, I walked down the narrow tree-canopied road in the direction of noises that sounded like shovels at work. The humidity was so high that water dripped off the leaves in a slow, steady beat, and I began to sweat profusely. Luckily, I had brought a case of Gatorade, which would be gone by the end of the second day.

After a short walk, I came upon the excavation crew, whose members were busy digging about half a dozen pits scattered among the trees. The site is open only one month of the year, and Al uses a large group of volunteers to make the best use of the limited time and resources. Many volunteers come from long distances for "digging vacations" on the cutting edge of archaeology, and some of them come back every year to uncover new artifacts.

Before long, I located Al, a friendly bear of a guy (fig. 9.1) who loves

Fig. 9.1. Al Goodyear at work in the Lower Site, holding pre-Clovis artifacts surrounded by the ever-present plastic rain tarps. *Courtesy of the University of South Carolina*

to talk about Topper, and for good reason. While it is a fine Clovis site, it is also one of the few sites in the Western Hemisphere with strong evidence for pre-Clovis settlers of the New World (for details, see Goodyear, 1998–2004). When I first saw Al, he was wearing tall rubber boots, and except at suppertime, I never saw him without them. After all this rain, the site was very soggy, including the tarp-covered pits, and before I left, I came to envy anyone with rubber boots.

First, Al gave me the grand tour, starting at the Lower Site just below the top of a forested hill. "There's a loop of the Savannah River not far away," Al explained, motioning to the south, where I could see the faint sparkle of water through the trees, "and the Paleo-Indians went down to the riverbank to gather rounded flint rocks, or cobbles, and hauled them up here."

"Like a flint factory," I commented, thinking of the Buck Lake property.

Al nodded. "In Clovis times, the river was a lot closer, so it was easier to get the flint up here. In fact, not long before Clovis, this spot where we are standing might have been under water. We've found a Pleistocene terrace over there with signs of periodic flooding that went on for a long time," he said, pointing to the deep pit nearby called the Lower Site.

As we stood there, a slow drizzle began, and the overcast sky threatened to dump a lot more water on Topper. We were somewhat protected by a twenty-foot-square plastic tarp suspended high above us with ropes. Even so, the light breeze carried the misty rain around it, and it was falling fast enough to add to a two-foot-deep pool of rainwater on top of the roof tarp. The plastic sagged ominously in the middle, as it swayed in the breeze and strained at the ropes.

Thankfully, the light rain stopped as Al continued. "When we started digging into that terrace, the real fireworks began—that's where we found the pre-Clovis stuff." He gave me a big wide smile, but I could tell he was a little sensitive about this. Pre-Clovis sites are still controversial, although Al is helping to change that.

He indicated a line of cream-colored chert sticking out of the carefully excavated wall about six feet below the surface. "See those artifacts?" he said. "They are Clovis, and as long as I excavated only down to that line, I was okay. I was a happy, contented university professor waiting for retirement, and, of course, I was a confirmed 'Clovis First' believer. But a few years ago here at Topper, that all changed, including the happy-and-contented part." Al flashed his customary smile, although he was half-joking and half-serious.

He continued, "A few years ago, there was increasing talk about possible pre-Clovis sites, such as the Meadowcroft Rock Shelter in Pennsylvania and Monte Verde in Chile, and I wondered if there might be something here too. So one summer I decided, what the heck, let's dig down into the Pleistocene terrace to see what's there. Well, we found a bunch of flakes and some worked chert that looked like tools, and at first I thought they must have washed down the hill from the Clovis spots above. Trouble was, none of them looked Clovis at all. They were much cruder, and there were no bifaces in the whole lot."

I had been around archaeologists enough to know that bifaces are tools that have been flaked and worked on both sides and are characteristic of Clovis toolmaking. Before that, most ancient people used a flint-working style that produced microblades, which are small, thin, knifelike pieces of chert that did not require much chipping before use.

I asked Al, "You started finding microblades?"

He nodded. "Lots of them, but it took me awhile to tell anybody about it. It just upset all that I had believed up to that point, but the more I found, the more I couldn't ignore them. That's when I decided to report our findings. The tools were there. I couldn't pretend that they weren't, and I was sure they were older than Clovis.

"I knew I was going out on a limb, but little did I know how many of my fellow scientists would want to chop the limb off!" He laughed, and I did too. He and our group were both working at the frontiers of science. Laughter is good therapy out on the frontier.

"The good news," Al concluded, "is that more than a few of my colleagues are coming around. I've had them out here, and they've seen the microblades. If someone had found them in Siberia, there would be no argument that they are older than Clovis, but because they are in South Carolina, it takes a little longer. Nevertheless, I'm hopeful. I believe many scientists accept, at least privately, that people were on this continent maybe 16,000 to 20,000 years ago, long before Clovis."

Al had to break off the tour when several potential donors arrived for a visit. The University of South Carolina partially funds his work, but it is not enough, so he has to scramble for private funding. His goal is to secure enough funds to build permanent shelters over the excavations to protect them from the rain and to make it easier to work year-round without rubber boots and raincoats.

WATCHING MY STEP

While he was gone, I went up to Topper's Upper Site (which was associated with the Clovis era) to collect samples; there several volunteers offered to help me. All around the pit, I saw Clovis artifacts sticking out of the wall in a clear, narrow line, just as the excavators had found them.

As with every test at each new site, I never knew what I would find with the magnet, and I was even less certain here, since Topper is 700 miles away from Gainey, the nearest site. I knelt down on the ledge, cautiously keeping my feet away from the edge, as I dug the magnet out of my pack. Sweat rolled down my forehead into my eyes, making it hard to see, so I found it by feel alone.

After wiping my eyes and glasses, I started to hold the magnet out toward the wall, but I hesitated with my hand in midair. Frozen, I stared at the wall as the full importance of that test swept over me. Our Event theory predicted that we would find plenty of magnetic grains in the Carolinas, maybe more than elsewhere. But what if there were none here? It would mean something was seriously wrong with the theory. Had I wasted my time flying to South Carolina? As my head spun with possibilities, I realized something else. Before I came to Topper, I had explained to Al that we had uncovered radioactive signatures at other sites and that we expected to find one here too. Finding that could help him date his site, especially the deeper layers, and he seemed particularly interested in that aspect of our theory. So what if there was no radioactivity here, and no magnetic grains?

Noticing that the crew was staring curiously down over the bank at me, I got back to business. After rubbing the magnet gently over a long stretch of wall right above the line of Clovis artifacts, I brought the magnet close to my glasses. Wiping sweat from my eyes, I saw a thick line of black pepper-like grains sparkling in the cloud-filtered daylight.

I couldn't have been more excited if I had seen diamonds hanging off the magnet! I held it up for the others to see. They gathered around, squinting hard at that thin line of iron grains. To them, the grains were hardly visible, and probably not very impressive, so I told them, "We're pretty sure these grains came from Canada . . ." Pausing for maximum impact, I finished with, ". . . at thousands of miles per hour." A few eyes got wider at that. "They probably arrived in less time than you can finish lunch," I added. Murmurs of surprise rippled around the group as they realized they had been sifting for days through sand laced with this stuff from Canada.

Next, I moved around the ledge, testing more areas near the artifacts. It was the same everywhere: the layer was loaded with them. Then I tested at the top of the wall, which had much less. Last, I tested about two feet below the Clovis, and there was much less there too. So Topper was just like all the other Clovis-era sites.

As Tony, the supervisor, watched me work, I pointed out to him that the magnet was a very fast and inexpensive way to find the Clovis layer, if they were ever unsure. He got the point right away. In fact, he mentioned that they were having a problem locating the Clovis layer in one of the nearby trenches and asked me to check it a little later. I agreed gladly.

TESTING FOR IRON WITH MAGNETIC SUSCEPTIBILITY

The next thing I did was to use a magnetic susceptibility (MS) meter, a device that allowed me to estimate the amount of iron in the wall. This gives us information similar to what the magnet tells us, and it is much faster. I had a lot of wall area to cover in a short time. Beginning six feet down at the bottom and working my way up, I took readings about every inch and recorded them. I did a quick graph of the results, which matched our expectations perfectly (fig. 9.2). According to Al, the entire sequence spanned about the last 40,000 to 50,000 years, with the biggest peak precisely where the line of Clovis artifacts ended, 13,000 years ago. According to our theory, the Event occurred at that time, peppering North America with countless trillions of hot iron pellets, all traveling faster than military missiles. It was not a happy time in South Carolina.

UNKNOWN PARTICLES IN THE SEDIMENT

My last objective at the Upper Site was to get sediment samples to take with me. Later, I planned to run a battery of tests, including testing the soil for radioactivity. After I prepared my tools, I looked uncertainly at the pristine

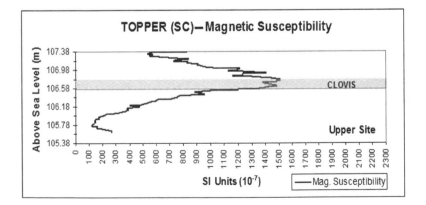

Fig. 9.2. Magnetic susceptibility, just like the magnetic grains, peaks in the Clovis layer.

walls of the excavation pit; I had to cut holes in one of them. Using a small hand auger, I cut out three-inch-wide by three-inch-deep plugs of sand from the wall and bagged the samples. Moving rapidly down the wall, I dug out double plugs about every eight inches. Stepping back, I surveyed the wall. It was a mess. Oh well, I rationalized, it's all in the name of science.

While filling the bags, I had noticed what looked like small, shiny bits of charcoal. Thinking about the charcoal at Murray Springs and the Buck Lake site, I was excited that we were also finding a widespread link to fires at Topper. Little did I know that those small pieces were not charcoal. They were something far more important—an extraordinary new type of particle, one that would be among the most important finds at any of our sites. At first, they stumped even a retired top NASA scientist, who had never seen anything like them before. Later, you will see that those tiny new grains contained vitally important clues to solving the riddle of the Event.

SIMPLE TESTS TO LOCATE THE CLOVIS LEVEL

Next, Tony and I went over to a shallow exploration trench that the crew was deepening. They had expected to find the Clovis layer close to the surface, but they were down nearly two feet without having found a single flake of chert. Tony was worried that they would be digging for hours only to find no artifacts. I hoped that my magnet and MS meter would give them some clues.

Running my magnet down the wall, I found a thin band of magnetic grains clustered about halfway down. Next, I swept the MS meter across the wall and found the same thing. The readings were very high in the middle, equal to the last pit we were in, and then decreased above and below it. Based on that, I suggested to him that he was already beyond the Clovis layer and that it contained no artifacts at that spot. To test the new idea, he stopped that trench and continued with another one nearby that was not as deep. Within a few minutes, the diggers began to uncover Clovis flakes at exactly the right depth, including part of a beautiful deeply fluted Clovis spear point. Tony was delighted and so was I, as we both realized that those two simple tools, a $30 magnet and a $2,000 handheld meter, had saved a lot of digging.

Later, he and I told Al about the test and discussed the implications of using these inexpensive, easy-to-use tools in any excavation that needed to find the Clovis layer, especially when no artifacts were visible. The Event appeared to provide a widespread marker for 13,000-year-old sediment layers. So far, everywhere we had been we had found magnetic grains, magnetic spherules, and radioactivity levels that peaked in nearly the same few inches of stratum. We suspected that might be true across most of North America. If so, these tools would be very helpful to archaeologists all over North America, and most likely in Europe as well.

PRE-CLOVIS AT THE LOWER SITE

After that, Al took me back down to the Lower Site, where he had made some of his most exciting discoveries. Pointing to an alcove excavated into the far wall, he said, "See that chunk of chert over there? That's one of the cores we found (figure 9.3, at arrow). Long ago, somebody sat around that block, probably with it between his crossed legs, while he whacked the core with a hammerstone to make microblades. It still looks just as it did when we found it."

The campfire at the Buck Lake site flashed into my head. A Clovis knapper (someone who chips stone tools) had been sitting around a core there in much the same way as at Topper.

"It gets better," Al said, jolting me out of my reverie. "Come on. Let's take a closer look." He strode off into the pit and across to the alcove with the enthusiasm of a kid showing his new toy, except that his "toy" was a large chunk of rough flint. He had already pointed out the Clovis layer to me, and I could see that this flint block was about two feet deeper than that;

Fig. 9.3. Lower Site alcove. The dots show the Clovis level, and the arrow points to a large pre-Clovis chert core. The knapper sat there knocking off flint flakes, which he used to make other tools.

this core was thousands of years older than Clovis. Because of that, it had to be one of the rarest chunks of flint on the planet. I was very impressed, to say the least.

Hopping up on the ledge, Al said excitedly, "See that?" as he pointed to several piles of flakes around the core. "Those are the rejects, the spoils. He broke those off—couldn't make anything out of them, so he left them there. They've been there for about twenty thousand years, until we dug them up. They weren't making bifaces from that core; they were making microblades. No need to do any armchair talking about pre-Clovis; you're looking right at it." I bent down for a closer look, being careful not to touch the core.

"But that's not the only one we found," Al declared with even more enthusiasm as he hopped off the ledge. I followed him to where he over-looked the deepest part of the twelve-foot-deep pit, filled with three rubber boot–clad volunteers. "See that one?" he asked, pointing to just above the waterline where an even larger chunk of chert poked out from the wall. I nodded. They had left it in place while they dug deeper. He continued, "That one is the same thing, a microblade core."

Instantly, I compared it with the upper core. The second one was more than six feet deeper and much, much older. "Do you know the age of it?" I asked.

He shook his head. "No, that's the frustrating part, but we know what it is older *than*. Do you see that gray-black streak just to the left of it? That's a campfire. Now, some of my critics claim it's just a forest fire, but it's only about an eighteen-inch-diameter circle. If they're right, it could be in the *Guinness Book of Records* as the smallest forest fire ever." He chuckled. As often as Al seemed to complain about being on the pre-Clovis frontier, part of him clearly enjoyed it.

"When we realized what we had down there," he continued, "I flew in the best radiocarbon man I could find, Tom Stafford, so he could test it. He couldn't find any ^{14}C in it at all—it was radiocarbon-dead. So it has to be more than fifty thousand years old. Maybe a *lot* more."

The implications of that were staggering. If true, then people were in the New World more than 30,000 years before most archaeologists accept that they arrived. Were they ancestors of today's Indians? What was their lifestyle? There are few answers, I realized, looking down at a single flint core and a small smear of charcoal. But it is clear that Al will keep digging for more answers.

Finally, Al looked at me and said, "Now, if that twenty-thousand-year-old core higher up there put me out on a limb, then that one by the campfire put me out so far, I can't see the trunk anymore."

We both had a good laugh at that, and I understood completely. He finished with a classic Al Goodyear line, a fitting summary for all he has endured at Topper: "As far as the skeptics and my limb are concerned, when they heard fifty thousand years, they dropped their axes and fired up their chain saws!"

SAYING A QUICK GOOD-BYE

After I finished and as I was saying a heartfelt good-bye to Al and the crew, a crash-and-boom, full-scale South Carolina thunderstorm began. Along with the others, I hastily grabbed my gear and raced to my car, feeling a surge of thanks and relief that the weather had given me such a temporary two-day break at Topper. While there, I had barely seen the sun once, only dark rain clouds, but the rain had let up when I arrived and waited to resume until I left, giving me the time to make important discoveries.

STARTING THE TESTING

And the discoveries were not over. When I returned home and began to test the samples, I found major surprises. One regarded the radioactivity: As at our other Clovis test sites, there were major peaks of radioactivity in the Clovis layer in both test pits, and we expected that. But when I tested both the Upper and Lower Sites, I found a second, deeper peak in each one. They were unexpected, although our theory predicted a small chance that we might find them.

At that time, we did not know exactly what elements caused these deeper peaks, but based on NAA and PGAA tests just coming back from Murray Springs and Blackwater Draw, we saw that radioactive isotopes of the elements uranium, thorium, and potassium-40 most likely caused the spikes. Still, we could not explain exactly how they got there. One peak was at 13,000 years ago, but the other one was at about 34,000 to 44,000 years ago, at the same time that the radiocarbon calibration curve makes its wildest climb. That point is where the ^{14}C corrections go from nearly 0 years to 8,000 years, the largest correction in the record, as we saw in chapter 2.

As I stood there holding one of those mildly radioactive bags of Topper sand in my hand, it looked just like all the other bags. Yet if this information was correct, I was holding star-stuff in my hands—part of the ancient land surface that had been exposed to one of nature's most violent spectacles: a supernova. But could we be sure? That would remain a mystery until nearly the end.

GOOD NEWS FOR AL GOODYEAR

The tie-in for Al was that, according to our theory, the lowest radioactive levels should date to about 45,000 years ago. In the Lower Site, the lowest radioactive level is only a short distance above the pre-Clovis campfire, providing confirmation that the campfire was very old.

After I e-mailed Al the results of my tests, he was very happy and called to invite us to give a talk at the "Clovis in the Southeast" conference in Columbia, South Carolina, describing our theory along with what we had found at Topper.

During the same call, Al gave me quite a surprise. It had begun earlier at Topper when I described our theoretical details and mentioned that the hot, high-speed particles would have had dire consequences for the Clovis people, possibly causing a serious population decline. I told Al about the 1,000-year cultural gap at Blackwater Draw and Murray Springs as suggestive evidence for such a population decline in the Southwest, and I asked if he had seen such a decline in the Southeast. He said he had not and did not believe there had been one. Apparently, in the Southeast, well-dated sites provide good evidence for continuous occupation. While he admitted that the Clovis style of point had disappeared, he felt that the later styles of points were just an evolution of the Clovis style that had been made by the same people.

When he called me this time, he had found something new. As he wrote in a follow-up e-mail, "I've thought more about what you said about there being some lethal effect on people and animals at 13,000 years ago, and I'm noticing a big drop in the incidence of spear points dating from right after that time. Our conference here may be able to show a serious decline in human population during that time interval in the Southeast. You may well have an explanation for it."

As I read those words, once again my spine tingled. Earlier, as Al and I had stood looking down on the Clovis-era artifacts sticking out of the wall at Topper, I had only a few meager clues to the colossal tragedy that had unfolded there. During my time at Topper, I was annoyed by the intermittent rain and lightning, but after reading Al's e-mail, I thought about what those ancient people had to go through. When the Event happened, some of them may have been sitting around a chert core, making new points at Topper, or keeping the fire going to ward off the bitter cold. Suddenly, a violent rain of blazingly fast, searing-hot debris crashed down on them from the northern sky, instantly shredding the pines and undergrowth like shrapnel, dropping the mammoths where they stood, and ending the Clovis way of life. I shivered involuntarily, thinking, *I'll take a heavy South Carolina rainstorm over that any day.*

NEW OR CORROBORATING CLUES TO THE MYSTERY

- At the Topper site in South Carolina, the Clovis layer had a peak in magnetic particles, just as at all other Clovis sites.
- At the Topper site, the Clovis layer had a peak in radioactivity, just like all other Clovis sites.
- Both Upper and Lower test pits at the Topper site had a radioactive peak about 41,000 years ago, matching the ^{14}C anomaly.
- Black-glass mystery particles were found at the Topper site. They were almost exclusively in the Clovis layer.
- Al Goodyear provided evidence for a catastrophic population decline after Clovis times.

10

THE MYSTERY OF THE CAROLINA BAYS

BAY COUNTRY

I unbuckled my seat belt on the flight to Raleigh-Durham, North Carolina, and tugged my briefcase from under the seat in front of me. Pulling out a thick folder, I reviewed my photos of the Carolina Bays, some unusual lakes and swamps located, as you might guess, in the Carolinas. They are elliptical and huge, as you can see from figure 10.1, which shows two bays superimposed over Manhattan for comparison. Someone called them "bays" long ago not because they look like bays in the ocean, strangely enough, but rather because bay trees often grow in them. Most have a slightly raised rim,

Fig. 10.1. Most bays are huge, as you can see by comparing an actual 1.3-mile-long double bay from North Carolina with the Manhattan skyline. *Composite photo.* *© 2005. Royalty-Free/Corbis; used by permission*

and they are aligned with one another, with their long axes pointing toward the Great Lakes. Along the coastal plain bordering the Atlantic, there are tens of thousands, if not millions, of them, varying in size from about fifty feet across to nearly seven miles in length. I had only seen aerial photos of them, so I was eager to see them in person for the first time. From the aisle seat, I leaned over to see if any were visible through the window. Nothing—we were still flying above the thick gray clouds that stretched to the horizon.

DISCOVERING THE BAYS

After airplanes came into general use, a timber company began the first photographic aerial survey of South Carolina in the 1930s, and for the first time, photographers had a bird's-eye view of thousands of lakes and wetlands in North and South Carolina. The new view of the lakes called Carolina Bays was surprising—they looked very different from the air, as shown in figure 10.2.

The most peculiar thing the photographer noticed was that nearly all the bays had raised rims, and that they frequently intersected other bays, covering them like giant overlapping footprints, one on top of the other. No other known lakes, swamps, or bogs on the planet display these raised, overlapping rims. In addition, all the lakes appeared to be elliptical or oval-shaped, and surprisingly, they all showed an identical northwest-to-southeast alignment. Another oddity: most were very shallow, and the sandy rims were highest on the southwest end. Incredibly, these strange lakes covered two-thirds of the surface of one county in North Carolina.

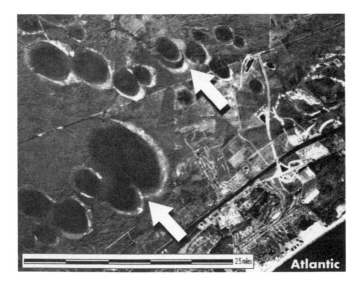

Fig. 10.2. This aerial photo of Myrtle Beach, South Carolina, shows clusters of overlapping bays (arrows), which usually appear on the ground as shallow-rimmed bogs. *Source: USGS*

The pilot and the photographer could not believe their eyes, since no one on the ground had ever noticed the peculiar bay formations. What force of nature could have created such massive elliptical lakes? The photographer thought no one would believe what they were seeing, but he had the proof in his camera—photos of tens of thousands of Carolina Bays.

Those photos instantly captured the public's interest, and newspapers carried sensational stories about the bays. Curious investigators spread out over the eastern seaboard and found that the bay lakes extended from Georgia to Virginia, although most are in the Carolinas. Later, other researchers discovered that there are occasional bays extending as far as Alabama, Florida, and New York. The total bay range is shown in figure 10.3. Altogether, bay lakes cover nearly 100,000 square miles of the Atlantic coast, an area larger than the combined size of North and South Carolina. Most bay researchers estimate that there are at least half a million of these lakes, though some estimates go as high as 2.5 million.

Throughout the book, we frequently use 3-D landscape images, called DEMs, short for *digital elevation model* (fig. 10.4). They are derived from U.S. Geological Survey (USGS) aerial photos and from space-shuttle measurements taken by NASA. Using DEMs makes it easier to see the bays clearly, since they remove the trees and underbrush that hide many bay features in conventional photographs.

For all the DEM images in the book, we increased the vertical contrast

Fig. 10.3. Bays range from Alabama to New York State. *Data from Prouty (1952)*

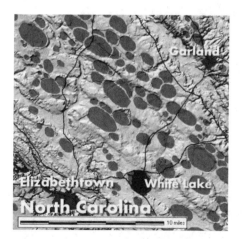

Fig. 10.4. DEM (digital elevation model) image of clusters of hundreds of Carolina Bays in Bladen County, North Carolina. Note that many bays overlap neighboring bays. *Source: USGS*

(elevation) multiple times to make it easier to see the subtle features of the bays. The resulting contrast makes it look like there are deep valleys in North Carolina; in reality, the area is quite flat. In addition, for many regular USGS aerial photos in this book, we selectively increased the image contrast to make the bay rims and other features more readily visible.

THE CONTROVERSY

After discovery of the bays, an immediate dispute began over how they might have formed—arguments that have not been resolved more than eighty years later. Some scientists theorized that powerful winds scooped the lake beds out of soft sand. Others proposed—seriously, it seems—that gigantic schools of spawning fish created them long ago when the area was beneath the waters of the Atlantic. The greatest controversy arose in 1933, when geologists Melton and Schriever proposed that the bays formed from multiple impacts by a giant iron or stone meteorite that broke up into thousands of pieces.

While the bays do look similar to craters on other planets, and though the impact theory fits many of the facts, the impact theory has several major weaknesses. First, no one could find meteorite fragments or other classically accepted signs of a meteorite impact, and if the bays were craters, they were far too shallow. In a typical meteorite impact found on Earth, if the crater is about eight miles across, it is usually about one mile deep. By contrast, bays that are eight miles long are only about fifty feet deep. Because of issues like this, many scientists rejected the impact theory in favor of terrestrial explanations, even though those do not explain the facts very well.

For years, the controversy raged between two camps of scientists, one supporting an extraterrestrial explanation and the other arguing for common terrestrial explanations, and neither side could disprove the claims of the other. In 1955, Frey, an early bay researcher, stated that there was "no adequate evidence substantiating any one particular theory to the exclusion of others." Forty-four years later, in 1999, bay researchers May and Warne said much the same thing: "More than 60 years of intense study and debate have yet to resolve the origin of the Carolina Bays." Unable to explain the formation of the bays, most researchers moved on to other subjects or retired years ago.

SEEING MY FIRST BAY

In addition to visiting Topper, I had come to the Carolinas because of the puzzle of the bays. Our theory offered tantalizing suggestions for how the bays may have formed. It built on earlier theories yet was somewhat different. I planned to collect sediment samples and search for magnetic grains, spherules, and high radioactivity, just as at other sites, and I hoped to find new clues to the formation of the bays, although I knew it was a long shot.

BLADEN LAKES COUNTRY

After getting my rental car, I drove to Lumberton, North Carolina, my base of operations for scouting Carolina Bay country. I picked this area because it is centrally located between a cluster of bays in the Bladen Lakes area and another cluster northeast of town. In fact, hundreds of bays surround Lumberton. They are nearly everywhere in that part of North Carolina—the town was even built over some of them.

Even though it was midafternoon, I was eager to see my first bay, so after checking in at a motel, I took off to the Bladen Lakes area, a state preserve containing a large cluster of bay lakes. Protected from development, these are some of the best-preserved bays in North Carolina. After slogging through the sloppy mud at Topper, I realized that the centers of the wet and soggy bays might be inaccessible, whereas I thought I could reach the higher and drier rims more easily.

I also reasoned that since all bays have rims, those rims might be a key part of the mystery of how the bays formed. For traditional meteorite craters, impact evidence is often contained in the rims, so I thought that if the bays were craters, then the rims might hold the answer. In my review of other researchers' work, I found that scientists had concentrated on the centers of the bays, and for reasons unknown to me, few had worked on the rims. I thought that perhaps I would find some new clues there.

After deciding to take samples only from the rims, I had to locate prospective bays with good rims, but I was limited to finding those with roadways cut through them, so that I could take samples without having to dig a lot. By carefully checking photos and the 3-D DEM files, I found about a dozen bays that looked interesting, and I typed their latitude and longitude coordinates into my GPS.

As I neared my first destination in Bladen County, the GPS screen counted down the distance. My excitement grew as I approached my first Carolina Bay. First, half a mile, then +950 feet . . . next +420 . . . then +240 . . . I was getting close . . . +110 . . . +65 . . . almost to my first bay . . . −35 . . . −140. *What!* I had driven past it.

Turning around, I saw absolutely nothing that looked like a Carolina

Bay. There was just the slightest hint of a very wide bump, maybe 150 feet wide and a foot high, that came from the woods on one side, crossed under the road, and disappeared into the woods. Thoroughly confused, I rescanned the area, and while the sand nearby was a little whiter than surrounding areas, other than that, there was nothing unusual. *This thing is a bay rim?* I was baffled. *It's almost invisible.* No wonder they couldn't see the things until they got into the air. *Oh, well,* I thought. *Maybe the next bay will be easier to see.*

It wasn't—and nor were the next eight. At each site, I was barely able to make out the faint outlines of the bay, and without the GPS, I never would have known.

As the sun began to set, I drove back to the motel feeling very disappointed. I wondered whether I would find even a single rim that was high enough to take samples. Was this a wasted trip?

That night, I was even more careful searching for rims with the DEM files on my laptop. The key, I decided, was to find the rims that were the tallest, steepest, and easiest to see. After a lot of searching, I found some likely prospects and entered them into the GPS. The next morning would tell if this was going to be a successful trip.

LOCATING A GOOD RIM

Just before setting out near daybreak, I recalled something unusual from the night before, that very white sand. In the literature on the bays, I had read that very often the southeast rims of the Carolina Bays are made of white sand. In fact, the only white sand that I had seen so far was in the rim areas that I had found with the GPS. That seemed more than a coincidence, since white sand is rare in the Carolinas. Except for the rims, all the sand that I had seen was either tan or reddish. Could the white sand have something to do with the formation of the bays? It was another clue.

Another clue about the sand came from a paper by Sharitz (2003), who mentioned that researchers had found a one-foot-thick white-sand layer beneath a thick organic layer at the bottom of Thunder Bay in South Carolina. The very deep layer was just above the tan hardpan-sand layer and just below the organic peat layer, suggesting that it had formed very early in the evolution of the bays, maybe at the very beginning. The black peat layer reminded me of the black mat at other Clovis sites. Others had reported fine whitish gray clay at the bottom of a large bay.

The case for a connection got stronger as my trip continued. Even though the tourist brochures touted "white-sand beaches," I could find no beaches that looked like those bay rims; the ones I visited were a light tan and not pure white like the bays. In fact, while driving many hundreds of miles across North and South Carolina and Georgia, the only white sand that I ever saw was in the rims of Carolina Bays. Even stranger was the fact that almost invariably, the white sand was thickest and sometimes only at the southeast end.

HOW TO MAKE SAND TURN WHITE

If the white sand was the product of bay genesis, then what could have caused tan sand to turn white? Two things typically do that. The first is chemical action, such as when acid etches quartz, removing the reddish iron-stained surface to reveal the white quartz sand beneath it. The other is very high temperature, usually above 1,500 degrees F, which burns off the reddish iron impurities from the surface of the quartz.

White sand and acid? Searing heat? Either of those solutions seemed difficult to believe, and the acid was the least plausible. High temperatures were more likely, since a huge explosion of some kind could have created the temperatures necessary to turn the sand white.

According to the impact theorists, that is what one would expect of an impactor arriving at a low angle from the northwest. The force of the impact would superheat the sand instantly, turn it white, and blast it out toward the southeast.

All that was still theoretical, and we lacked hard evidence either way. We thought, however, that no matter what had happened, the white sand was one clue that might help solve the riddle of the bays.

FINDING A RIM AT LAST

The next day, I used the GPS to locate the first few new bays, but I had no luck locating good rims. Then, as I rounded a curve in the road on the way to the next bay, I saw a large, promising bright-white sand rim up ahead. As the GPS counted down the distance, I headed right toward the northeast rim of Salters Lake, a Bladen County bay, shown in figure 10.5. As I arrived, I saw with great relief that I could get good samples from this one. I was ready to get started.

At all the previous Clovis-era sites that we had tested, we found a predictable pattern. First, there was a layer where Clovis and the megafauna still existed. Next, we found a layer from which the Clovis people and the megafauna were gone. The evidence for the Event appeared at the boundary

Fig. 10.5. The arrow points to the first rim I sampled at the northeast corner of Salters Lake. Notice the bright white rim sand, which is too nutrient-poor to support much vegetation. *Source: USGS*

of these two layers, as indicated by significant spikes in magnetic grains, spherules, and radioactivity. If the bays formed 13,000 years ago during the Event, then that same pattern should hold for them, and I would be able to find similar peaks near the tops of the bay rims.

ARCHAEOLOGY AND THE BAYS

Before I left for the Carolinas, I had wondered if the Carolina Bay archaeological record supported our theory. So far, all of the other Clovis-era sites I had visited involved water, and we reasoned that if the bays formed at or after the time the Clovis people disappeared, there should be little archaeological evidence inside the bays.

If the bays had been around prior to Clovis times, however, we would expect to find extensive evidence of the Paleo-Indians having used the water-filled bays. The thick layers of peat and aquatic fossils in today's bays show that they contained water often throughout their history, so if they were around prior to 13,000 years ago, they should have attracted now extinct animals and Paleo-Indians. Yet our research could find no reference to scientific excavations that had recovered from inside a bay even a single bone of a mammoth or other extinct megafauna. Al Goodyear indicated that he did not know of any either.

To check further about Carolina mammoths and Clovis artifacts, I contacted Dr. Mark Brooks, with the University of South Carolina, who has done a lot of research around the bays. Mark made much of his research available and confirmed that he knew of no mammoths that had ever been found in a bay. He has studied a small number of bays in South Carolina along the Savannah River, and he concludes that some of them formed long before 13,000 years, possibly as much as 100,000 years ago. He may well be correct, but some might not be true bays. Since Carolina lakes come in all sizes and shapes, some certainly would be older than 13,000 years. We maintain that not all round or even elliptical lakes are impact craters, just most of them that have distinct raised rims all around.

Regarding Clovis, he told me about one site where they found a single Clovis point in a bay rim that a construction company was demolishing. They had to do "crisis archaeology," as at Blackwater Draw, so it is hard to know the exact situation before they arrived, but Mark is confident that the point came from the lower portion of the rim, and he concludes that the Clovis people used that bay. But they found only a single point, which is not much evidence, nor is it completely clear whether Clovis people actually left it in the rim.

At another site, Mark and his colleagues (Grant et al., 1998) reported finding 1,700 Native American artifacts in and around Flamingo Bay, but only dating to hundreds of years after the Clovis era ended. They found no evidence that Clovis people had used the bay, in spite of scientific dating by Ivester and his coworkers (2002) showing that the bay rims have been around for 108,000 years. Many other Carolina Bays show only post-Clovis occupation, a fact that is easy to explain only if the bays did not exist when the Clovis people were alive.

There is only one other report from the South Carolina Paleo Point Database (Charles and Michie, 1992) about a Clovis spear in a private collection that was "found on a Carolina Bay," although it is unclear exactly what that means. Other than that, we found no references to Clovis points from bays, meaning that out of many thousands of points found along the Atlantic coast, only two seem weakly associated with a known bay.

More evidence for the lack of Clovis usage surfaced when we came across a scientific paper by Anderson and colleagues (1998), who had cataloged and mapped the distribution of Clovis points by county. When I looked at the distribution for North and South Carolina, I was startled to see that the counties with the most Carolina bay lakes have among the least number of Clovis points. This was completely opposite to all other known Clovis sites that were near water. The Clovis people favored lakes and streams because they attracted animals, and, of course, because they needed a daily supply of water themselves. It makes no logical sense that the areas with the most water should have the fewest people.

Although lack of evidence is not positive proof, the question to ask is what explanation best fits the facts. If most bays formed at the same moment that all the mammoths and Clovis people disappeared, then that would explain both the lack of points and the absence of mammoth bones.

COLLECTING THE SAND

Getting my gear, I was eager to test my first bay. On the cut bank where the rim was exposed, I scraped away the surface debris to uncover several layers, including two feet of pure white rim sand near the top.

Running a quick test with the magnet, I saw that the pattern matched our other sites and that the magnetic grains peaked in the white layer, as predicted. There was one other substantial peak below the upper one, but it was smaller.

At first, the multiple peaks puzzled us. Since we believe the Event created the bay rims within just a few seconds, however, this was to be expected. It is plausible that magnetic grains would spread throughout the rim, even though they might be higher in some areas, such as near the top, which is the case. This is because, according to our theory, the grains would have drifted down last and, therefore, would be near the top of the rim.

When I ran the test for hollow spherules, I found just one major peak of them in the entire rim, and it was near the top. It also matched the location of the largest peak of the magnetic grains. It is logical that the hollow spheres would be near the top, since no matter what they were, they are the lightest of all the particles.

TEN BAY RIMS

After Salters Lake, I drove across the state line to the next group of bays, near Marion, South Carolina. At one remote bay with no local name, which I called M31, I found that the magnetic grains and hollow spherule charts looked nearly identical to those for Salters Bay, even though they were fifty miles apart.

Altogether, I stopped at or drove past thousands of Carolina Bays, yet I found only ten suitable for taking samples. Those ten were dotted across the landscape from the middle of North Carolina into South Carolina, almost to the Georgia border. The most distant two were about 250 miles apart, and all of them covered a collective 1,500 square miles. In spite of that considerable separation, the test results for magnetic particles were remarkably similar for every bay. If the bays had formed at different times, the results should vary widely, but they do not. This evidence supports our theory that all the bays formed at one time from the same cause.

FOAMED BLACK GLASS

After driving south to near Marion, South Carolina, I followed my GPS coordinates to reach a promising bay, but when I turned onto the forest road, I had to stop suddenly. It had been raining off and on for days, and the road ahead of me was an almost continuous series of muddy puddles. Faced with miles of slippery, rough road, I turned the car around and headed back for the motel, discouraged about reaching a dead end. Surely, I could find some bays that were more accessible, I thought, resolving to recheck the maps.

Just then, that "luck" feeling hit me again. Against all logic, I had a hunch to keep going down that remote forest road, even though I risked becoming hopelessly stuck in the mud. Luckily, after about a mile of continued driving, the dirt road became good again, and the mud disappeared.

After about half an hour, I used my GPS to find the unnamed bay, which I designated M33, and I began digging away in the bank collecting samples. Suddenly, I heard an odd humming noise. Puzzled, I looked up over the bank, where, to my astonishment, I saw a golf cart driving through the trees beyond the bank. The golfers saw my head suddenly pop up over the top of the sandbank and were as surprised as I was, but they kept going. Scrambling up the loose sand of the bank, I saw among the trees one hole of a previously hidden golf course, which could be accessed by a different road. Apparently, the course architect thought that white sandy bay rims would make good golf traps, so he or she integrated them into the course. Undoubtedly, neither the architect nor the golfers realized how genuinely unusual those sand traps were.

That brings up one of the more disappointing things that I realized while I was there: the bays are disappearing at an unprecedented rate. Probably 60 to 80 percent of them have vanished, mostly in the last 100 years, casualties of our relentless expansion across the landscape with subdivisions, malls, and golf courses. Within another 100 years, the only bays left may be in the few state preserves that protect these unusual sites.

After the golfers played through at my bay rim, I returned to collecting samples. From top to bottom, I dug out several pounds of sand from each five-inch to ten-inch interval and put them into labeled bags. Then, while digging out a trowel full of sand about halfway down, as I emptied the sand into the bag, a small, dark, one-inch piece of something rolled off the trowel. When I picked it up, it was very light and was rough to the touch. After shaking the sand off it, I saw that it looked like a piece of brittle, black sponge (fig. 10.6). Thoroughly puzzled, I washed it with water from my drinking bottle to get a clearer look. It was full of holes like a sponge, and when I turned it in the sunlight, it glittered like glass. Trying to break off a small piece, I found it to be very hard.

Was it charcoal? I wondered. The intense heat of a firestorm (or more accurately, a conflagration) can do odd things to wood, including forming superheated explosive gases that can char an entire mature tree in seconds.

Was it obsidian, or volcanic glass? None that I have ever seen was full of holes like that. And there are no volcanoes in or near the Carolinas, so that seemed unlikely. I felt a surge of excitement when I considered the last option. What about meteorite impact glass? That was possible, since the glass did appear to be melted (fig. 10.7), but I had never seen meteoritic glass with as many large holes. This was the oddest stuff from any of our sites, and I sensed that it was a vital clue in our search for answers into what happened 13,000 years ago.

Wondering if there was more of it, I began sifting through the bank until I had collected a small handful of the stuff. Later, when I weighed it all, I found that it made up an amazing 2 percent of the sediment, meaning that every hundred handfuls of sand would produce about two handfuls of this amazing black glass. The distribution closely matched the magnetic particles, with a peak in the upper part of the rim, although the black glass was found in smaller quantities throughout the rim.

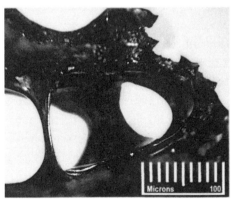

Fig. 10.6. Odd spongelike piece of black glass at about 50× magnification. It is about a half inch long.

Fig. 10.7. Another piece of black glass at about 90× magnification. It appears to have been melted, bubbled, and stretched.

Suddenly, it hit me that I had seen something similar at Topper—the pieces that I thought were charcoal. They were much smaller, but they had the same black, shiny look. Could it be that these things were in all the bays and at other sites as well? My mind raced with possibilities, questions, and eagerness to retest all the sites that I had visited already. This could be a major development.

As I sat there holding that unusual black glass, it suddenly dawned on me. A hunch had led me to that spot where I held a strange new clue in my hand. Less than an hour earlier, I had already turned the car around to leave, and if I had ignored that intuitive flash, I might never have stumbled on the mysterious black glass. It was a crucial breakthrough, as you will see later, that would help tie the formation of the bays to some well-dated 13,000-year-old Clovis sites.

MORE ON THE GLASS MYSTERY

When I returned home, I pulled out the boxes of sediment from all the other Clovis-era sites. There were tiny pieces of easily overlooked black glass from Gainey (in Michigan), Murray Springs (in Arizona), the Buck Lake site (near Edmonton, Alberta), and the Topper site (in South Carolina), and it was only in the Clovis layer, not above or below it. I could find none at Blackwater Draw (in New Mexico), St. Mary's Reservoir (near Calgary), or Lake Hind (in Manitoba). Altogether, the glass distribution spanned the continent from Arizona to Alberta to Michigan to the Carolinas, and it appeared to have fallen 13,000 years ago.

The only difference between previous test sites and the bays is that at the known Paleo-Indian sites, I found the black glass only in the very narrow Clovis layer. At the bays, it was all throughout the bay rims, although it invariably peaked near the top.

All of the evidence fits our theory that the rims and bays formed all at the same instant. In support of that, Ivester and coworkers (2003) dated two bay rims at 11,300 and 12,630 years ago using OSL (Optically Stimulated Luminescence), a technique for determining the last time sand was exposed to sunlight. We used the same technique to date two different levels of M33 Bay rim sand that included the black glass. Dr. James Feathers and Jack Johnson of the Luminescence Dating Laboratory at the University of Washington reported that the "highest age (11,400 ±6,100 years) is close to the age of Clovis, and given all the data, it is unlikely that the rim sediment tested can be older than Clovis." This means that, for the bays, the entire rim most likely dates to Clovis times. For a while, we struggled to unravel the mystery of the black glass, since there were so many puzzling aspects to it. If it was from an impact, how could we explain the unusually shallow craters of the bays? In addition, we were stumped by the absence of any pieces of meteorites. Could those glassy fragments be the only pieces that remained after the impact? A meteorite made of black glass? If so, it was a very odd object indeed.

To solve the problem, we needed help. Dr. Ted Bunch, the retired chief of exobiology at NASA's Ames Research Center and now with Northern Arizona University (NAU), had worked with us on the embedded mammoth grains, and he offered to look at the black glass. When I sent him a piece of it, he called me to say, "This is the darnedest stuff I have ever seen."

Ted tested the glass samples with the NAU microprobe, a device that can determine the precise elemental composition of unknown substances. When he finished, we were all astounded by the results. It was not regular silica glass; in fact, it had almost no silica in it at all. It contained carbon, oxygen, and very little of anything else. In Ted's opinion, it formed from material that came to this planet from somewhere else, and neither he nor we knew exactly what it was or how it had gotten here.

We ran more tests on the carbon glass. Those tests finally revealed the truth (which we will discuss more thoroughly in future chapters). In the end, with Ted's help in solving the riddle of the black glass, we believe we have finally solved the mystery of the bays and now know the last key details of the enigmatic Event.

NEW OR CORROBORATING CLUES TO THE MYSTERY

- There are possibly millions of Carolina Bays that look just like meteor craters.
- The raised rims of the bays overlap just like the rims of lunar and Martian impact craters.
- Some bay rims are made of pure white sand that is unlike any other sand in the Carolinas.
- The rims have Clovis-like peaks in magnetic grains, radioactivity, hollow spheres, and charcoal.
- There is almost no evidence indicating that the Carolina Bays existed prior to Clovis times.
- Exotic foamed black glass was found only in bay rims and in many other Clovis layers.

TAKING STOCK OF THE EVIDENCE

By this point, we had collected a lot of information, and as the full picture took shape, we began to look beyond pure science for more information. When we did, we found remarkable and widespread corroboration for our theory in the myths and legends from more than fifty diverse ancient cultures around the world. These cultures, most of which had no contact with one another, told astonishingly similar stories of an immense global catastrophe that shattered their way of life and nearly destroyed all their people. According to their stories, the disaster almost miraculously spared a few people, who were able to resettle the planet. Those survivors told and retold stories of the great calamity and passed them on to their descendants. Most important, those people were not just survivors; they were eyewitnesses as well, and their accounts describe many details about the Event that we can substantiate with hard scientific evidence.

By combining the legends with the scientific evidence, we think we now have an overall picture of what happened. That picture is still evolving, though, and undoubtedly some of our interpretations are incorrect. In spite of that, we think the broader details are accurate, with multiple lines of supporting evidence, although we might not be able to elucidate the finer details just yet. We still do not understand many things, so it may take years of discovering and analyzing more evidence to get the full picture.

In the next section of the book, we present our overall picture, just as we think it happened. As you will see, that time in history, from 41,000 to 13,000 years ago, truly was a time of trial by ice and fire.

PART TWO

THE MAIN EVENT

11

WHAT REALLY HAPPENED?

A COSMIC CATASTROPHE

The narrative that follows is based on the evidence we uncovered in part 1. Other details are drawn from descriptions by Hiroshima atomic-bomb victims, from evidence of the dinosaur extinctions, and from the collision of the Shoemaker-Levy comet with Jupiter. The rest is fact-based speculation, since we do not have all the answers yet for what happened. The entire narrative is possible and plausible, however.

[Authors' note: The following vividly describes an almost unimaginable cosmic catastrophe. If you want to skip the graphic details, just go to the summary at the end of the chapter.]

41,000 YEARS AGO

The Flash of Radiation

It was warm that night in Southeast Asia, too warm for a fire, but the tribe built one to roast the two deer they had killed that afternoon. At the edge of the rock shelter in the foothills of the mountains, some of the tribe members stretched out against rocks and others dozed while the meal cooked. They were tired; they had trotted for dozens of miles after trapping the deer.

Suddenly, high in the sky, a star exploded soundlessly. The colossal flash filled the night sky, shining brighter than the full moon. Instantly alert, some of the startled hunters grabbed their spears and jumped to their feet. As the shocked tribe looked up, the brilliant light in the sky wavered and shimmered. Larger now than the moon, it radiated glittering spikes of multicolored light in all directions. Around the frightened tribe, the nighttime landscape lit up under the eerie flickering glow, causing shadows to shift, flow, and merge as if demons crawled among the bushes. Fearful, the hunters edged backward toward the shelter.

The stellar rays damaged oxygen in the atmosphere to create ozone. Like the smell after a strong thunderstorm, ozone filled the air with its unmistakable and pungent odor.

Colliding with atoms of the atmosphere at nearly the speed of light, the deadly radiation launched cascades of particles that lit up the atmosphere with weird electrical effects. Multihued auroras shimmered across the Southeast Asian skies in unprecedented displays, and nearly nonstop lightning flickered from cloud to cloud or leaped earthward. Thunder rumbled nearly continuously as the tribe watched the strange lightning flash across the horizon.

Skilled as warriors, the men scanned the electrified sky looking for danger, but they confronted no earthly enemy—the true menace was beyond their senses. And it was unstoppable. For ten seconds, the intense radiation shot through their bodies and brains, and it was too much for some. A few dropped to the ground moaning; others quickly pulled the fallen ones into the shelter, but none of them would recover. The rest who were outside the cave moved quickly under the overhang, but most of them would not survive either. Because of the high altitude, the only ones that made it through were those in the cave who were shielded from the direct glare of the exploding star.

The Aftermath

After the initial flare, the light continued to grow brighter over the next few weeks, and it remained visible in the daytime for the next six months, appearing in the sky as a second sun or moon. When the supernova and the moon appeared together, there was no normal nighttime—the landscape was as brightly lit as a perpetual dawn. Gradually, the giant glow faded over many years, until it was no longer visible.

The radiation pulse was over quickly, but the effects were not. Europe, Africa, and the Americas were hidden from the supernova's direct glare and were mostly unaffected. However, in Southeast Asia and Australia, which bore the brunt of the exposure, the powerful rays sliced through leaves and flesh, damaging or killing animals and plant life all over the region. Afterward, temperatures plunged, putting even more stress on the survivors. As vegetation wilted and turned brown across vast areas of Australia, the unstable weather touched off frequent lightning strikes, setting fires that raged unchecked across the devastated environment. Besides creating ozone, the barrage of radiation also formed nitrates and cooked up a brew of other chemicals that rose into the sky, so that most of Australia and

Southeast Asia was covered by a lingering thick pall of dark smoke and brown chemical haze.

The radiation was lethal for many animals, but those that survived were not much better off. Lack of food doomed many, and before long, nearly all Australia's largest animals disappeared, including huge kangaroos and giant bearlike animals called wombats. Smaller animals were affected too, and soon after the massive blast of cosmic rays, many species of lizards, snails, fish, birds, and frogs disappeared forever. Australia suffered through a major extinction, and by the time it was over, almost no animal weighing more than a few hundred pounds was left alive. Most of the rest of the planet was spared for a while.

The surviving people did not do well either. Within the first week or two, the radiation sickness began—nausea, lack of appetite, apathy, and hair loss. Population levels plummeted as sterility rose, and even when there were births, genetic damage was high.

Most mutations were deleterious, with a high mortality rate, but other mutations were beneficial. From that great catastrophe, the survivors brought new genetic variations to the human gene pool, and after that time, human brain size began to grow larger. Gradually, from then on for thousands of years, a flurry of intense human progress occurred, as complex speech, new hunting technologies, art, and music appeared or expanded and went through rapid change.

It was also a time of increasing cosmic impacts. The pulse of radiation illuminated comets and asteroids orbiting the solar system, and when it struck the supernova-facing side of comets, the heat was intense enough to vaporize small amounts of their frozen compounds. As jets of gases escaped from the surface of giant comets and asteroids, they gently nudged these bodies into Earth-crossing orbits. Over time, the Earth and moon suffered increasing levels of bombardment from space. Then, after thousands of years, the barrage dwindled, and conditions on Earth slowly returned to normal. But there was more to come.

34,000 YEARS AGO

The Shock Wave

Unlike the initial explosion of the star, the arrival of the first supernova shock wave went unnoticed by those on Earth. There were no fireworks, no flashes in the sky—nothing to announce it. Even so, the countdown to trouble started when the leading edge of the wave passed silently across the border of the solar system.

Across the Northern Hemisphere, inhabitants may have noticed more "fireflies" in the sky. The profusion of shooting stars striking the upper atmosphere was the result of the rain of cosmic ions and grains, rich in iron and titanium along with radioactive thorium, potassium, and uranium. The native people interpreted the shooting stars as a bad omen, and they were right. The flux of particles damaged the ozone layer, and although no one noticed, levels of harmful radiation began to creep higher. Over time, ultraviolet light and increasing levels of silent cosmic rays began to injure plants and animals, especially the weaker ones.

Occasionally, the hypervelocity cloud kicked loose new waves of comets and asteroids, which, along with the dust and debris in the fast-moving cloud, produced impacts on Earth or aerial bursts in the sky. Most of the incoming debris was small and did little damage, but sometimes the objects were large enough to penetrate the atmosphere. Slamming into the ground, the oceans, or the ice caps, they exploded violently, sending out high-velocity microscopic shrapnel that was lethal to animals, plants, and people. Sometimes waves of debris collided with the Earth, producing multiple explosions, which devastated the herds of mammoths, bison, and horses beneath. Big Ed, the mammoth, was in one of those herds. The incoming grains were high in toxic metals and radioactive elements, so that some animal species began to suffer from radiation sickness and heavy-metal poisoning. The second round of extinctions was under way.

16,000 TO 13,000 YEARS AGO

The Debris Wave

As the third wave of supernova star-stuff moved into our system 16,000 years ago, its touch was mostly lighter than dust in the wind, yet that was just enough pressure to nudge giant comets into new paths. Some of those dust-and-ice giants slowly swung around toward the inner system to take aim at Earth. Although they were fluffier than newly fallen snow, they were far from harmless. Over time, firmly in the grip of the sun's gravity, some of those enormous dustballs headed straight toward Earth at 70,000 miles per hour.

In addition, the incoming supernova cloud was not uniform. There were lethal lumps in it—thick clumps in the wave that had pulled together due to the cloud's weak though persistent gravity. Even though lighter than Styrofoam, some lumps were many miles wide. There were tens of thousands of them, and many were headed directly toward our planet.

The Impact

Around noontime on a cold late-winter day about 13,000 years ago, a band of a dozen fur-clad Clovis hunters trotted effortlessly and silently across the icy countryside of the American Midwest. Off in the distance to the north, jagged half-mile-high cliffs of blue-green ice towered above the frigid landscape; near them, swiftly moving milky glacial streams cut across the land, making cross-country travel difficult.

Life was harsh for those Paleo-Indian hunters along the edge of the ice, where cold high-speed winds howled across the landscape and where only the hardiest plants and animals survived. There were only open stands of sparse, stunted pine and spruce, carpeted in between with patchy mosses, lichen, and tough grasses. Shallow, ice-glazed pools and lakes dotted the landscape, as small herds of mammoths grazed the open grassy areas and mastodons roamed the thin forests. Along with them, there were oversized predators—saber-toothed tigers and packs of dire wolves.

The People, as they called themselves, gathered whatever they could find for food. Mostly they hunted smaller animals, but sometimes they stalked mammoths. That day, the tribe was not thinking very much about mammoths. They were uneasy about the bad omens in the sky. Over the last few days, several stars had been growing larger and brighter, until finally they could be seen even in the daytime. It was the duty of the tribe's shaman to watch the stars carefully for signs indicating changes in the weather, but he had never seen stars grow like that. The shaman did not consider it a good sign for the hunt, and he was right. Their world was about to change.

Yelling in alarm, one of the men pointed a trembling finger at the crescent moon. As the group stopped and stared at the sky, a huge, silent fireball engulfed the moon's dark upper-left limb. The explosion was brief but violent. Convinced that the gods were angry, the men and women ran hastily toward shelter in a series of small caves cut into a rock overhang. It was a critical choice that shielded them from the worst of the trouble that was racing toward them. At that moment, the densest part of the debris cloud of dustball comets began to collide with the solar system. Angling in from the northern sky, the lumpy cloud of cosmic debris slammed sideways into the plane of the planets. The tribe could not see it, but enormous silent explosions were ripping across Mars, Venus, and the other planets and their moons; the worst of the system-wide onslaught continued for many hours.

At the same time, although invisible to those hunters, giant cosmic dustballs were plunging soundlessly into the flaming body of the sun. Instantly, writhing plumes of sun-stuff exploded as massive flares, some of them heading directly toward the Earth. Within minutes, high-energy solar particles were buffeting the upper atmosphere, sending colorful auroras twisting and arcing across the daytime sky.

The shaman had never seen these sky-signs before, but, trying to hold down his rising fear, he knew they were bad. As the tribe nervously huddled in the rock shelter, he hurriedly recited incantations and tried to intercede with the gods, but it was too late.

Scrambling backward like a crab, one of the tribesmen near the outside shouted in terror as he tried to squeeze farther under the overhang. Wedging his way inside, he repeatedly stabbed a finger at the sky, where an array of glowing blue-white comets now stretched from horizon to horizon. Growing larger every second, they streaked into the atmosphere, each one lighting up brighter than the sun. One dustball comet was more than 300 miles wide; others were nearly as big. Shimmering fiercely, the largest fireball was too brilliant to watch. It cast shadows that shifted rapidly across the back of the cave as the comet moved, forming silhouettes of the tribe that twisted over the rock wall like drunken dancers.

Heated to immense temperatures by its passage through the atmosphere, the lethal swarm exploded into thousands of mountain-sized chunks and clouds of streaming icy dust. The smaller pieces blew up high in the atmosphere, creating multiple detonations that turned the sky orange and red across the horizon. Beyond sight of the tribe, the largest comet crashed into the ice sheet, instantly blasting a gigantic hole through the ice into Hudson Bay. Within moments, other comets exploded over Lake Michigan, northern Canada, Siberia, and Europe, as every northern continent took direct hits.

The dazzling flash of the nearest explosion temporarily blinded some of the tribe. Others closed their eyes tight, but even then, the fierce light stung their eyes. Some held their hands up as shields, but the intense light illuminated their flesh with an orange glow, eerily outlining the bones in their hands. Most of them curled up and buried their heads to shut out the painful brilliance, but their skin began to blister from the intense glare and heat. Hastily, they pulled cloaks and blankets over their heads, mumbling frantic prayers to the angry gods.

When the ground shock waves from the impacts arrived, the earth shook violently for a full ten minutes in great rolling waves and shudders. Rocks broke loose, clattering down from the outside cliff face, and fell from the roof over their heads. Screaming in pain, some people were injured by the falling rocks. Nearby trees shook and swayed violently before toppling over. Short, narrow fissures opened and snaked across the rocky field in front of them. The stream beyond the caves cascaded into the newly opened cracks and disappeared. Choking clouds of dust rose to obscure the view, as the People huddled in their shelter, certain that they were going to die.

The Blast Wave and the Bubble

Within seconds after the impact, the blast of superheated air expanded outward at more than 1,000 miles per hour, racing across the landscape, tearing trees from the ground and tossing them into the air, ripping rocks from mountainsides, and flash-scorching plants, animals, and the earth. Nearly the only living things to survive close to the blast had hidden underground, underwater, behind hills, or inside some other type of shelter. Millions of unlucky animals were out in the open.

By the time the blast wave passed over the Clovis tribe, it had slowed, but it still was traveling four times faster than tornado winds. It shook the ground just as the earthquakes had done. Fiery hot, windborne sand and gravel pelted the walls of their hiding place and ricocheted around them like bullets. Within moments, the flying debris cascaded off the outside cliff face and piled up around the opening of the shelter to the height of a man, almost closing the entrance and sealing them in. That probably saved their lives.

Across upper parts of North America and Europe, the immense energy from the multiple impacts blew a series of ever-widening giant overlapping bubbles that pushed aside the atmosphere to create a near vacuum inside. After the outer edge of the closest bubble passed over the Clovis band, the wind speed dropped, and air pressure fell precipitously, making it difficult for the survivors to breathe the thin air. In the back of their cave, the People gasped for breath as their bodies became starved for oxygen. Each labored breath of the superhot dusty air seared their lungs.

Behind the expanding edge of the bubble, Earth was stripped of the protective shield of the atmosphere and was at the mercy of a different barrage. The enormous blast ejected tiny, fast-moving grains in all directions through the thin air. Some went sideways to lodge in trees, plants, and animals; others went up and came back down. With almost no atmosphere to slow them, they fell faster and faster, hitting Earth at hundreds of miles per hour. At the same instant, large atoms blown out of the turbulent sun and high-speed galactic cosmic rays streamed unhindered through the empty bubbles. Traveling at several percent of the speed of light, the radiation bombarded the planet like microscopic bullets, speeding deep into exposed flesh and bone.

In their headlong retreat toward shelter, some of the tribe had dropped their spears. As they stared out in disbelief, the invisible particle bombardment made their abandoned spears twitch and vibrate on the ground. Millions of high-speed microscopic grains peppered the exposed flint surfaces and blew tiny holes in the wood. At the same time, tiny speeding pellets ripped unseen through trees and plants, shredding and stripping off leaves and twigs, which danced around eerily on the ground, as if demons were trying to pick them up.

The particle onslaught struck mammoths and larger animals that could not find shelter, lodging in their tusks or horns and sinking deep into their eyes and flesh. Some animals stampeded in terror, while others dropped where they stood, unaware of the invisible particles that had hit them.

Before long, the outward push of the shock wave slowed and stopped, and then the vacuum began to draw the air backward. As the expanded atmosphere rushed back toward the impact site, the bubbles collapsed, sucking white-hot gases and dust inward at tornado speeds and then channeling them up and away from the ground. Climbing high above the atmosphere, some of the rising debris escaped Earth's gravity to shoot far out into space like a dusty geyser, while the rest flowed out as a reddish brown mushroom cloud that flattened out for thousands of miles across the upper atmosphere. As the expanding cloud blocked the sun, darkness engulfed the areas near the impacts.

Some of the dust and debris lifted by the powerful updraft was too heavy for the atmosphere and began drifting and crashing back to earth. Still superhot from the blast, it gave off a powerful lavalike glow, reheating the still smoldering ground and air as it fell. In places, temperatures, which had fallen after the fireball passed, quickly rose dozens of degrees again. Landing on top of the continental ice sheet, the hot particles melted holes and pits into the top of the ice. Suddenly liberated, meltwater coursed off the ice sheet in all directions.

The raging updraft through the hollow bubbles created an equally powerful downdraft of frigid high-altitude air, traveling at hundreds of miles per hour. With temperatures exceeding 150 degrees F below zero, the downward stream of air hit the ground and radiated out from the blast site in all directions, flash-freezing within seconds everything it touched. Some of the animals that had survived the initial fiery shock wave froze where they stood, while others survived for only a few minutes more. The howling, frigid blast turned trees and plants into brittle ice statues. The rapid temperature fluctuations meant the end for millions of plant and animals—and it was not over yet.

The Earth Shakes and Burns

The impacts and shock waves triggered enormous earthquakes along various fault lines from the Carolinas to California and shook some dormant volcanoes awake in Iceland and along the Pacific coast. Erupting with furious activity, they spewed hot lava across the landscape, releasing dust, sulfur, and noxious chemicals into the atmosphere and adding to the already heavy cloud cover.

The impacts, the blast wave, and the eruptions started thousands of ground fires wherever there was enough fuel to feed the flames. In some cases, the fires went out quickly, as the high-speed winds and oxygen-poor air snuffed them out. Other parts of the landscape, tinder dry from the winter freeze, burned with fierce intensity for days following the impact. Fast-moving, wind-driven wildfires formed spiraling tongues of raging flames that twisted for thousands of feet into the air, and the wild inferno raced through the forests faster than birds and animals could flee. The roar of the fires shook the ground, and the fierce heat blew apart trees like bombs, exploded rocks like shrapnel grenades, and set off steam explosions whenever the fast-moving fire front jumped across frozen ponds and streams. When the fires had finally burned themselves out, there was little left besides smoldering stumps and telltale charcoal strewn across the continents.

The cold climate supported more sparse grass than trees, and the extensive Ice Age grasslands burned intensely and quickly. Afterward, the ash from the grass fires washed away in the steady rains, leaving little evidence behind. As those grasslands went up in smoke, the main food supply of millions of mammoths, horses, camels, and bison disappeared within minutes to hours, leaving hundreds of thousands of square miles stripped of plant food and covered with black charcoal and white ash. For many parts of the north, as far as the eye could see, the ground was either burning or already black and charred.

The eruptions and continental fires lifted additional tons of ash and soot into the atmosphere, further darkening the sky. Along with that dust, millions of tons of dangerous cometary chemicals drifted high up into the sky, only to float back to earth later. In some places, the air was too toxic and oxygen-depleted to support life.

Earth's Magnetic Field Flickers

Shocked by the thunderous impacts, Earth's magnetic field flickered briefly, causing the magnetic poles to wander crazily across the planet. The north magnetic pole briefly approached the equator before it recovered. As the field wavered and weakened, Earth became even more exposed to incoming cosmic rays.

With the magnetic field oscillating, animals that navigated by the field became lost and confused. Unable to find their way, tens of thousands of turtles, whales, and porpoises beached on shorelines and became stranded. Millions of birds, trying to flee the explosions, used the wavering field to navigate in the wrong direction and perished.

The bombardment of the sun continued to create massive explosions that blasted solar material toward the Earth and the moon in seemingly endless waves. With less protection from our magnetic field, life on Earth was pummeled by nearly continuous deadly solar radiation.

The Carolina Bays

In the split second that the bottom half of the giant dustball-comet crashed into the ice of Hudson Bay, it vaporized on impact and exploded upward violently, shattering the upper portion of the impactor and spewing pieces of the comet across the continent. At the same instant, the impact blew apart nearly 200,000 cubic miles of the glacier, sending the icy debris hurtling through the air or skipping across the landscape. Arcing swiftly away from the multiple impact sites, a rain of incandescent debris and chunks of steaming ice showered down across most of North America, Europe, and Asia.

People and animals many hundreds of miles away from the north saw the bright flare of the massive explosions and felt the ground shake. Those that looked up saw the incoming sizzling clouds of debris hurtling toward them through the daytime sky in total silence. The dangerous chunks traveled far faster than the speed of sound, so no one heard them coming.

Within minutes, the massive low-flying lumps crashed into the Carolinas and the eastern seaboard, exploding into fireballs and gouging out the Carolina Bays. In some areas along the coast, the thickest ice bombs leveled nearly all the landscape for hundreds of miles and torched entire forests. Other giant flying lumps exploded to form shallow craters across wide stretches of Nebraska, Kansas, Texas, Oklahoma, Arizona, and New Mexico. Within thirty minutes of impact, some ice had fallen as far away as California and Mexico, more than 2,000 miles away.

Pieces of flying ice and debris both large and small fell on nearly every section of the continent from the Atlantic to the Pacific and from the Gulf of Mexico to the Arctic Ocean. A similar barrage blanketed large parts of Europe and Asia, and some pieces reached as far as Africa and the rain forests of South America, although Australia was spared. More than one-quarter of the planet was under siege.

The dustball-comets were high in carbon, and so was the burning vegetation. The massive explosions melted the carbon and lifted it high into the atmosphere or carried it in the flying chunks landing across the Northern Hemisphere. As it came back down, it littered the landscape with millions of tons of small chunks of black glass, carbon spherules, and fine carbon dust. Trapped within the glass were minute particles of star-stuff, with chemistry unlike anything on this Earth.

Surging Ice

As the largest dustball-comet collided with the North American glacier in Hudson Bay and blew a hole completely through the ice sheet, it sent high-velocity meltwater surging under the ice. At the same instant, it cracked the ice and sent it skidding out through Hudson Strait, releasing hundreds of thousands of fractured ice chunks into the North Atlantic as icebergs. Caught by the powerful ocean currents, some of them eventually drifted as far away as Europe, Africa, and Florida.

Along the southern edge of the ice sheet in North America, fires were the greatest danger, but the ice was a serious problem by itself. The meltwater surges lifted and floated large sections of the ice, causing monolithic ice blocks to slide southward along hundreds of miles of the ice front. Moving nearly as fast as a horse can run, the blocks plowed over the forests, shearing trees off at stump level and burying meadows that were close to the ice sheet.

The surging meltwater sluiced through the soft sediment under the glaciers, carving hundreds of thousands of spindle-shaped drumlins across three continents. Jolting ahead like a relentless bulldozer, the ice pushed up huge piles of glacial rock and gravel moraines, until the meltwater finally subsided and the wild ice ride was over.

Slides and Tsunamis

Sailing through the air, thousands of ice chunks and clouds of slushy water splashed down into the Atlantic, exploding with colossal detonations. The multiple concussions triggered immense underwater landslides along the continental shelf of North Carolina and Virginia, releasing thousands of cubic miles of mud. In turn, that unleashed 1,000-foot-high tidal waves that raced away from the coast at 500 miles an hour. Aimed out to sea, the waves had little effect on the North American coast, but they were headed straight for Europe and Africa.

Things were quiet in Europe and North Africa for a while after the last of the impacts, but the damage was extensive, and no part of the continents was spared. Tens of millions of animals lost their lives all across those continents, and large areas of vegetation were flattened or still burning. Europe seemed safe in the aftermath.

Some people had survived the impacts in Europe, and nine hours later, a few survivors in Ireland came out of hiding to forage for food along on the seacoast. The only warning of trouble came just minutes in advance, when the tide withdrew at an alarming rate, suddenly exposing hundreds to thousands of feet of offshore mudflats. Startled and fearful, the people turned and ran, but it was too late.

Nearly 100 feet high and moving at 400 miles per hour, 1,000-mile-long megawaves suddenly rose up from the ocean to surge across the shorelines of Europe and Africa. Rushing hundreds of miles inland beyond the coasts, they devastated everything in their path and obliterated all remaining human coastal settlements. Nearly everyone living along the shores of western Europe and North Africa perished instantly.

With its momentum spent, the churning water paused briefly and began its rush backward to the coast. As it did, the swirling muddy water carried with it rocks, smashed trees, and the battered remains of plants and animals, pulling them all back into the Atlantic and the Mediterranean.

During this surge, the immense force of the crashing waves triggered offshore African slides, sending a second round of megawaves racing back toward North and South America. The people in the Americas suffered little damage from the initial tsunamis, but they were not so lucky with the ones returning from Europe. This time, the monstrous waves caught unsuspecting survivors on the Atlantic coastlines of both continents. Hundreds of people disappeared under the churning 100-foot waves that rolled miles inland across what had once been fertile lowlands.

This second round of giant waves triggered the largest slides off the mouth of the Amazon River in South America. A third round of mountainous tsunamis raced back off to smash into North America, Africa, and Europe once more. This time, it mattered little—no one was left on the coasts to see the waves coming. For more than a day, the reflecting waves ricocheted back and forth across the Atlantic, growing smaller with each transit. Finally, with their energy spent, the deadly waves faded away.

If there had been anyone still alive along any coastline around the Atlantic, they would have seen a startling sight in some places: burning water. The churning waves and underwater slides exposed giant deposits of frozen methane off North America and Europe, and after the pressure was removed, the methane flashed into gas and bubbled energetically to the surface. The firestorm and falling hot rocks and particles ignited some of the rising gas plumes, so that here and there miles-long tongues of orange-blue flames flickered and danced across the sea surface. For weeks, the sea burned or boiled with escaping methane.

Rain and Snow

Within minutes after the impacts, the subzero air and the rising water vapor combined to produce heavy snow and sleet as the supersaturated atmosphere dropped its burden, causing snow to fall as far away as Mexico, the Caribbean, and North Africa. Gradually, in the south, the snow turned to rain, which continued day after day, stretching into weeks, and then into months. It did not fall heavily, as in thunderstorms, but steadily as a slow drizzle. Rivers and streams swelled beyond flood stage and remained there for months.

You might think that the rain was good—a cleansing rain to put out the fires and wash the land clean, but there was a dark side. Millions of tons of noxious chemicals fell with the rain. Combining with hydrogen, they formed a toxic brew of acid rain—hydrochloric, sulfuric, nitric, and carbonic acids. In places, they etched the rocks, ate holes through leaves, and burned the flesh of living animals.

Even though the surviving people found shelter from the blistering rain, they still had to drink. But the water was laced with acid and traces of arsenic, formaldehyde, cyanide, and toxic metals—not enough to kill the strongest ones, but enough to make them ill.

Meltwater Flooding

All the combined heat turned millions of tons of ice into water and sent it coursing off the continental glaciers to pool in the existing glacial lakes. As the lakes quickly filled and overflowed, their ice dams failed, sending monstrous floods roaring on to the next lake. As each flood poured into the lake below it, that lake's dam failed in turn, creating a cascade of ever-growing floodwaters raging toward the Mississippi, into the St. Lawrence, and out through all other outlets into the oceans. In what is now Washington state, a dam broke in a huge glacial lake, sending floodwaters more than 800 feet deep cascading to the Pacific through narrow clefts in the mountains, sweeping away topsoil, trees, plants, people, and animals, and etching giant grooves into the solid rock walls.

From every existing river and stream, frigid freshwater flooded into the Atlantic, Pacific, and Arctic Oceans. The sudden pulse of melted ice caused sea levels to rise many feet within a few weeks, inundating the lowlands around the world. Relentlessly, day by day, thousands and then millions of square miles of once verdant grasslands and forests slipped beneath the rising ocean waves, never to be seen again.

The Gulf Stream and the Climate

When the underwater slides began, thousands of cubic miles of mud, sand, and gravel slid right across the path of the Gulf Stream and the other deep ocean currents that make up what is called the "ocean conveyor" in the Atlantic. The immense force of the slides rerouted the conveyor sideways, disrupting its flow.

Simultaneously, near-freezing freshwater from the floods acted like a lid on the North Atlantic, slowing down the conveyor and then stopping it completely. With the North Atlantic deprived of warm water, within days temperatures began to fall around the North Atlantic from the northeastern United States to Canada and on to Europe. The stalled conveyor, coupled with sunlight-blocking dust and clouds, was too much for the climate to overcome. Within days or weeks after the impacts, continental temperatures rapidly fell to well below freezing, and a brutal Ice Age chill once again spread across the land. Temperatures remained low for more than 1,000 years during a time called the Younger Dryas.

Algal Blooms and the Black Mat

By the time the waves grew quiet and the raging winds subsided, additional tens of millions of animals had perished, including most of the people living in the Northern Hemisphere. Slowly, the survivors struggled to return to a normal existence. Within months, the landscape began to stabilize, but it still looked like a barren, blackened moonscape in many areas. The impacts, firestorms, and tidal waves had devastated large tracts of the most fertile land along the oceans and river tributaries.

Even though the worst had passed, the trouble was not over. It was still too dark and cold for most plants to grow well. The surviving animals had little to eat, and starvation overcame many of them during the next few months.

A few plants did very well during the troubled times, however. Known as "disaster species," some of them were the most primitive plants, which could live well under difficult conditions that hampered other forms of life. One such species was freshwater algae, which underwent explosive growth. With almost no remaining predators to keep them in check, the algae feasted on the rich mix of chemicals in the environment, some of which are fertilizer for algae. As other plants and animals decayed, they freed iron, nitrogen, sulfur, and other nutrients that spurred on the algae to ever-more-frenzied growth. Dense blue-green mats of algae choked the ponds, streams, and rivers, and as the algae died, they drifted down to clog the bottoms. All over the continents, thick black mats formed. Their explosive growth sometimes filled the lakes and ponds with deadly nerve toxins, which unsuspecting thirsty animals drank in large quantities. Within hours, many animals died alongside the seemingly life-giving waters.

For nearly 1,000 years, algae controlled the lakes and waterways, until their abundant food supply eventually dwindled and their predators returned. Finally, balance was restored, and the algal blooms ceased.

A New Beginning

In spite of all that had happened, most of the Clovis tribe survived in their rock overhang. Across the continent, most people were gone, but a few other small bands made it through. There was no pattern to the survival—the tragedy skipped over some people to strike those standing next to them. The survivors could only believe that the hand of the Creator had spared them that day, but they still faced a devastated world.

In Australia, Africa, and South America, far away from the northern blasts, most people made it through, even though the Event altered the global climate. Over the course of the next 1,000 years, they gradually migrated to fill the void left by the incredible catastrophe that had overtaken the planet.

The surviving animals struggled out of hiding to scratch around for food, and hardy seeds began to germinate again. The animals that fared best were small, and they were mostly omnivores or scavengers, able to feast on the huge banquet of animal and plant remains. The large plant-eaters fared worst in the aftermath; they had the largest appetites and the hardest time satisfying them. Being big, they were also the most visible targets for the hungry human survivors. Almost all conditions were against them, so before long the remaining megafauna in the north dwindled and disappeared.

By the time the effects of the catastrophe had subsided —maybe months, years, or decades in total—the collision had irreversibly altered the planet. The old Earth was gone, destroyed by an exploding star's invisible radiation, by white-hot fire from the sky, and by floods raging across the Earth. A new world was born from the mud and ashes.

The new children who entered the world to replace the missing of course knew nothing of the Event. Their parents were determined, however, that their offspring should know about the Great World Fire and the Great Flood, and that they remember to honor the Creator who had spared them. Grandfathers and grandmothers told stories of the Event to their grandchildren, who passed them on to their own children. For many generations, when children heard stories of the Event for the first time, they sat wide-eyed and enthralled. They were amazed that such things could occur.

Eventually, many generations passed, and the children came to doubt the old stories. Surely, they thought, the Old Ones just made them all up. These terrible things could never really have happened.

SUMMARY OF THE MAIN EVENT

41,000 years ago a supernova exploded close to Earth.

- The burst of radiation caused widespread extinctions in Australia and Southeast Asia.
- Much of the human race perished in and near Southeast Asia.
- Human genetic mutations led to larger brain size, fostering art, music, and a burst of creativity.
- Being shielded from the explosion, the other continents were affected very little.
- For about six months, the supernova was bright enough to be a second sun or moon.

34,000 years ago the first shock wave of the supernova buffeted the Earth.

- Radiation increased and small ions and particles bombarded Earth.
- There also were increased comet and asteroid impacts.

16,000 years ago the second shock wave of the supernova arrived.

- As with the first shock wave, radiation increased and small ions and particles bombarded Earth.
- As well, there were increased comet and asteroid impacts.

13,000 years ago multiple impacts of cometlike objects hit the Northern Hemisphere.

IMPACT-RELATED EFFECTS:

- Similar objects bombarded Mars, the moon, the sun, and probably the other planets.
- The impacts touched off a series of major solar flares.
- The blast waves spread across North America, Europe, and northern Asia.

- The explosions launched ejecta across the Northern Hemisphere, carving the Carolina Bays and other basins.
- The shock of the impacts triggered major earthquakes and volcanic eruptions.
- The intense heat caused continent-wide firestorms.

WATER- AND ICE-RELATED EFFECTS:

- Hundreds of thousands of cubic miles of ice vaporized, collapsing parts of the ice sheets.
- Fast-moving subglacial meltwater surges carved out millions of drumlins.
- Rapid melting of the ice caps raised sea levels quickly, inundating coastlines worldwide.
- The explosions triggered nearly a dozen massive underwater slides.
- The slides produced enormous tsunamis that ravaged coastlines around the Atlantic.

CLIMATE-RELATED EFFECTS:

- The underwater slides released frozen methane, a climate-altering gas.
- The impacts produced vast amounts of water vapor that fell as rain and snow for weeks.
- The fires produced carbon dioxide, a climate-altering gas.
- The slides and meltwater flooding halted the ocean conveyor in the Atlantic.
- The combination of climate effects triggered a 1,400-year return to the Ice Age during the Younger Dryas, followed by rapid warming.
- The water vapor, soot, and debris in the atmosphere created a long-lasting thick cloud cover.

BIOSYSTEM-RELATED EFFECTS:

- Darkness, cold, and fires destroyed most vegetation in the Northern Hemisphere.
- Millions of animals perished from combined causes. Large-sized species went extinct.
- Disaster species flourished. Algae underwent explosive growth, producing the black mat.
- Human populations began to grow explosively.

PART THREE

THE EVIDENCE

12

SCIENTISTS AND STORYTELLERS

MORE ANSWERS

It is overwhelmingly clear that a mass extinction took place between about 41,000 and 13,000 years ago. In part 1, we followed the clues to uncover what happened, and in part 2, we presented our overall theory about what probably transpired. In this section, part 3, we will look at the evidence in more detail. In addition to reviewing the scientific research by our group and by others, we will hear from eyewitnesses to the catastrophes—the Native Americans, the ancient Greeks and Persians, and other ancient people who lived through them. Of course, the witnesses are long gone, but we still have handed-down versions of their stories to describe what these people saw.

THE EYEWITNESSES

Presently we will hear from two Native American tribes who were eyewitnesses, but first, let's put their stories in context.

Some cultural researchers suggest that all the early peoples, including Native Americans, greatly exaggerated the stories that exist today. This is undoubtedly true to some extent, since storytellers tend to embellish stories, and the oldest stories have had the most time for embellishment. Furthermore, the storytellers intended to entertain and instruct, not to present literal history. Because stories of a global catastrophe are nearly universal among many unrelated cultures, however, some researchers assume that they accurately reflect the broad outlines of an actual ancient shared experience. "Something" happened long ago, although many details may be lost.

These two stories below and many of the rest that follow describe a long period of steady rain with immense flooding. Although some critics maintain that the storytellers merely exaggerated some large local flood event, the stories seem to describe something very different. Every culture periodically experiences occasional huge floods, but they are normal nevertheless, even though they are certainly catastrophic for anyone in their path. But one would not expect the survivors of such floods to claim that the world was flooded nearly to the mountaintops, and that almost everyone drowned, as most of these ancient stories do. Modern native peoples who experience catastrophic floods do not make such claims. Something extraordinary must have happened long ago.

WORLDVIEW OF THE ANCIENTS

To comprehend the cultural stories, we need to understand the mind-set of native peoples. It is clear that they did not think the way most people do in our logic-oriented Western civilization. Typically, native cultures think in more symbolic or dreamlike terms.

An example of the symbolic style of thinking comes from the actual reports of the Native Americans who witnessed the Battle of the Little Big Horn. It was a wide-ranging battle, so some of the eyewitnesses saw parts of it but not other parts. Later, when asked about the details of the battle, the Indians gave widely varying accounts. For example, when asked if Custer was there, some said that he had not been there, while others said he was. What accounts for these contradictions? According to the Indian mind-set, if a person did not see Custer, then, to that person, he was not there. This suggests that many ancient stories were intensely personal accounts of something that the original storyteller witnessed. When we read these stories, we must recognize that they are subjective, emotional accounts of an enormously catastrophic event, not scientifically objective reports.

Similarly, when ancient peoples spoke of "the world," they meant from horizon to horizon in their own area. Most had trading networks and knew the world was large, but they had no concept of India, China, and Europe, prior to Columbus. So if they said that "the world flooded to the highest peaks," they really meant that their local world flooded to the highest spot around them. They did not mean that India, China, and Europe had flooded, and they were not claiming that every mountain in the world was inundated, only the ones nearby. In addition, they probably exaggerated somewhat.

Another difference is that ancient peoples used very different words to describe forces of nature or attributes of people. Whereas we might simply say that someone is clever and devious, the Indians might say he became Coyote. They did not mean he had physically turned into a coyote, a notoriously sly and clever animal; they meant it in a symbolic way. Most of their stories are like that, containing symbolic or allegorical references. To understand the references, we have to translate them just as we would any other foreign language, and we have to be careful that much is not lost in translation.

As you read the stories that follow, it will be perfectly clear to you that the Native Americans and other ancient cultures had a unique worldview. Disappointingly, most Westerners automatically assume that our way of thinking is superior to theirs, but in reality, it is just a different way of viewing the world. After all, ancient peoples were quite successful in their world, thinking just the way they did. If they had not been, we would not be here today—we are their descendants.

Now, let's look at one story typical of the ones we uncovered that, we believe, refers to the time of the Event, 13,000 years ago.

THE SURVIVORS: THE LAKOTA NATION OF NATIVE AMERICANS

This story, which describes a worldwide fire and deluge, along with giant animals that sound very much like descriptions of Ice Age animals, very closely matches the scientific evidence we have uncovered. The story tells of Thunderbirds in the sky, and some people say that one of the Lakota tribes, the Blackfoot, has a sacred ceremonial pipe made of "thunderstone," a rare piece of an iron meteorite. Corroborating the events of the story, tribal elders say that the thunderstone fell from the sky after the Thunderbirds created it in a terrible lightning storm (from W. Taylor, 1993). Many other tribes use the Thunderbird symbol to indicate both the destructive effects of lightning and the impact of meteorites.

This particular tale comes from the Brule Indians, another member of the Lakota Nation; it tells of a time long ago before our present world came into being.

Battle with the Giant Animals

In the world before this one, the People and the animals turned to evil and forgot their connection to the Creator. Resolving to destroy the world and start over, the Creator warned a few good People to flee to the highest mountaintops. When they were safe, the Creator sang the Song of Destruc-

tion and sent down fierce Thunderbirds to wage a great battle against the other humans and the giant animals.

They fought for a long time because the evil humans and the animals had become very powerful, and neither side could gain an advantage. Finally, at the height of the battle, the Thunderbirds suddenly threw down their most powerful thunderbolts all at once. The fiery blast shook the entire world, toppling mountain ranges and setting forests and prairies ablaze. The flames leapt up to the sky in all directions, sparing only the few People on the highest peaks. It was so hot that the world's lakes boiled and dried up before their eyes. Even the rocks glowed red-hot, and the giant animals and evil people burned up where they stood.

After the Earth finished baking, the Creator began to make a new world, and as the Creator chanted the Song of Creation, it began to rain. The Creator sang louder and it rained harder until the rivers overflowed their banks and surged across the baked landscape. Finally, the Creator stamped the Earth, and with a great quake the Earth split open, sending great torrents of water surging across the entire world, until only a few mountain peaks stood above the flood, sheltering the few People who had survived.

After the waters cleansed the Earth and subsided, the Creator sent the surviving People out to populate the new world, our world today, warning them not to fall into evil, or the Creator would destroy the world again. As the People went out over the land, they found the bleached bones of the giant animals buried in rock and mud all over the world. People still find them today in the Dakota Badlands.

RETOLD FROM ERDOES AND ORTIZ, 1984

Fig. 12.1. Brule warrior. This photo and most of the dozen that follow in the "Survivor" sections were taken of actual Native Americans in the late 1800s. *Source: Library of Congress*

COMMON ELEMENTS OF THE STORIES

That story and the ones in future chapters come from many different cultures. Each one adds different nuances to the overall picture of the catastrophes, and although the stories are different, the basic imagery is remarkably similar. There are seven common threads in the ancient cultural stories in this book, and even though not all threads are in every story, each story has most of the following:

- The Creator warns of trouble.
- Almost everyone ignores the warnings.
- The few people who listen take action to save themselves and others.
- Fire, stones, and/or ice soon fall from the sky.
- Thick clouds form, heavy rains fall, and flooding begins.
- Many people, plants, and animals perish.
- Some survive to rebuild and repopulate the new world.

We propose that these stories are from the witnesses to a terrible disaster 13,000 years ago, but they are more than a simple chronicle of that disaster; they are just as much a story about survival. After all, these stories came from the ones who survived, and they tell us how they did it: by listening to the warnings and taking the appropriate action. Today, we face the same threat from the skies that they did. It has not gone away. The question is this: will we learn anything from our ancient ancestors? Or will we have to stumble blindly through a similar disaster with equally grave consequences?

THE SURVIVORS: THE OJIBWA TRIBE

This story comes from a tribe from northeastern North America, the Ojibwa, part of the larger Algonquin tribal group, who also called themselves "the First People." Their traditions go back to the Ice Age. What follows is one of their oldest stories, and one of the least symbolic of the stories that have survived. It clearly corroborates most of the scientific evidence we have uncovered, and even attributes the effects directly to the passage of a glowing star with a long tail. Ominously, it also warns that the star will come back again one day.

⭒

"Long-Tailed-Heavenly-Climbing-Star"

Once long, long ago, Chimantou, the Great Spirit, visited the Ojibwa tribe, who lived near the edge of the Frozen Lands. Chimantou warned them that a dangerous star was about to fall and urged them to hurry to the bog to cover their bodies with mud. Most People did not recognize the Great Spirit, however, and made fun of Chimantou. "Do not listen. That man is just a crazy person," they said, laughing. "Cover ourselves with mud! Ha!" they said as they went on their way and paid no more attention to the Great Spirit. Only a few hurried to the bog as Chimantou suggested.

Before long, when the sun was high, the day suddenly grew brighter. The People all looked up in panic and someone shouted, "Look! A second sun is in the sky!" The new star was growing larger, brighter, and hotter as it hurtled toward them. It became so bright that they had to shield their eyes.

The People who had not covered themselves with mud ran for shelter in terror, but it was too late. The star flew down to Earth and blanketed the world with its long, flowing, glowing tail. Tall trees burst into flame like giant torches, lake and rivers began to boil, and even the rocks glowed and shattered from the heat, as terrible fire swallowed up the entire world.

Then suddenly, when the heat was the greatest and the People in the bog thought even they would surely die, the star climbed back up and moved away from Earth.

After the world cooled down, the mud-covered People cautiously came out of the bog to look around. Stunned, they saw that the world had changed completely. In all directions, all that remained were smoldering, blackened trees and scorched grasslands. The People who had not listened

Fig. 12.2. An Algonquin-Fox warrior. *Source: Library of Congress*

to Chimantou had perished, along with all the giant animals. Only their skeletons remained.

The People were afraid and did not know what to do, until Chimantou came to them and said, "Put aside your fear. The star is gone for now. Go out and multiply, for this new world is yours. But if I come to warn you another time, do not forget to listen, because Long-Tailed-Heavenly-Climbing-Star will surely come back again to destroy the world."

RETOLD FROM CONWAY, 1992

Next, we examine how all the many clues and facts fit together, and to accomplish that, we have to explore a supernova. We will start by traveling hundreds of light-years out into the Milky Way galaxy, and then we will come back again.

13

THE OMINOUS "GUEST STARS"

QUESTION: Were the Clovis people aware of comets and supernovae? Is it possible that they saw the supernova shock wave or the impacts coming?

No one knows if the Clovis people knew about heavenly bodies, because they left no written or artistic record, but it seems likely. Most people today whose survival depends on animal migrations and growing seasons have a keen awareness of the motion of the sun, moon, and stars. Clovis people certainly witnessed eclipses, supernovae, meteorite falls, and the passage of comets, and they explained those events according to their worldview, even though that is now lost to us.

In the modern world, many people know nothing about supernovae, but chances are most know their astrological birth sign. Therefore, in some sense, they connect to the heavens. In addition, most people have probably seen a comet at least once during their lifetime. That would be even truer for the Clovis people, who spent much more time outdoors than most of us do today.

So while we know nothing about the Clovis people's understanding of supernovae, we do know a lot about what modern people have thought about them. To put our supernova story into a cultural setting, let's look at a bit of the history of supernovae.

OUR ANCESTORS AND SUPERNOVAE

From time immemorial, people have observed the skies in an effort to divine the future through the motion of the stars. Ancient observatories like Newgrange and Stonehenge in Great Britain, El Karnak and Nabta in Egypt, Gaocheng in China, Angkor Wat in Cambodia, Machu Picchu in Peru, Bighorn Medicine Wheel in Wyoming, and Chaco Canyon in New Mexico were designed to follow the movements of the stars and planets. Early people placed great importance on astrology both to plan their daily lives and to predict the future. In many ways, modern interest in astrology parallels this tradition as we still try to understand the universe in terms of the motions of the stars.

Today's Native Americans, who believe they descended from the Clovis people, left occasional references to comets and supernovae chipped into the rocks of the American Southwest. For example, Chaco Canyon, New Mexico, served as a ritual center and observatory, and the builders arranged its walls to align with the solstices and other important celestial events.

We know that a supernova occurred in 1054 CE, as reported by the Chinese; today we find its remnant body in the Crab nebula. According to contemporary accounts, the star flared suddenly and remained so bright for a few years that viewers could see it even in the daytime. It was also visible in North America, and archaeologists at Chaco Canyon believe Native Americans recorded the event there in the pictograph shown in figure 13.1. Today, nearly 1,000 years after the star exploded, astronomers using powerful telescopes still can see filaments of star-stuff streaming away from the explosion center at an incredibly high velocity equal to half the speed

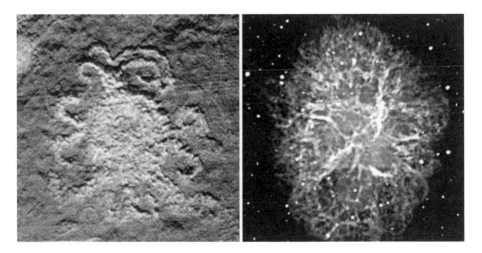

Fig. 13.1. On the left, this Chaco Canyon pictograph, made by Native Americans around 1054 CE, looks remarkably similar to the filament-laced Crab nebula supernova from the same time. *Sources: NPS, Chaco Canyon (left); Jeff Hester and Paul Scowen (Arizona State University) and NASA (right)*

of light. The pictograph in Chaco Canyon seems to show those same filaments, which most likely were much more visible in 1054 CE.

The earliest written records of the sky come from China, Korea, and Japan. The Chinese recorded star motions as part of the daily histories of the dynasties, although most of their interest was in the fixed, predictable stars, which they used to make their astrological predictions. They described several kinds of temporary stars; "bushy stars" and "broom stars" were the names they gave to comets, whose tails had two distinct appearances to the Chinese, and they called meteors "flowing stars." They viewed comets and meteors as particularly menacing, no doubt because some of them had crashed into the Earth and the people on it. Some think the Chinese legends of dragons are descriptions of fiery comets or meteorites, which looked to the Chinese like giant fire-breathing dragons gliding ominously through the night sky. So is this wariness of comets because the Chinese thought they were real dragons? We cannot be sure that "dragon" was not just a metaphor. Perhaps their uneasiness was merely from the disruption of the harmony of the night skies. Or maybe it is part of the cultural memory of the catastrophic Event long ago when dragon-comets attacked out of the northern sky. The last type of temporary star they called a "guest star"; that was their name for a light in the sky that suddenly appeared but stayed fixed in the constellation in which the ancient astronomers first noticed it. Of course, today we know that the "guest stars" of the ancient accounts are novae and supernovae, the explosions of stars at the end of their lives.

MODERN SUPERNOVAE

Descriptions by ancient astronomers of where these stars appeared in the sky are often sufficient for modern astronomers to locate their remnants. Today astronomers routinely discover supernovae in distant galaxies, and these explosions are brighter than the entire starlight from the galaxy in which they occur, as in figure 13.2. Most modern supernovae are discovered by comparing before-and-after pictures of the sky to look for a bright star that wasn't there before.

In February 1987, at the Las Campanas Observatory in Chile, an astronomer named Ian Shelton saw something strange in the telescope, and he went outside to look at the same sky his telescope was imaging. He saw a new, bright star that he recognized immediately as a new supernova. Quickly, he sent a telegram to astronomers around the world announcing the first visual discovery of a supernova since 1604.

Shelton's new supernova, SN1987A, occurred when a supergiant star,

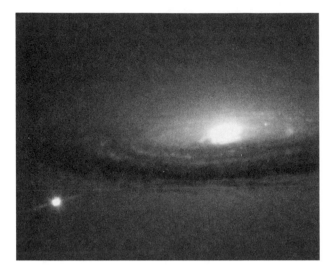

Fig. 13.2. The immense power of a supernova is obvious in this image; supernova SN1994D, a single star, briefly shone nearly as brightly as the entire nearby galaxy. *Source: High-Z Supernova Search Team, Hubble Space Telescope, NASA*

containing twenty times the mass of our own sun, exploded violently and blazed with the initial power of 100 million suns for several months following its discovery. The supernova still exhibits a very unusual set of rings puffed out of the explosion like smoke rings, the innermost of which is composed of brilliantly glowing gas. Just before the star exploded, it ejected millions of tons of slow-moving gas, but after the explosion, the fast-as-light radiation and swiftly moving ejecta caught up with the gas cloud. When it did, the gas became luminescent, displaying almost the same colors as neon bulbs.

THE DANGER TO EARTH

During the past 1,000 years, we know that at least six supernovae occurred within 20,000 light-years (6,100 parsecs) of Earth, an area that represents only a small part of our Milky Way galaxy. As many as fifteen supernovae might occur in our galaxy each century.

Historically, we may expect a supernova closer than 815 light-years (250 parsecs) every 100,000 years. This is within the range at which the Event supernova occurred. As we will see later, that fact has important implications for life on Earth.

QUESTION: Has anyone ever seen a supernova that blew up very near the Earth?

According to the above statistics, supernovae near the Earth are infrequent, and no one has seen a very close one in recorded times. However, a few ancient cultural stories describe seeing a twin sun in the sky. This is exactly the way a close supernova would appear. We saw earlier in this chapter that some supernovae in distant galaxies briefly give off more light that the entire rest of their galaxies, but normally, they are visible only with powerful telescopes. If a type II supernova (see below) occurred very close to Earth, it would rival the moon in size and brightness for many weeks. In addition, the supernova would be visible during the day, dimming as the radioactive isotopes produced in its remnant decayed.

THE SURVIVORS: THE ATAYAL

This story comes from the Atayal tribe in Taiwan and tells of a time when Earth had twin suns, one bright and one dimmer. The area is relevant, since we think that deadly radiation of the Event supernova hit Asia and Australia the hardest about 41,000 years ago, although the supernova would have been visible worldwide. The Aztecs in Central America tell a story of two suns (or moons) that is somewhat similar to this one from the Atayal.

In this story, we see a very good description of the aftereffects of the Event supernova. Tribal members saw what appeared to be a second, moon-like sun, which was bright enough to be visible in the daylight. In addition, the radiation was strong enough to wither grass and kill trees. That effect, along with climate changes, such as severe drought, probably led to many wildfires across the land.

The Atayal and the Twin Suns

Long ago, when the Atayal tribe first came across the water to Taiwan, there were twin suns in the sky; the normal one was large and yellow and the other was small and blue. Sometimes the normal sun came up in the day and the other shone brightly at night. When this happened, the birds stayed up all night singing and the animals ran around making noise just as if it were daytime. When the suns came up together, the weather became so hot that the crops withered, leaves fell from the trees, and no one could go out in the scorching heat. The People could not sleep or get enough to eat, so they became very angry at the two suns.

Exhausted and hungry, the tribe called a meeting, at which the head-person said, "We must do something about the two suns if we are to have a normal life. Who has any ideas?"

One man suggested that they climb up to Heaven and put out the second sun like a candle, but no one had a ladder that long. One woman said someone should fly up like a bird and pour water on the second sun, but the headperson asked, "Are you going to put on wings and fly up there?" She shook her head and said no more. No one else had a solution to the problem of the two suns.

Finally, a young boy spoke up, "I have been practicing with my bow and arrow every day. I think I can shoot it down." All the older hunters laughed at the brash boy. One of them said, "You cannot even hit a monkey in a tree. How can you hit the sun, which is farther than any monkey?"

Indignant, the boy insisted, "I will hit it!" But nobody would listen to him. As they left the meeting, most of them still laughed about the boastful boy who wanted to shoot down the sun. Nobody thought he could do it, except his father.

The next morning, the father and his son packed their belongings for the long trip to the highest mountain on the far end of the island. The father said, "On the tall mountain, we will have the best chance to hit the sun and bring it down." Putting his arm around his son, he said, "I believe you can do it." With that, they set out, enduring many long days of baking heat, along with sleepless nights, vicious bandits, wild animals, and raging streams.

Finally, they arrived on top of the tallest mountain during what was supposed to be nighttime, when the blue sun was shining brightly. The boy did not want to rest, so he took out his bow and arrow and aimed carefully at the small blue sun. He pulled the string back as far as it would go and let the arrow fly. The arrow landed squarely in the round body of the sun. The sun shuddered, and then boiling blood poured out of the wound. The boy and his father had to get out of the way of the huge falling drops of blood, which splashed across the land, setting fire to trees and grass and even scalding the rocks. Then, in the same way that the light of life disappears from a dying animal, the life-light disappeared from the sun, and it went dark.

For the first time, the boy and his father clearly saw the moon and stars all over the night sky. They laughed and embraced each other and danced in the moonlight. Then they lay down and slept well for the first time in years.

RETOLD FROM A TRADITIONAL ATAYAL TRIBE LEGEND

Now, let's see what other trouble a supernova might have caused.

14

EXPLOSIONS IN THE SKY

QUESTION: You claim the Event was highly destructive. How could a far-away supernova cause all that destruction on Earth?

To answer that, we need to examine how supernovae happen. All stars either explode or burn out eventually, and, in fact, our sun, a yellow dwarf star, has its origin in the cloud of debris from long-vanished supernovae that produced all the elements necessary to create the sun, our planets, and even life itself. That star-stuff makes up everything on our planet, including you.

The birth of the sun began more than 4.5 billion years ago, when a cloud of mostly hydrogen gas, along with many other elements, collapsed of its own weight to form the star we call our sun. The force of gravity pulled the hydrogen atoms together so closely that nuclear fusion reactions ignited, turning hydrogen to helium and releasing a great amount of energy into space. Since then, the sun has continued to burn hydrogen smoothly and reliably, providing Earth with the warmth and energy necessary for our evolution and survival.

So far, our sun has used up half of its hydrogen, but there is no need to worry about running out; it will take billions more years for the sun to use nearly all its hydrogen. Then as it uses the last remaining hydrogen near its core, it will begin to burn helium and other elements, causing it to swell to a radius 40 percent larger than it is now and become twice as bright. At that time, if any people remain on Earth, they will need to worry; life will begin to become extremely uncomfortable. About one and a half billion years later, the sun will have swollen into a red giant, three times its current size, looking like a big orange disk. We'd better have left the solar system before then, because the temperature on Earth will be more than 100 degrees warmer, and the seas will begin to boil.

About 250 million years later, the helium-burning sun will have grown to 100 times its original size, occupying half of the sky, and it will be 500 times as bright as it is now. The surface of the Earth will melt, and the sun will become so hot that the remaining helium will fuse into carbon, generating a massive explosion that will send a third of the sun flying out into space. After that, the sun will continue burning its contents into heavier elements, becoming a planetary nebula, which constantly throws its outer layers into space as a dense solar wind.

Finally, after burning and blowing off gas, the sun will become a white dwarf, with about half of the mass of our current sun squeezed into a ball the size of the Earth. Its nuclear fuel exhausted, it will begin to cool down, becoming a black dwarf with the ashes of Earth and the other planets circling silently around it in the dark of space.

MEDIUM-SIZED EXPLODING STARS

The death of our sun may sound like a spectacular event, but larger stars disappear in even more spectacular explosions called supernovae. These types of stars often form a binary- or multiple-star system with a companion star similar to themselves. In fact, our sun is unusual in that it has no apparent twin, so there is a good chance that it has an undiscovered companion, sometimes referred to as Nemesis. If so, some scientists propose that it could trigger extinctions if and when it appears periodically.

For most stars with twins, one of those stars will die and become a white dwarf after billions of years. When it does, its partner will become a red giant, but before then, something unusual often happens. In a binary system, frequently the white dwarf's powerful gravity pulls gas from its red giant partner, forming a disk of hot gases. In the process, the white dwarf gains mass, so that when it reaches 1.4 times the mass of its partner, the gravitational forces trigger uncontrolled fusion of carbon, nitrogen, and silicon into a radioactive form of nickel. This causes the white dwarf to explode in a "type Ia" supernova, totally consuming the star and sending its companion flying off into space. The supernova radiates 10 billion times the light of the sun for several weeks, and the ejected atoms and neutrons recombine to form the heavier stable elements. Because type Ia supernovae are formed when a specific mass is attained, they are all very similar, with a common luminosity that allows them to be used as a "standard candle" whose relative brightness is proportional to their distance.

TYPE II, THE LARGEST EXPLODING STARS

The cores of stars more than five times as massive as our sun burn in a layered structure like an onion, with heavier elements up to iron, the most stable of all elements, at the center. They burn brightly and rapidly, exploding into type II supernovae in only a few million years after their birth. During the time before such a star explodes as a supernova, it regularly blows off gas in a strong stellar wind.

The massive heat released from these large stars balances the force of gravity that is trying to squash them. When the stars try to burn iron, they are no longer able to generate any more nuclear energy (or heat), and the center of the star begins to collapse rapidly. The iron-burning stage lasts only a fraction of a second as the star shrinks to the size of the Earth. When it does, a rapid-fire series of cataclysmic events begins.

MULTIPLE SHOCK WAVES

In a type II supernova, the dense core collapses within less than a second, becoming tightly compressed. The dense core, which contains only about a few percent of the star's total mass, collapses so quickly that it leaves behind the outer layers. Those layers then crash down onto the core in a "reverse shock," heating everything to 5 billion degrees and generating more energy than the star did in its entire previous existence. This energy expands catastrophically in a massive explosion or "forward shock," sending most of the star's mass hurtling into space in a giant shell-like shock wave, as shown in figure 14.1. The remnant core is so tightly compressed that the protons and

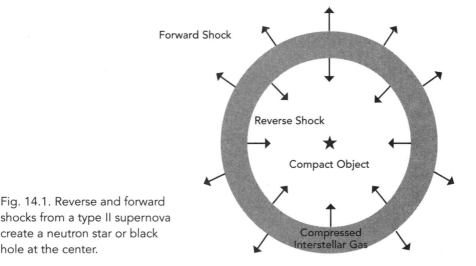

Fig. 14.1. Reverse and forward shocks from a type II supernova create a neutron star or black hole at the center.

the electrons are crushed together to form neutrons, and the core becomes a neutron star, or even a black hole, invisible to the naked eye and with an extraordinarily high density.

As the explosion blasts the star's material into the interstellar medium, it undergoes nucleosynthesis, producing all the elements, including uranium. Hurtling through space, those elements eventually form new solar systems like our own.

Unlike type I supernovae, which are observed in all galaxies with equal probability at any location, type II supernovae are restricted to the dense, star-packed regions of the arms of spiral galaxies, such as our Milky Way. Because type II supernovae are so violently explosive, they present a grave and severe threat to every star system close to them. The sun and Earth move around in the galaxy independently of other star clusters, with the result that for the last few million years, we have been passing through a galactic arm. This motion has brought us into one of the most dangerous supernova-producing regions of our entire galaxy.

THE EXPANDING SHELL

After a type II supernova explodes, most of the mass of the original star expands as a bubble into space at enormous velocities, typically 6,200 miles per second (10,000 km/s). As it expands, it sweeps up gas present in the interstellar medium from the former star's solar wind and the interstellar debris of stars long dead. The material swept up and carried along by the ejecta of the supernova may actually be greater than the mass of the original star.

Later, the shell of ejecta cools and becomes thinner as material toward the front of the shock wave slows down and faster material behind it catches up. The shell may also begin to distort as it encounters different densities of gas and varying magnetic fields in the interstellar medium.

After several thousand years, the supernova remnant enters the "snow-plow" stage, when the edge of the shock wave has become slower and denser and plows through space at a constant velocity, much like a snow-plow, eventually dispersing into open space.

This is the current model scientists use to describe the propagation of material from a type II supernova, but it must be viewed as a theoretical exercise since little experimental evidence exists to support it. Detailed investigations of supernovae are relatively new to science, and it is likely that supernovae vary significantly depending on the local interstellar environment and the dynamics of each supernova.

QUESTION: You claim the expelled supernova debris forms cometlike clumps, yet the expanding shell is just an ultralow-density cloud of elements. How can you explain the clumping idea?

While it is tempting to imagine that the supernova produces an evenly dispersed shell that expands uniformly in all directions like a soap bubble, becoming increasingly less dense, that does not seem to be the case. Supernova explosions tend to begin, not from the exact center of the core, but off-center, producing unequal forces of expansion. Because the explosion is unbalanced, it sends the remaining neutron star careening off from the explosion at high velocity.

The same thing happens to the gases in the body of the star, according to calculations by a group of astrophysicists at the University of Chicago and by the Space Telescope Science Institute (STSI). Jim Truran (personal communication, 2005) said the Chicago group's analysis showed that the type 1a explosion produces blobs of debris that shoot away from the blast in much the same way as the remnant core (fig. 14.2).

One of the researchers, Michael Shara of STSI, said, "Based on these observations, our previously standard view of what nova shells should look like may be fundamentally wrong. The [prevailing] view is that a nova

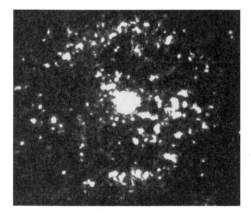

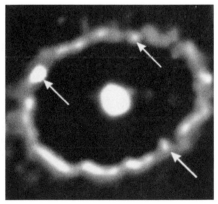

Fig. 14.2. T Pyxidis, an exploding dwarf star, produced thousands of huge blobs of material rather than the expected smooth shell (hubblesite.org/newscenter/ newsdesk/archive/releases/1997/29). *Sources: Shara, Williams, and Zurek (Space Telescope Science Institute); Gilmozzi (European Southern Observatory); Prialnik (Tel Aviv University); and NASA*

Fig. 14.3. Bright balls are produced as a high-speed shock wave from SN1987a plows into a gas ring at more than 1,000,000 mph. The collision heats the gas ring, causing some regions to thicken and glow (brightest ones at arrows). *Sources: P. Garnavich (Harvard Center for Astrophysics), et al.; NASA*

explosion is the same in all directions, with debris traveling at the same speed, so that a fairly smooth cloud is formed. Instead, we've found this myriad of individual knots [blobs]" (from Shara, *Astronomical Journal* 114 [1997]: 258).

This new observation is consistent with atomic and dynamite explosions, in which lumps of debris with highly variable density fly off unevenly in many directions. In addition, as the shock wave from the exploding star expands, the asymmetric dust cloud surrounding it may be compressed, producing much denser regions of gas. This has been observed in the explosion of SN1987a, shown in figure 14.3. The bright regions in the ring are interpreted as thicker lumps of debris heating up and glowing due to collision with the shock wave.

Whenever the expanding cloud passes close to another star system, the gravity of that system tends to concentrate the cloud into clumps. Also, even though the attraction between the supernova dust particles is weak, gravity will aggregate them slowly, in much the same way that scientists theorize it causes comets and asteroids to form in our own system. In support of this, recent scientific studies of some comets, such as Halley's and Tempel-Tuttle, which was hit by NASA's Deep Impact probe, show that they are just ultralight dustballs of loosely aggregated material, weakly held together by gravity. As the supernova cloud travels millions of miles through space, the same thing should happen to it: the cloud tends to clump together in some regions while leaving voids elsewhere that contain almost no debris. If so, then we propose that our solar system was bombarded by the thick clumps in just such a region of the debris wave. Even though those lumps had the same consistency as Styrofoam cups, they were miles wide. Imagine being hit by a hundred-mile-wide Styrofoam cup.

HOW MANY SUPERNOVAE ARE NEARBY?

David Green, from the Cavendish laboratory in Cambridge, has cataloged 231 supernova remnants in the Milky Way galaxy, and about 77 percent of them appear to be simple shells of ejecta. The distance from Earth to forty-seven supernova remnants from the catalog has been measured, and they range from 980 light-years to 48,000 light-years away. Older or more distant remnants are too dim to observe, so we see far fewer distant remnants than actually exist, and, in fact, we see none from the far side of the galaxy. The supernova remnants that Green observed all lie within the distance of recent historical supernovae, so assuming there are six supernovae per 1,000 years, the 231 remnants would represent the historical supernovae of the past 40,000 years. Of those 231 supernovae, some would have been extremely close to us. One of them may have led to the Event.

THE EFFECTS ON EARTH

We have seen that supernovae are powerful events that occur frequently, in geological terms, near Earth. These explosions are massive by any yardstick, and the consequences to Earth would be dire. For example, imagine the burst of gamma rays and neutrinos irradiating half of the Earth in a few seconds, exposing the inhabitants to a Hiroshima-like radioactivity. It wouldn't end there, however. After the violent detonation of a star twenty or more times the size of our own sun, the remnants would come hurtling toward us through space at enormous velocity to crash into the sun and every planet and moon in the solar system. The scars on our solar system would be long lasting, as our evidence for the Event shows.

THE SURVIVORS: THE ARAWAK

There are only a few cultural stories that could be descriptions of the conditions caused by the supernova, but when those traditions still exist, they are the same from Europe to Asia to the Americas. Typically, the pattern describes three phases of worldwide catastrophes:

- fire, just as we would see from the supernova radiation
- ice, such as would form as a result of supernova-induced climate change
- floods, caused by the shock wave and large impacts on the ice sheet

We would expect to see descriptions regarding the first, fire, because the cosmic rays would have killed many plants within a few days after the irradiation, leading to widespread wildfires. Describing the end of the First World by fire, the next story comes from the Arawak, a Caribbean tribe who were the first to meet Columbus in 1492.

Fire from the Sky

Ages ago, the Creator became impatient with all the evil in the world and decided to destroy it and create a new one. Looking around the Earth, the Creator could find only one righteous family that deserved to live. Appearing to them one day, the Creator told them, "Go dig a large pit, cover it with logs, and pile sand over the top. After it is done, seal yourselves up inside the pit for protection."

The family recognized the ominous tone in the Creator's voice, so without any questions, they quickly began to dig the pit. Just as they finished sealing themselves inside, the Creator sent a terrible rain of fire and hail down from the sky to destroy the world.

Inside their pit, the ground shook so violently from the concussions that they were afraid the walls might collapse on them. Huddled in the middle of the pit, they heard the roaring and crackling of the fire as the forests around them burst into furious flames. The air in the pit rapidly grew warmer and warmer until it was almost too hot to breathe. The family began to fear that the pit was not deep enough to protect them, but soon, the noise stopped and the roof of the pit grew cooler to the touch.

After a while, the family came out to find a changed world. The fire had scorched the Earth as far as they could see in every direction, so that almost every living thing had been destroyed. Some animals and a few other People survived to rebuild the world.

RETOLD FROM BRETT, 1880

Fig. 14.4. Gulf Coast Indian.
Source: Library of Congress

WHAT THE EVIDENCE SHOWS

- The largest and most dangerous supernovae occur in the galactic arms of the galaxy.
- We are passing through the danger zone of one galactic arm right now.
- When a supernova explodes, it ejects massive amounts of stellar debris in shock waves.
- A supernova produces two explosions; one creates a forward shock wave and the other forms a reverse wave.
- The explosions create lots of radioactive uranium and thorium, just as we find in the Event.

We know that there were many supernovae near Earth in the past, and we know about where all of them occurred. Let's see if we can find the "smoking gun."

Let's see if we can find the one that caused the Event.

15

GIANT BUBBLES IN SPACE

QUESTION: Okay, so you find evidence on Earth for a supernova, but is there any evidence for it in space?

To find evidence for a supernova, let's begin by considering where we are located in the galaxy. When you look at the night sky, the stars appear randomly distributed, except for a bright streak across the sky; that streak contains the stars of our own Milky Way galaxy. Many of the stars we see outside that streak are not actually stars at all; they are distant galaxies just like our own.

In that streak, which is the plane of the galaxy, the center appears darker or even empty of stars, but that is only because dust obscures most of the visible light from the Milky Way. Many millions of stars along the galactic plane hide in that dust, which hides the center of the plane from view, but we can infer from our knowledge of other galaxies that we live in a spiral galaxy. Earth may be vitally important to humans, but by galactic standards, it is an insignificant planet orbiting an insignificant star, located far away from the center of the Milky Way out in the galactic "suburbs."

The Milky Way is composed of a system of spiral arms radiating from the center of the galaxy, and scientists believe that the arms formed because stars near the center of the galaxy orbit the center faster than do those farther out, and as the galaxy rotates, it appears to wind up like strands of spaghetti on a fork. This differential rotation compresses the gas in the arms sufficiently to initiate the formation of stars.

In between the spiral arms, there also are many stars, but these tend to be very old stars, which, like our own sun, wander in and out of the arms in their travels around the galaxy. This wandering is caused by cosmic currents in the galaxy that are just like currents in the ocean. The Atlantic Ocean always remains the same shape, but the water, fish, and seaweed all move around somewhat independently inside it.

Much brighter, more massive stars called type O and B dominate the regions within the arms, where they are constantly forming and dying. Following a "live fast and die young" lifestyle, these stars are the short-lived supergiants that burn fiercely and then erupt in a blaze of glory as powerful type II supernovae. This means that the most powerful supernovae occur most frequently inside the spiral arms, which, consequently, makes the spiral arms the most dangerous places to be. Unfortunately, and as alluded to earlier, our sun is currently passing through the Milky Way's densely packed Orion arm, a hazardous region filled with many exploding and about-to-explode supernovae. Astrophysicist Narciso Benítez and colleagues from Johns Hopkins University have estimated that about twenty supernovae have occurred near Earth during the last 10 to 20 million years. Some occurred very close to Earth, and we think one of those was the precipitating cause of the series of catastrophes that befell Earth beginning approximately 41,000 years ago.

THE BUBBLES NEAR US

Closer to home, the Earth is surrounded in space by bubblelike features, which, unlike normal areas of the galaxy, are nearly empty of gas and dust. Most bubbles are highly active areas that give birth to giant stars, which go on to die as type II supernovae. In fact, it appears that the gigantic explosions of these supernovae are responsible for sweeping the bubbles clear of gas and dust. The outer edges of these space bubbles are the areas where the expanding shock waves are pushing back the gas and dust that surround the supernovae. The collision of the expanding supernova shells with the existing dust causes the interface to heat up and become optically incandescent, so that when astronomers look at those expanding shells, they see them as vividly colored bubbles, an innocuous image that masks the violent events occurring there.

The sun lies in the middle of the Local Bubble, a young, relatively small bubble. A series of recent supernovae created our bubble, unleashing multiple explosions that blew a tubelike chimney completely through the galactic disk out beyond the plane of the galaxy on each edge. The enormous blasts pushed gas away from the area around us, creating a near-perfect vacuum. In the bubble, a human-head-sized volume of space contains, according to NASA, only a trifling four atoms of gas, 1,000 times fewer than found in the regular galaxy. In addition, those atoms are up to 100,000 times hotter than the atoms in normal Milky Way gases.

Evidently, the sun was very close to the action when the last supernova went off, and that supernova must have exploded nearly due north or south of our solar system, as aligned with Earth's poles. Clearly, Earth was in the direct path of the blasts, since we are in an empty part of the bubble. This is solid proof that at least one or more massive stellar shock waves rolled through our solar system not too long ago, and all life on Earth has felt the effects intermittently over the last several million years. The supernova eruptions most likely are not over, but we think the last one in the Local Bubble occurred 41,000 years ago. Recent supernova shells have been observed to expand to one-third the size of our local bubble in less than 2,000 years, so it may very well be that the shell of our 41,000-year-old supernova has disappeared into the boundaries of the Local Bubble.

QUESTION: The bubble provides evidence for supernovae in our region of the Milky Way. Beyond that, is there any other evidence for the Event, such as a shell or the remnants of the star?

A TELLTALE PULSAR

If a type II supernova exploded near Earth about 41,000 years ago, the remnant, either a neutron star or a black hole, may still be nearby. Or it may no longer be close, because supernova explosions can eject remnants at high velocity, as shown in figure 15.1, causing them to hurtle off to distant parts of the galaxy. Even if they *are* nearby, these remnants are very small and usually dark, so they are very hard to see.

If the remnant becomes a neutron star, it is likely we can observe it as a pulsar, which emits huge bursts of energy. We believe that the catalyst for the catastrophes which began 41,000 years ago is the pulsar Geminga (see fig. 15.2) in the constellation Gemini, which, as noted by Ramadurai (1993), may have produced distinct peaks in supernova-related beryllium around the Earth about 50,000 to 30,000 years ago. At about 500 light-years' distance, Geminga is the closest pulsar to Earth, and it emits a pulse of X-rays and gamma rays at precise 0.237-second intervals, but it gives off no visible light or radio waves, making it invisible to optical telescopes. Evidence suggests that Geminga was much closer to Earth in the past, perhaps closer than 100 to 150 light-years, which is well within the danger zone for supernovae.

Astronomical images of Geminga indicate that it is moving very rapidly through space, as often happens to remnants expelled from a supernova. The pulsar displays an X-ray bow shock wave with twin tails that stretch

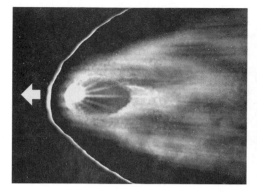

Fig. 15.1. Some supernova remnants move incredibly fast, as in this illustration of a neutron star in the constellation Sagittarius. Named the Mouse nebula, it travels at more than 1.3 million miles per hour, forming a bow shock wave in front. *Sources: NASA and the Harvard-Smithsonian Center for Astrophysics*

Fig. 15.2. Is this the actual image of a killer? This image shows two supernova remnant pulsars, the Crab (right) and Geminga (left). Geminga is our candidate for the cause of the Event. *Source: NASA, Compton Gamma Ray Observatory*

2 billion miles away from the twelve-mile-diameter, supercompact pulsar. Imagine the remains of a star being jammed into the same twelve-mile space occupied by many small cities, and still being able to emit vast amounts of X-rays and gamma rays across the galaxy.

Some scientists, however, such as Leonardo Pellizza and colleagues (2005), think that Geminga exploded much farther away. They estimate that the parent supernova occurred 290 to 780 light-years (90 to 240 parsecs) away from Earth, which still would have been close enough to cause great damage. This distance, however, is based on Geminga's calculated age of 342,000 years, and this age, in turn, was calculated according to the rate that the gamma-ray pulses are slowing. But these calculations may be off by a considerable margin, because pulsars sometimes undergo periodic "star quakes," also known as "glitches," when a pulsar suddenly slows down. Alpar and coauthors (1993) have presented arguments that Geminga is indeed a "glitching pulsar," in which case it would be considerably younger than previously thought, and it could have been considerably closer to Earth when the deadly explosion took place.

Ellis (1995) used measurements from ice cores and ocean cores to suggest that Geminga did indeed explode much more recently and closer to Earth, and therefore is responsible for the apparent shock wave that passed through our system about 41,000 to 34,000 years ago. This span, as you may recall, includes the dates that we have on the mammoth tusks, which contain hypervelocity iron particles that one would expect a supernova to eject. If Geminga was the culprit, then the explosion most likely struck Earth more or less from an equatorial direction, but because of Earth's various wobbles through space, the effect certainly could have reached the Alaskan and Siberian mammoth herds.

One more line of reasoning connects Geminga to the catastrophes discussed in this book. The bubbles from known historical supernovae reach a diameter of 120 light-years in 1,800 years, an average expansion rate of 10,000 km/s, by which time the expansions had slowed to about 620 miles per second (1,000 km/s). At that rate, the leading edge of the shock wave would travel from the moon to Earth in just under seven minutes. That is blindingly fast; for comparison, it took the Apollo astronauts about three days to travel the same distance. In addition, the leading edge of the bubble would pass by Earth in 7,000 years, and the entire shock wave would expand to 350 light-years' width after about 40,000 years. From our discussion above, do you recall the Local Bubble, that immense supernova-created chimney that cuts through the galactic plane? Amazingly, that figure, 350 light-years, is nearly the exact size of our Local Bubble! The time fits, along with the age and the width of the Local Bubble, so maybe Geminga was indeed the catalyst for the Event.

THE SURVIVORS: THE HOPI

There are few ancient stories with descriptions of what the supernova may have been like, such as the Arawak story of destruction of the world by fire. The Hopi are another tribe that has a clear tradition of fire having destroyed its First World. There are also only a few stories about what would have happened next: the bitter cold that set in as Earth descended into the deepest part of the Ice Age. One of these stories comes from the Hopi as well. Perhaps the continuity of their mythology is due to their fierce resistance to assimilation into another culture and their struggle to maintain their own identity through all their troubles.

This story clearly depicts events that would occur in the aftermath of the supernova, when temperatures would plummet and the Earth would descend into an ice age. In addition, a supernova would affect the location of Earth's magnetic poles, although not the poles of its axis of rotation. Some scientists, however, suggest that oblique impacts by large comets or asteroids might cause the Earth to wobble on its axis of rotation, as described in this story.

☞

End of the Second World

After the destruction of their former Earth, the few surviving People cautiously emerged from the hole in the Earth into the newly created Second World. Everywhere they saw beautiful mountains, spruce trees, crystal waters, and all the natural abundance they needed. Before long, they multiplied and prospered, spreading over the entire planet, even to the other side of the Earth.

For a long time, the People were happy they had been spared, and they remained grateful to the Creator, but gradually changes occurred. No longer satisfied with what they had, they began to trade and barter with other tribes to acquire rich clothing and ornaments. Before long, some People began to belittle those who had less than they did. Year by year, the People became more contentious and more isolated from each other and from the Creator. Then, finally, no longer content with trading, some People decided just to take what they wanted, so wars broke out between the tribes. Before long, warfare was everywhere, and the People became crazed with violence. Those who spoke against war were ridiculed or killed for their peaceful words.

One day, the Creator appeared to the few peaceful People who were left. "There is no hope for this world, which has been wrecked by the Evil Ones," the Creator told them. "I will have to cleanse it again. Therefore, I

Fig. 15.3. A Hopi girl. *Source: Library of Congress*

have asked the Ant People to open their world to a few of you, just as they did at the end of the First World. Go quickly while you have time." The People crawled for safety through the ant doorway into the Earth.

The Creator then spoke to the twins, one at each of the Earth's poles, who kept the Earth spinning properly. In a loud booming voice that shook the sky and mountains and oceans, the Creator commanded, "Leave your posts, and move away from the poles. This World is finished." As the twins moved away, the Earth began to wobble and spin like a drunken man. Mountains toppled over and fell into the seas. The oceans washed back and forth, splashing over the land. The frigid winds from the North and South Poles roared out over the middle lands, and the Earth spun out into the cold depths of space, where it froze solid. People and animals on the surface were frozen into statues of ice. Only the chosen few were safe and warm deep down with the Ant People. And so ended the Second World, two worlds before ours.

RETOLD FROM FRANK WATERS, 1963

WHAT THE EVIDENCE SHOWS

- Over a few million years, twenty supernovae exploded close enough to Earth to cause damage.
- Our sun is right in the middle of the Local Bubble, a recent supernova creation.
- That we are in the Local Bubble means that at least one supernova shock wave passed us recently.
- The Local Bubble is the width expected for a 41,000-year-old supernova remnant.
- Geminga, a near-Earth pulsar, is the X-ray-emitting remains of a recent supernova.
- Evidence suggests Geminga is the right age and distance for triggering the Event.

Let's recap. Have we just seen an X-ray photo of the cosmic catalyst that precipitated Earth's catastrophes? Do we have the smoking gun for the mammoth extinctions in the dense, burned-out core of Geminga? It seems like it, but we cannot be sure yet that it was Geminga. We need to find more evidence.

16

RADIATION FROM
THE HEAVENS

QUESTION: It is hard to imagine that tiny particles or rays from a supernova could travel millions of miles to Earth and have any catastrophic effect. How is that possible?

You don't notice it, but supernovae bombard your body every day with invisible particles and radiation. Traveling at nearly the speed of light, one atom crashes into each square centimeter of Earth's atmosphere around every second, creating a shower of particles. These projectiles are cosmic rays, and supernovae in our own galaxy created the vast majority of them. They travel through space for many thousands of years, darting about the galaxy guided by the shifting magnetic fields they encounter in their travels. Most of these atoms are hydrogen, but helium, carbon, iron, and all the other elements in our solar system make up the cosmic rays, which arrive from far away in pretty much the same abundance as they occur near Earth. Most of the low-energy cosmic rays (<1,000 MeV, or million electron volts) actually come from our own sun's solar wind, while most of the others come from beyond our solar system. The highest-energy cosmic rays have energies greater than 10^{20} MeV, the energy of a well-thrown baseball, except that we cannot see the "pitcher," and, in fact, the true origin of these high-energy cosmic rays is still a mystery.

What happens when a cosmic ray strikes the atmosphere? Fortunately, our atmosphere is so thick, with a density of 14.7 pounds per square inch, that even the most energetic cosmic ray can't make it to Earth. Instead, each cosmic ray smashes apart atoms in our atmosphere (mostly nitrogen, oxygen, and argon), producing a shower of protons, neutrons, pions, muons, electrons, and gamma rays, which in turn break up other atoms. The collision produces a cascade of particles, many of which ultimately reach the ground, where they can be detected easily. The cosmic-ray barrage produces neutrons that bombard us each second at sea level at a rate of one neutron per 11 square inches (71 square centimeters). This means that, for example, two neutrons are passing through your hand every second, although, of course, you don't feel them. Your DNA might, however, as we will see later. In most places on Earth, our main exposure to radiation comes from these cosmic rays. If you live at higher elevations, your exposure to cosmic radiation is greater; and flying in airplanes, the exposure is high enough to be a health concern for flight crews. The most serious danger from cosmic rays is to astronauts, who often go to the edge of the atmosphere and beyond, where they suffer much greater exposure. One of the enduring problems with future long interplanetary missions is that it is difficult and costly to add enough shielding to protect astronauts, who will suffer continuous exposure to high levels of deadly radiation.

COSMIC RAYS AND EARTH'S WEATHER

Cosmic rays from supernovae may also play an important role in the weather. Scientists believe that cosmic rays commonly trigger lightning by creating an ionized path through the atmosphere for a lightning discharge to follow. There is also evidence of a correlation between the cosmic-ray rate and low-altitude cloud formation. Nigel Marsh and Henrik Svensmark (2001) found the connection, as shown in figure 16.1. Increased ionization

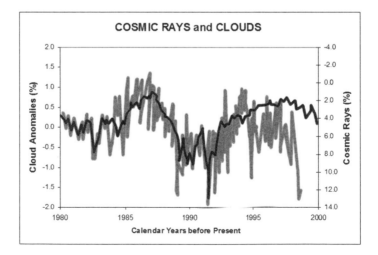

Fig. 16.1. Cosmic rays (black) have tracked nearly perfectly with low-level cloud cover for the last twenty years, indicating that increased radiation makes more clouds, which alters the climate. *After Marsh and Svensmark (2001)*

from cosmic rays seems to facilitate the formation of atmospheric molecules like sulfuric acid (H_2SO_4) into small droplets, which act as condensation nuclei for water vapor. Regarding the effects on climate, the chain of events works like this:

- Nearby supernovae produce more cosmic rays.
- More cosmic rays cause more water vapor to condense into clouds.
- More clouds block more incoming sunlight. The weather gets cooler.
- More clouds form larger and more violent storms. The climate becomes colder.
- More storms produce more rain and snow. Temperatures drop.
- More snow reflects more light energy to space. The atmosphere turns colder.
- More cold water and ice enter the ocean. The climate grows even colder.

With that scenario, it is clear that supernovae can alter the climate, although there are several ways besides supernovae by which cosmic rays can increase. If the Earth's magnetic field weakens, that allows more cosmic rays to hit our atmosphere, so the same cooling should happen then. In addition, whenever the sun enters a phase with very few sunspots, our cosmic-ray rate increases. This is because the sun's plasma no longer protects the Earth, and more radiation leads to global cooling. These other processes, however, do not produce as severe or sudden cooling as does a supernova.

COSMIC RAYS FROM SUPERNOVAE

Our focus is in the effects of cosmic radiation from supernovae, since many of the cosmic rays that strike Earth originate in the exploding stars in our local neighborhood of the galaxy. Because this region has had more than its share of supernovae in recent millennia, Earth's cosmic-ray rate is probably higher now than in the distant past, although the rate does vary over time.

This obvious link between recent cosmic rays and climate makes one wonder about the connection to long-term climate. Is it possible that changes in cosmic radiation affect the multimillion-year cooling cycles on Earth? It does, according to the work of Shaviv and Veizer (2003). They compared the cosmic-ray rate to Earth's climate over the last 500 million years and found an amazingly strong correlation, as shown in figure 16.2. Every time cosmic-ray rates increased, the climate got colder, and every time cosmic radiation decreased, the climate warmed up in perfect unison.

Notice also that most of the major extinctions (black bars) along the

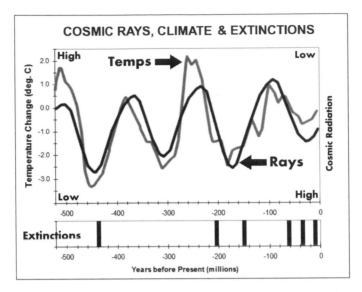

Fig. 16.2. The black temperature line, calculated from cosmic-ray flux, tracks perfectly with actual climate (gray line) for the last 500 million years. Six of the nine largest extinctions (lower bars) also line up with the increased radiation. *After Shaviv and Veizer (2003) and Raup and Sepkoski (1984)*

bottom line up quite well with the coldest lower half of each valley in the chart. Only a few (shorter gray ones) occurred during the warmer times with low cosmic-ray counts. This suggests that changing levels of cosmic radiation directly influence rates of extinctions.

MASS EXTINCTIONS

In 1980, Luis Alvarez and a group of other scientists discovered evidence for the extinction of the dinosaurs by a huge impact event, which researchers eventually linked to a hidden crater on the Yucatán Peninsula. Before their research, no one had found a strong connection between impacts and extinctions. Not long after that, in 1984, Paul Raup and Jack Sepkoski proposed that the dinosaur extinction was only one of about ten large extinctions that seemed to follow a regularly recurring pattern at about 26-million-year intervals. Of the ten, the five largest occurred about 65, 210, 245, 364, and 440 million years ago, as marked in figure 16.2. For most extinctions, at least 20 to 60 percent of all species vanished in a geological instant, but the one 245 million years ago was particularly catastrophic; an incredible 90 percent of all life disappeared from the planet.

The theory of regularly spaced extinctions is still controversial, but scientists are in widespread agreement that the extinctions did take place. So what caused them? Comets or asteroid impacts are well linked to several of the extinctions; massive volcanism is linked to others; and the reasons for the rest are currently unclear. Recently, however, Adrian Melott and a group of astronomers (2004) proposed that the second-largest extinction, 440 million years ago, when two-thirds of all species died out, may have been caused by a supernova. They theorize that gamma rays destroyed the Earth's ozone layer, allowing ultraviolet radiation to sterilize most of the planet.

Many scientists accept that a major cosmic impact happened at the same time as the dinosaur extinction 65 million years ago. Surprisingly, a supernova may have been involved then too. It is well known that a major climatic cooling event occurred back then, and in the 1970s, Russell (1977) proposed that a supernova caused the climate change and contributed to the extinctions. As evidence, Russell described an enormous, rapidly expanding shell of interstellar hydrogen, called Lindblad's Ring, which is about 18,000 miles in diameter and is part of our Local Bubble. If the ring was the remnant shell of a supernova, the center of the supernova explosion would have been dangerously close to Earth 65 million years ago, leading Russell to conclude that the supernova played a crucial role in the extinction of the dinosaurs, including causing the impacts.

More recently, a smaller extinction occurred that, according to Benítez and colleagues (2002), might be linked to supernovae in the Scorpio-Centaurus association, or Sco-Cen, a group of young stars that may be responsible for the Local Bubble. Benítez and his team found that about 2 million years ago, some stars in this group might have passed within 130 light-years of Earth, close enough for a supernova to bathe us in cosmic radiation and other particles. When these researchers realized that the stars had been so close, they searched for evidence of an explosion, and they found it. In deep ocean cores, researchers had already found abundant iron-60, one of the isotopes produced by supernovae. The layers dated to 2 million years ago, the exact timing of the extinction. With this evidence, Benítez and his coauthors concluded that a supernova had killed off millions of plants and animals at that time.

Those are three examples of possible connections between supernovae and some of the largest known extinctions; almost certainly, there are more yet to discover. Supernovae are one of several mechanisms that can produce extinctions, but they are one that is mostly overlooked—partly because they happen so far away and are less easily studied than large craters, climate change, and volcanic eruptions. Yet they may be the triggering mechanism that set in motion the more obvious effects.

SUPERNOVAE AND NITRATE

Melott's group, which studied the oldest supernova-related extinction, suggests that the supernova radiation would have caused nitrogen to oxidize and build up in the atmosphere as brown smog. If the same thing happened with our recent supernova, it should appear in the record. Nitrate levels also should rise when impactors crash into Earth's atmosphere.

One of the longest ice cores retrieved from Greenland is the GISP2 core, which scientists tested for nitrate, or NO_3. Living plants produce this compound and so do human activities, but, interestingly enough, the compound also has a link to clouds. Mayewski and Legrand (1990) found that nitrate may be an indicator for polar stratospheric clouds, and, as we saw above, cloudiness has a connection to increased cosmic radiation. Making an even closer link, Dreschhoff and Zeller (1998) suggested that nitrate might be deposited in the ice core during severe solar flares, which may have a link to the incoming supernova debris wave, as we will see in a moment. The last connection is from the impacting bodies themselves: when any high-velocity object enters Earth's atmosphere, it "burns" nitrogen to produce nitrate, among other nitrogen compounds.

This gives us three links between supernovae and nitrate, so it should show up in the ice-core record. In figure 16.3, you can see that it appears at all three times mentioned in our theory—13,000, 34,000, and around 41,000 years ago. The 13,000-year peak is the strongest, showing up as

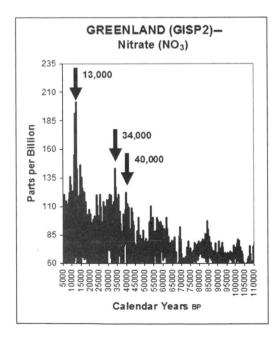

Fig. 16.3. Nitrate levels from the GISP2 ice core in Greenland. Notice that during the Event, amounts are nearly double and are the highest in the last 110,000 years. *Data from Mayewski et al. (1997)*

a very distinct, sharp peak that is about triple the normal level during the entire 110,000-year record. The next two oldest peaks correspond to the supernova's shock wave (34,000) and radiation pulse (41,000).

But is it possible that the peaks are just a normal rise due to climate? Maybe, since nitrate values tend to peak when the climate gets warmer, as it is today. We can check that possibility in the GRIP core, another ice core from Greenland, since it spans the previous interglacial around 130,000 years ago, when climate was much like it is today. Surprisingly, nitrate levels in the last interglacial were among the lowest in the entire 154,000-year record. Therefore, it seems unlikely that the high nitrate results from our warmer climate. Now that we know this, it is hard to explain away that peak as being unrelated to the supernova and impacts, since there is nothing like it previously. Instead, it provides reasonable evidence for the Event, since nitrate has many connections to supernovae. More of it can wind up in our atmosphere due to supernovae-produced cosmic rays, to impacts, and to solar flares, all three of which are part of our scenario. Is it any wonder that nitrate levels 13,000 years ago were higher than at any other time during the last 154,000 years? It was an extraordinary time.

CAUSE OF EXTINCTIONS

Past extinctions almost all show multiple events that occur simultaneously, or nearly so, to produce widespread die-offs—one event, no matter how severe, is rarely enough. If you mention *dinosaurs*, most often a listener thinks of "asteroid crater," although that was only one event in a cluster of serious problems, many of which were interrelated. This is no less true of the Ice Age extinction.

We do not believe or suggest that just the supernova or just the impacts caused the extinction of the megafauna—it was more complicated than that. Some percentage of plants and animals perished from each of the following causes, although some are more important than others:

- supernova radiation, both directly and through genetic damage
- high-velocity supernova particle bombardment
- impact blast wave and heat
- impact-related flying debris
- fires, both directly and through destruction of the food supply
- toxic chemicals and heavy metals in the air and water
- event-related climate changes, directly and by destruction of the food supply (chill theory)

- epidemics induced by radiation and by ecosystem destruction (ill theory)
- predators, both human and animal, that hunted the surviving animals (overkill theory)

As you can see, the ill, chill, and overkill theories are compatible with the Event and almost certainly played a role. In addition, while it is tempting to think that the word *extinction* means all the animals died on the same day, that was not the case. Small, isolated populations most likely hung on for hundreds or thousands of years before disappearing.

THE CONTINUING EXTINCTION

So far, we have been talking mostly about an extinction that ended 13,000 years ago, and we have touched on extinctions that occurred millions of years ago. Since these are in the distant past, you may wonder whether we need to be concerned about them.

Leakey and Lewin (1995) might have an answer for us. In their book *The Sixth Extinction,* they describe a time in which 50 percent of all life disappears from the Earth, a time not in the distant past, but right now. This current extinction is already the most extensive one in the last million years, and by the time it runs its course, some believe it will join the other five as one of the six worst in Earth's history, with the disappearance of more than 50 percent of all species, possibly including our own.

Most of this sixth extinction is apparently caused by human overpopulation, overhunting, overfishing, overindulgence, and overexploitation. There is more to the story than just blaming humanity, however—the Event played a major role in our current ongoing extinction as well. How, you might ask, do our modern troubles relate to an Ice Age Event long ago? To answer that question, we need to step back to see the bigger picture.

AFTERMATH OF THE EXTINCTION

When people use the term *extinction,* they mean that many living species vanished. That is just part of the equation, however. Another side is that some species survived. In all past major extinctions, with ecosystems out of balance, many of the surviving species experienced explosive growth. This is what happened 13,000 years ago, when an unusual mix of conditions created favorable conditions for the human species.

First, there were the new genes. Spurred by genetic mutations that produced a burst of creativity and technological resourcefulness, humans became even more efficient predators. The Event killed off many competing predators, and skilled hunters decimated many surviving ones, ensuring that more people survived. The same resourcefulness made humans better at finding food in every way, and consequently human populations expanded rapidly.

Second, a benign climate, due partly to the greenhouse effect of the impacts, fostered larger human populations. The warmer climate, coupled with humans' newfound resourcefulness, fostered the invention of agriculture, freeing humans from a nomadic lifestyle. The development of better housing, clothing, and weapons all allowed human populations to increase.

Third, increasing populations led more people to live together in villages and towns, where the division of labor allowed a larger pool of skilled talent to develop. This fueled an almost constant technological boom in many fields, producing, among other things, pottery making, metalworking, and writing.

All that may seem positive, except that the burgeoning population, initially fostered by the extinction Event, contained the seeds of many of our current troubles. When overpopulation occurs in any species—whether it is rabbits, locusts, lemmings, or people—a host of problems comes along with it, including epidemics, starvation, extreme aggression, ecosystem destruction, and scarce resources, every one of which is a major pressing problem in our society today.

THE EXTINCTION SEQUENCE

An extinction sequence comprises the following stages:

- A major catastrophe leads to the disappearance of some species.
- These disappearances lead to the overpopulation of some surviving species.
- Overpopulation leads to devastating depopulation.

This equation has held true for every past extinction event. In the current sequence, we have passed through the first two stages as a species but not the last stage, depopulation.

As in all past extinctions, overpopulation is producing all the ills for humankind suffered by other overpopulated species. Just as happened with the explosive algal blooms that created the black mat after the impacts, our species has run amok to spread across the entire planet in uncontrolled growth. Eventually, the algae came under control and returned to sustainable levels, but we have not done so yet.

That brings us to the third part of the extinction equation: depopulation.

In any species, radical depopulation is a classic, well-known result of the problems we face. In similar situations for other species, disease, predation, conflict, and starvation have always managed to "solve" the overpopulation problem. We don't see the severity of the problem clearly because we are *inside* the extinction sequence, but if we do not control population growth by choice, painful forces eventually will do it for us. The question is whether we will do it by choice or have it imposed on us.

Today, most of humanity's current problems, including starvation, warfare, pollution, global warming, and AIDS, result from or relate to the severe overpopulation that has occurred in the aftermath of the Event. Past evidence tells us that after a catastrophe has disturbed the ecosystem, some species go extinct *long after* the initiating events are over. That threat of extinction still looms over humanity today, and it is not an idle threat. It has happened many times before.

THE SURVIVORS: THE PERSIANS

Here is another story similar to the Hopi tale about the cold that would have occurred after the supernova irradiated our planet with cosmic rays. It is from Persia, in the Middle East, a long way from the Hopi homeland. Although the irradiation from the supernova would have been brief, the climate effects would have spread around the globe. It is also a story of overpopulation and one solution for it.

⚶

The Very First Winter

Yima was the first mortal that the Creator spoke to, so he was appointed to run the Earth, and he did so for 900 years with great success. During that time, there was no disease or death. Neither was the weather too hot or too cold, and everyone got along very well and prospered. Since there was no death, the people, animals, and plants quickly became so plentiful that after 300 years, Yima had to make the Earth bigger. Using a magical golden ring and a golden dagger from the Creator, whose name was Ahura Mazda, he quickly accomplished the enlargement. Everything went along well on the newly larger planet for another 300 years, when they ran out of space again. Yima was forced to make Earth larger once more, but after another 300 years, it was the same problem.

Yima complained to Ahura Mazda, who told him, "I do not want you to enlarge Earth any more; it is already big enough. We will just have to have fewer people. I will create winter and ice, and that should solve the problem."

Ahura Mazda said to Yima, "To survive the winter, you must immediately build a Vara [a city-sized enclosure] with lots of storage room. Then, gather two seeds each of all sorts of cattle, birds, dogs, horses, plants, and trees, and store them all there. When you are finished," Ahura Mazda told Yima, "take your family and a thousand couples of your own choosing into the Vara and close the great doors behind you. As long as you stay there, you will be safe." After Yima collected all the seeds, he shut the doors, as instructed, and his family was protected.

Outside, the snow began to fall and the rivers froze solid. Ahura Mazda sent frost that killed the plants and snow that buried the trees. Before long, the entire world was locked in the grip of winter, a season it had never seen before, and everyone who was left above ground perished from the cold. Meanwhile, in the Vara buried under the deep snow of the North Pole, Yima and the others lived the happiest of lives.

RETOLD FROM FRAZER, 1919

WHAT THE EVIDENCE SHOWS

- Supernovae produce cosmic rays that produce clouds that cool the climate.
- A supernova may have caused Earth's second-largest extinction 440 million years ago.
- Flares and impacts create nitrate. Ice cores show nitrate levels were high 13,000 years ago.
- Climate became the coldest in 130,000 years right after the supernova 41,000 years ago.
- The current or "sixth extinction" is a continuation of the Event that began 41,000 years ago.
- The extinction sequence: *extinction* produces *overpopulation,* which results in *depopulation.*
- Human overpopulation is the source of or a major contributor to all humanity's problems.
- If the extinction sequence holds, we face impending depopulation.

Now, let's shift to a different aspect of the Event, the Carolina Bays, those mysterious shallow holes in the coastal plain of the Carolinas.

17

THE CAROLINA BAY CRATERS

> QUESTION: Many scientists believe the bays formed, not from impacts, but from wind and water action alone. What evidence is there supporting the impact theory?

One of the strongest arguments supporting the impact theory is this: the bays seem to be unique on the entire planet, and whatever process formed them is not creating any new bays. In other words, if common agents like wind and water had formed the bays (instead of an impact), then wind and water should still be producing more bays at this moment somewhere on vast land areas of our planet. However, we want to point out that nearly all researchers agree that wind and water have modified the shapes and rims of existing bays; the only disagreement is whether wind and water *created the initial bay depressions.*

BAY CHARACTERISTICS

The impact theory of the bays is controversial, and many scientists prefer various terrestrial explanations, which we will review in due course. Before we go further, we need to specify the exact definition of a bay to make sure everyone is talking about the same thing, since there are many round, oval, and irregular lakes along the Atlantic coast that are not bays. Several scientists, such as Prouty, Thom, and Frey, have spent years studying them, and, they generally agree that bays have the features listed below. The references are from Eyton and Parkhurst (1975).

- *Elliptical shapes.* Bays are usually elliptical or egg shaped, and they tend to become more elliptical with increasing size.

- *Raised rims.* A true bay has a distinct raised rim around most of the perimeter, and many have full, continuous rims.
- *Low centers.* The center of a bay descends below the surrounding landscape, and the strata beneath the bay are undistorted.
- *Overlapping rims.* Frequently one bay overlaps another bay without distortion, and the rim of the overlapping bay nearly always remains intact.
- *Clusters.* They occur often in groups, clusters, and chains.
- *Common alignment.* In any given area, bays are arranged in the same general compass direction; in the Carolinas, the long axes typically point northwest.

MAJOR BAY THEORIES

With the above characteristics in mind, let's review some theories for formation of the Carolina Bays. The most widely accepted theories are:

Wind-water theory (D. W. Johnson, 1942, one version; and Thom, 1970). The wind-water theory is that Ice Age floods and winds created depressions, or "blowouts," which filled with water, forming lakes and ponds. As strong winds blew across the lake surfaces, this caused eddy currents that eroded the banks to create the elliptical bay shape. During times of drought, strong winds blew out loose sand from the dry lake beds to form the distinctive rims.

Solution theory (Johnson, 1942, alternate version; Legrand, 1953). Johnson and Legrand propose that groundwater or springs flowed through areas of loose sand, dissolving the clay and various water-soluble minerals, such as iron and limestone. After that process removed a substantial volume of minerals, the earth subsided to form depressions that filled with water to become the Carolina Bays. Over time, wind and water reshaped the depressions to form the rims.

Impact theory (Melton and Schriever, 1952; Prouty, 1952; Eyton and Parkhurst, 1975). This theory has several variations, but in the latest, an incoming comet or meteorite blew up high in the atmosphere to the northwest of the Carolinas. The resulting shower of pieces either landed on or exploded over the Atlantic coast to create the bays. Although the impact or explosion created the original depressions and rims, wind and water altered both over time, and they continue to do so.

SOLVING THE OVERLAP PROBLEM

Although scientists have argued each of these theories back and forth for decades, and although each explains some features of the bays, there is no consensus. But any successful theory must explain one particular bay characteristic: the overlapping rims. Overall, researchers estimate that 10 to 20 percent of studied bays overlap another bay (Thom, 1970), meaning that there are more than 50,000 overlapping bays on the Atlantic coast. As we will see, of the three major theories about bay formation, the impact theory is the only one that adequately explains the overlap of the bays.

The best explanation for the overlaps is that they result from the impact of multiple objects. After the first ones landed, others fell on top of them, creating the overlaps. Figure 17.1 shows that the bays overlap in nearly all compass directions, a fact impossible to explain with the wind and water theories, which propose that the strong bay-forming winds blew only from either the northwest or the southwest, varying according to the particular version of the theory. Either direction is insufficient to explain how the bays could expand backward for miles against the prevailing winds.

MORE ON THE OTHER THEORIES

We won't go into an in-depth discussion of the weaknesses of the other theories here. For those who are interested in a detailed comparison, however, see Prouty (1952), Johnson (1942), or Savage (1982). The last two out-of-print Carolina Bay books are available online, along with a wealth of other bay-related information, at Bob Kobres's Web site (http://abob. libs.uga.edu/bobk/cbaymenu.html).

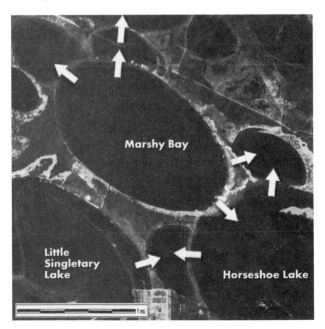

Fig. 17.1. Marshy Bay and a cluster of bays in North Carolina with overlapping rims. Notice the "fish-scale" effect at the arrows, where a top bay overlaps a middle one, which overlaps a lower one. Of the eleven bays, ten overlap others or are overlapped. Wind or water cannot create this effect, whereas an impact can. *Adapted from USGS*

What we will say is that we believe that neither the wind-water theory nor the solution theory can adequately explain several major problems with the bays. First, they cannot describe precisely how the action of wind, water, or solution could create such complex bay overlaps in which the rims remain intact. Second, and even more damaging, they cannot explain why such common, widespread forces as wind, water, and solution created the bays just once and then, mysteriously, stopped creating any more of them.

THE SURVIVORS: THE MATTAMUSKEETS

Modern scientists are not the only ones to believe that an impact caused the bays. The first were the Mattamuskeets, an Algonquin-speaking tribe that once lived around Lake Mattamuskeet, a large bay lake in northeastern North Carolina. It is part of a cluster of Carolina Bays, all of which have the classic characteristics of bays: they are elliptical craters with raised rims that sometimes overlap one another. Here is a retelling of a Mattamuskeet story that attributes that bay lake to a cosmic impact.

❦

The Princess and the Falling Stars

As the princess knelt before the altar, her mind turned to the plight of her dying People. It had been brutally cold for as long as anyone could remember, and there had been no summer that year and not much harvest. She chanted and prayed to the Great Spirit to help them through the winter season that had come very early again. Every day seemed colder than the last, and every week, a few more of her People weakened and fell ill.

After her chanting, she wondered if the Great Spirit was listening anymore. Then she heard a loud, high screaming noise and an enormous explosion, and then more blasts, one after another. Terrified, she raced out of her lodge. An immense bluish white steam cloud billowed high up into the noon sky above the crest of a nearby hill. At the same time, hundreds of falling stars streaked out of the sky and exploded in the wooded hills around her, shaking the earth so that she could not keep her footing. Fierce, orange-red flames and plumes of black smoke rose into the sky as the forest burned all around the camp.

But before she could run and hide, the star-fall was over. She waited for more to fall, but only the sound of the fires remained.

Before long, the rains began to fall. It rained night and day, and the rain put out the fires.

Fig. 17.2. The "Princess."
Source: Library of Congress,
Edward S. Curtis Collection

That afternoon, the princess and other tribe members left the camp to see what had happened. They found that the explosions had burned giant holes into the ground and that rain-swollen streams cascaded into them. The rains continued every day for thirteen moons, so that soon the craters were full. The giant hole closest to their camp became a huge fifteen-mile-long inland lake. They called it Mattamuskeet, the Lake on the Hill, and named their tribe after it.

Thanks to the falling stars, which brought much-needed water, they had survived. Out of seeming catastrophe came a new life.

RETOLD FROM BAREFOOT, 1995

PROBLEMS WITH THE IMPACT THEORY

Although the impact theory solves some major problems that the other theories cannot, it still must answer some remaining difficult questions. One of the main problems is with the dating. For the impact theory to be valid, we should find that all available bay dates are nearly the same, whether by radiocarbon or by a scientific dating method called optically stimulated luminescence, or OSL for short. That is not the case, however. Although there are only a limited number of radiocarbon dates available for bays, the dates vary widely within a range of 48,000 to 6,000 years. OSL and radiocarbon dates on bay rims range from 108,000 to 2,000 years ago.

So how can we explain that wide range? In summary, the bay impact theory proposes that tremendous impact explosions blew tons of material into the air and randomly jumbled together older and younger sediment in a geologic process called reworking, or redeposition, so that, for example, a 70,000-year-old leaf was left lying right next to a 17,000-year-old tree branch. Redeposition makes accurate carbon dating very difficult, since nothing is in its original position.

In his report in 1953 on Singletary Lake, a Bladen County bay, Frey complains about "redeposition confusing the record" (pp. 290–91). Second, he complains about the difficulty of dating the bays because "there are quite a number of . . . abrupt changes," by which he means that there are lots of missing sections of sediment. Clearly, the bays he worked on, which are typical of other bays, presented major difficulties for dating. Because of that, Frey, who did more work on the bays than nearly any other scientist, considered bay radiocarbon dates unreliable.

The same is true for OSL dating, a process that allows researchers to approximate the last time that quartz sand was exposed to sunlight. This is an important method for determining the age of a sandy surface that might include, for example, a Clovis spear point with no datable charcoal. The method works well when sand is deposited in a normal fashion, such as by wind action, but does not work well when rivers deposit jumbled sand of varying ages.

Likewise, there is a problem with OSL when an impact tosses all the quartz grains into the air and the sediment becomes seriously jumbled, with, say, 100,000-year-old grains landing next to 13,000-year-old grains. In that case, OSL dating is unlikely to be accurate; it returns only an average age of the oldest and youngest sand, making the layer appear considerably older than its actual age. In addition, radiation levels are crucial to OSL dating, and our evidence shows significantly increased levels of radiation in the Event layers, but not above or below them. Too much radiation in a thin layer can produce incorrect OSL dates.

Researchers have found this reworking problem in nearly all bays that they have dated. A good example of this is from Big Bay in North Carolina, where a large sand dune had blown in over the top of a bay after it formed. Ivester and colleagues (2003) dated the dune to around 74,000 years ago and the bay rim to around 11,300 years ago. The dune had blown over a deeper part of the bay that Brooks and coworkers (2001) radiocarbon-dated to about 48,000 years ago, but this layer was above an earlier layer that radiocarbon-dated to about 26,000 years ago. Here's a summary of what they found:

Dune covering the bay	74,000 years
Bay rim alongside the dune	11,300 years
Interior part of the bay	49,000 years
Deeper in the interior of the bay	26,000 years

The four dates are jumbled in an impossible order—the sediment could not have formed that way naturally, with the highest dune layer being 62,000 years older than the rim that ran under it. Something has to be wrong with the dates. Because the site was turned upside down by the impacts, the dates are upside down too. It is well known that impacts can invert strata, depositing older layers on top of younger ones, as is reported here. The 11,300-year bay-rim date (error range from 12,300 to 10,400 years) is near Clovis times and is the only one that actually reflects the true age of the bay; we believe the others represent pre-bay reworked sediment. The Clovis-era date is consistent with bay research by Riggs and coauthors (2001), as discussed below.

Our position about the dating is not meant to imply that the scientists did not accurately date the material; we accept that their dates are accurate. We propose only that some of their radiocarbon or OSL dates apply not to the age of the bays, but rather to the age of older material that was reworked during the impacts. We will not go into all the technical details for concluding that there are impact-related dating problems, but if you would like an in-depth discussion of the processes and the problems, please see our Web site at www.cosmiccatastrophes.com.

QUESTION: So, is there any way to get an accurate date for the formation of the bays?

Yes, there is research in North Carolina at Lake Waccamaw, a lake that also has a Native American connection. The tribe that lived near it was the Waccamaw-Sioux, or the Catawba, a name that some scientists translate as People of the Falling Star. Like the Mattamuskeets, the Catawba believed that, ages ago, a large fiery comet blasted a ten-mile-wide hole into the ground that filled with water to become today's lake. The lake, which is close to smaller, overlapping bays, has a distinct raised rim around it.

The scientific research on the lake was conducted on behalf of the state of North Carolina by Riggs and colleagues (2001), who made a comprehensive study of Lake Waccamaw, one of the largest bays on the Atlantic, and of nearly two dozen other bays nearby. They used extensive sediment cores, pollen studies, and about three dozen radiocarbon dates to determine the age of the Wando formation, a layer of sediment that they found beneath Lake Waccamaw as well as underneath all the other bays. Based on their data, the group concluded that, because several dozen bays are located on top of the ancient Ice Age river system and the Wando formation, Lake Waccamaw and the other nearby bays must be younger than the river. They estimated that the bays formed "during the cold climatic regime associated with the last glacial maximum, which occurred between 14,000 and 20,000 years ago."

This study is the most extensive to date for the Carolina Bays. If we accept the dates of this study, it is consistent with our theory that the bays formed about 13,000 years ago, well within the age limits of the river system on which they are found.

Another matching date comes from Ivester (2003), who studied the rim of Arabia Bay in southern Georgia and concluded that it was active 12,630 years ago, exactly at the time of the Event, and very close to the 11,300-year date of the rim at Big Bay, as discussed above.

WHAT THE EVIDENCE SHOWS

- The wind-and-water and solution theories cannot explain the overlapping rims of the bays.
- The wind-and-water and solution theories cannot explain why no new bays are forming.
- The impact theory can explain both of the above.
- Riggs concluded that the bays are 14,000 to 20,000 years old, close to our 13,000-year date.
- Two of Ivester's dates show bay rims to be about Clovis age.
- Older dates are most likely in error because of reworking during the explosions.
- Two Native American tribes have stories of fiery impacts that created bay lakes.
- Two Native American tribes also report rain for more than a year, coupled with severe flooding.

MORE QUESTIONS ABOUT THE BAYS

There are some remaining problems with the impact theory, such as why the bays are so shallow and why they are elliptical. To explore this issue more closely, we will have to travel millions of miles around the solar system to uncover clues that will solve these mysteries. Our first stops are Mars and the moon.

18

THE MARTIAN CONNECTION

In discussing the impact theory and its relationship to the creation of the bays, we need to define what we mean by the word *crater*. A crater is a depression formed by some type of explosion or by the impact of a meteorite or comet. By maintaining that the impact theory prevails in the creation of the bays, we are saying, in effect, that the bays are craters. To play devil's advocate, however, we will look at this issue from every angle.

QUESTION: How can the bays be craters when nearly all of them are elliptical? And how can the bays be craters when they are considerably shallower than almost any other known crater on Earth?

Unlike the bays, most accepted Earth-impact craters are round, and the depth-to-width ratio ranges from about 8 percent to 20 percent (Mazur, 2000). This means that if a bay such as Lake Waccamaw, which is about six miles long, is also a traditional crater, we would expect to see a crater depth of at least 2,500 feet. Instead, Waccamaw is a maximum of only 20 feet deep (Riggs, 2001), and the deepest known bay is less than about 50 feet deep. How can we reconcile these differences?

EARTH CRATERS VS. MARTIAN AND LUNAR CRATERS

To answer that question, let's look closer at meteorite craters, which are rare on Earth but very common on other planets. Scientists know that meteorites have hit the Earth much more often than they have hit the moon or Mars, which are smaller than Earth. There should be more craters here, but there are not. Why is that?

The reason is that Earth has a thick, dynamic atmosphere and lots of water, whereas other planets and our moon do not. Wind and water action tends to bury craters and to level the crater rims. So the simple answer to the bay rarity is this: there are no other impact features like them visible on this world because they have been erased. Earth impacts happen infrequently, and when they do happen, wind and water obliterate them quickly. The Carolina Bays show up only because they are relatively recent. Even so, wind and water will erase them all soon.

NOT-SO-RARE BAY-SHAPED CRATERS

There's another important fact to understand, which is that bay-shaped craters actually are not rare at all. The startling truth is that there are tens of millions of bays—it is just that they are on other worlds rather than on Earth. Because of the lack of water on most other nearby worlds, many baylike features appear there. As you will see in a moment, if some of the craters we see on the moon and on Mars happened to be on the Atlantic coast instead, everyone would call them Carolina Bays. They are identical.

Keeping in mind all the typical Carolina Bay characteristics, let's compare some lunar and Martian craters with the bays.

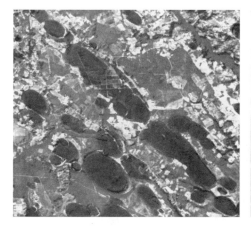

Fig. 18.1. *North Carolina*. Aerial photo of Big White Bay and other bays in a long chain across Bladen County. This string is about twenty-five miles long and more than a mile wide. It probably formed just like those on Mars and the moon, where a long string of ejecta from a larger crater created them. *Source: USGS*

Fig. 18.2. *Ganymede*. There is a nearly identical feature on Ganymede, one of the moons of Jupiter. The long chain is made up of thirteen separate overlapping craters that NASA thinks were created by the rapid-fire impact of a string of comet fragments. *Sources: Galileo Project, Brown University, Jet Propulsion Laboratory, NASA*

Fig. 18.3. *How the chains formed.* When an asteroid hit the moon long ago, it produced the huge crater at the upper arrow and blew out millions of tons of debris that formed starlike streaks radiating out from the crater (indicated by the long arrows). The ejecta formed long chains and clusters of smaller, elliptical craters at #1 and #2 that are nearly identical to the Carolina Bays. Each tiny crater in this image is actually many miles long. *Source: NASA*

QUESTION: Okay, but just how similar are bays to Martian impact craters?

To find the similarities between bays and Martian impact craters, let's do more side-by-side comparisons.

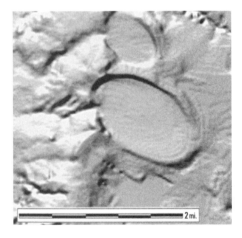

Fig. 18.4. *North Carolina.* This USGS DEM image shows Kings Bay and a smaller bay located northeast of Ivanhoe in Sampson County. They have all the classic features of Carolina Bays: they are shallow and elliptical, with encircling raised rims. Note how the impacts cut into the existing hills to the left. This is difficult to explain with the wind theory of bay formation. *Source: USGS*

Fig. 18.5. *Mars.* This large elliptical crater is located in the Margaritifer Sinus region of Mars at 25°N, 9.5°W. It has all the characteristics of a Carolina Bay. It is shallow and elliptical, with an encircling raised rim, and the smaller crater overlaps it (arrow), keeping its rim intact. *Source: NASA*

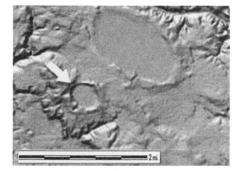

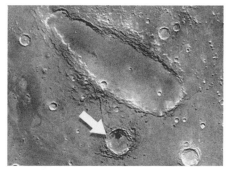

Fig. 18.6. *North Carolina*. This USGS DEM image shows two bays in Bladen County. Dismal Bay, the larger one, is elliptical. The other bay (arrow) is nearly round, as is somewhat common for smaller bays. This may be because smaller bays formed from impacts by smaller objects, traveling at a steeper angle. Those conditions tend to produce rounder craters. *Source: USGS*

Fig. 18.7. *Mars*. This miles-long, baylike elliptical crater is located in the Elysium region of Mars at 14°N, 180°W. NASA scientists have determined that the extreme elongation of this crater is due to a very low angle of impact of about 5 to 15 degrees. This confirms that to produce elliptical bays on Earth, the incoming objects traveled along a very low-angle trajectory. The smaller craters around this Martian crater came in at an angle steeper than 15 degrees, as did the smaller ones in North Carolina. *Source: NASA*

Fig. 18.8. *North Carolina*. Salters Lake, which looks like a giant bowling pin, is the bay we tested in Bladen County. As you can see, it overlaps two smaller bays. In this series of impacts, the incoming object that formed the largest bay landed on top of the other two bays probably only seconds after they formed. When the last object hit, the force blew out the rim to the northwest at the arrow, leaving a distorted section of rim between the two bays. *Source: USGS*

Fig. 18.9. *Mars*. Inside Newton crater at 39°S, 157°W, these overlapping, somewhat egg-shaped craters show the identical process at work. Two asteroids most likely landed one after the other, and as the upper one landed, the force of the impact blew the common rim out into the smaller crater. Also, notice the older, larger, and fainter crater, which is overlapped by the newer craters. *Source: NASA*

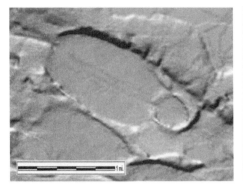

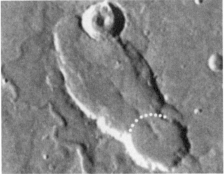

Fig. 18.10. *North Carolina.* This is a USGS DEM image of David Gam Bay in Bladen County, which looks a little like a giant's shoeprint with the heel to the right. The smaller "heel" bay with the intact rim appears to have formed after the larger bay. Such a feature is simple to explain by an impact but is almost impossible to explain by wind, water, or any other usual terrestrial process. *Source: USGS*

Fig. 18.11. *Mars.* This crater, which also looks like a giant's shoeprint, is in the Oxia Palus region of Mars at 26.5°N, 10°W. It shows an impact feature identical to the one shown in North Carolina, and here too a smaller crater formed on top of a larger one, destroying the common rim. In this case, the rim between them in the "heel" area is only faintly preserved (dotted line). *Source: NASA*

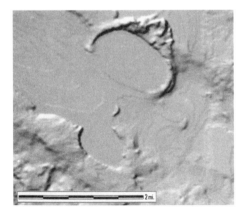

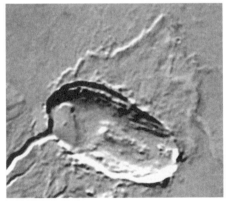

Fig. 18.12. *North Carolina.* A winglike pattern of ejected debris can be seen around Johnson Mill Bay. The incoming object struck the low ridge and blew the debris sideways about 3,000 feet, or half the width of the bay. But how can this have been caused by an impact from the northeast? Mars has the answer to this at right. *Source: USGS*

Fig. 18.13. *Mars.* This "butterfly crater" is named for its distinctive winglike ejecta pattern. Near 25°S, 97.5°W, it has a single visible wing, like the bay at left. The crater is about the size of the Hawaiian island of Oahu, and it was created when a very low-angle impactor kicked ejecta sideways rather than forward. This effect is much like that of a water-skier, who raises a wake to the sides rather to the front of the skis. *Source: NASA*

WHAT THE EVIDENCE SHOWS

- Baylike craters are rare on Earth, but they are very common on the moon and Mars.
- The only reason we do not see more craters on Earth is that comet impacts create shallow craters that quickly disappear due to the erosive action of the prevalent forces of wind and water.
- Elliptical craters are common on Mars and the moon. According to NASA, about 5 percent of all such craters are elliptical, amounting to more than 5 million on the moon and tens of millions on Mars.
- The Carolina Bays are shallow because they formed from low-angle impacts (5 to 15 degrees above horizontal), as do all elliptical craters on other planets.
- Nearly all crater chains of shallow, elliptical bays on Mars and the moon are low-angle, secondary impacts. This leads to a new understanding of the Carolina Bays: they also formed from the *ultralow-angle impacts of secondary ejecta*.
- The raylike pattern of ejecta on the moon and Mars always points back to the parent crater. For the Carolina Bays, this suggests a direction toward the Great Lakes and Canada.

What we have discovered on Mars and the moon has important implications for our search for the main impact crater on Earth. Now let's go back to the Atlantic seaboard.

19

THE MAIN CRATERS

QUESTION: Does evidence exist for the location of the main impact area?

Let's see if the Carolina Bays can give us some clues about where the main crater might be. Early researchers noticed that the axes of most bays in North Carolina are parallel. In trying to explain this odd feature, they proposed that an incoming swarm of meteorites created the bays, because meteorite fragments usually travel along parallel paths. As an extension of that theory, they thought that the long axis of each bay should be roughly parallel to others, no matter whether the bay is in the Carolinas, Georgia, or New Jersey.

Scientists must test all theories to see if they match the evidence, so Douglas Johnson set out to analyze the parallel-bay idea. He was an early researcher who developed one version of the wind-water theory, and he did not believe that impacts caused the bays. He surveyed eleven counties covering about 7,000 square miles along the North and South Carolina border, recording compass headings for nearly 400 of the largest bays. When he compared them at a county-wide level, he found that the long-axis compass headings of the bays point consistently to the northwest, as shown in figures 19.1 and 19.2.

Next, Johnson moved about 200 miles south, where he surveyed eight counties covering about 4,500 square miles along the South Carolina–Georgia border. There, he found that the average long-axis compass heading was very different; it pointed just 10 degrees west of due north, a difference of about thirty-five degrees (see fig. 19.3).

Johnson finished surveying nearly 400 large bays and published his results in a 1942 book, in which he concluded that a comet or asteroid could not have caused the bays, since they were not parallel, as the impact theory had proposed. Other researchers, however, such as Prouty and Eyton

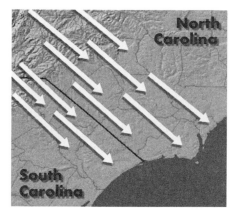

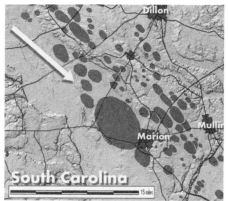

Fig. 19.1. Along the North and South Carolina border, Doug Johnson measured the compass heading of bays in eleven counties. All appeared to be parallel, coming from the northwest. *After D. W. Johnson (1942)*

Fig. 19.2. The parallel alignment is clear in this fifty-mile-long bay cluster near Marion, South Carolina. The heading is to the northwest, although individual bays vary. For size reference, notice that some of the bays are much larger than the cities nearby. Two large bays actually underlie parts of the city of Marion. This chain is just like ones on Mars and the moon.

Fig. 19.3. Along the Georgia–South Carolina border, Johnson measured the compass orientation of the bays in eight counties. There the headings were toward the north-northwest. *After D. W. Johnson (1942)*

Fig. 19.4. This USGS DEM image shows a site near Lakeland, Georgia. The bays in that area display an orientation that is almost due north, more so than Johnson found. Source: USGS

and Parkhurst, interpreted the same evidence in a very different way. They thought it actually supported a possible impact.

When he analyzed the bay angles, Prouty (1952) revised his impact theory to suggest that the asteroid had exploded in the air, causing multiple fragments that fanned out in a starburst pattern to create nonparallel bays. Such an object is called a bolide, and it frequently produces an intense aerial explosion without leaving a crater. The 1908 Tunguska event in Siberia was most likely a bolide that exploded about five miles above the remote Russian forests.

In 1975, deciding to explore the alignment of the bays further, Eyton and Parkhurst analyzed the compass heading of 358 additional bays inside and outside the Carolinas, including more in Georgia. They found that the most southerly bays pointed even more to the north than Johnson had found (fig. 19.4). With that, Eyton and Parkhurst had identified a clear pattern for the bays and an important clue: *the farther south you travel, the more the bays line up to the north.*

BAYS ALONG THE POTOMAC

When Eyton and Parkhurst went north to Virginia, they discovered bays there too, except that the axes of these bays tended to point almost east–west. The directional pattern was consistently different from that in Georgia, allowing them to conclude that *the farther north you travel, the more the bays line up to the west.*

In the north, there is a large cluster of bays just south of Washington, D.C., as shown in figure 19.5, but because these features are so highly eroded, it is difficult to be certain that they are bays. It is possible, but unlikely, that they are simply loops carved by ancient floodwaters from the river, rather than from an impact.

There are confirmed bays nearby, however, so it is plausible that these are giant bays. If so, then it is probable that other comet fragments landed squarely on the Washington, D.C., area, which is only thirty miles away. Even if the bays in this image were the closest impacts, the enormous explosions certainly would have had a catastrophic effect on the land around Washington. It happened long ago, so it seems safely out of mind. But the effects of such an event in this area would be catastrophic today.

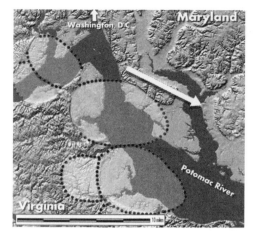 

Fig. 19.5. This USGS DEM image shows the nearly east–west orientation of some prospective bays along the Potomac River, a few miles east of Fredericksburg and about thirty miles south of Washington, D.C. The largest one is about nine miles long, spanning the entire river to cut into both the Virginia and Maryland shorelines.

Fig. 19.6. The ring centered on Elizabeth, New Jersey, is about fifteen miles in diameter, making it the largest bay ever found. It covers most of Staten Island and is about the size of all of New York City. The southernmost bay appears to have a faint double rim (lower arrow), which is an expected feature for the type of explosion that formed these bays. The compass direction is nearly west, as one would expect for the northernmost bays ever found.

BAYS IN NEW JERSEY AND NEW YORK

Prouty (1952) identified bays as far north as New Jersey, and our analysis of USGS DEMs and aerial photos has located more potential bays in that state, as well as in neighboring New York (fig. 19.6). These have all the classical bay features, such as raised rims, shallow center depressions, and overlapping rims. All are elliptical, although the largest one is only slightly so. As with other bays, these occur on soft, unconsolidated formations that would not have preserved these features for long, so they must be relatively young.

Most scientists believe that the rim around the Elizabeth, New Jersey, bay is actually glacial debris, called end moraine, which meltwater shaped into mounds and ridges when the last glacier retreated. This may be true, or it may be an impact, since it has every one of the usual bay features, and because it is more regular than the typical ice-sheet moraine. Such smoothly curved features are somewhat rare along old ice-sheet margins, but they are almost universal among bays. For comparison, in figure 19.6, the upper arrow points to a section of unquestioned moraine between Brooklyn and Queens in New York City. You can see that it appears to be much more irregular than the rim of the proposed bay.

In any event, the moraine theory would not explain either the other two overlapping bays or the fact that the other three almost identical, smaller bays are located miles beyond the accepted margin of the last ice sheet. Since they are much farther south, it seems unlikely that the glacier formed them, and it is more likely that they are very large eroded bays.

THE SURVIVORS: THE IROQUOIS

As you can imagine, these simultaneous miles-wide massive explosions would have devastated any area in which they occurred. Our research indicates that the immense force that formed the largest bay might have been more powerful than a large percentage of the world's current nuclear arsenal. When it fell during the Ice Age, however, there may have been only a few hundred people in all New York state. Most of them probably perished from the explosions that formed these bays. If those are bay craters in New York and New Jersey, the Native Americans who lived in that area probably saw them form.

The story that follows, first told long ago, is unusual in that it refers to extinct Ice Age mammoths about which the modern Iroquois should have no knowledge (and they are not the only tribe to have knowledge of Ice Age megafauna).

- The Ojibwa tribe referred to "giant animals" that caused trouble.
- The Brule told stories about large "water monsters" that lived near lakes and rivers.
- The Pawnee tribe said there were nonhuman "giants" that lived in the previous world.
- The Yuma tribe struggled with "large dangerous animals."

Did they all just make up these stories? Did they get the idea of extinct animals from Europeans, or maybe from fossil bones protruding from the earth? Or maybe they are exactly what they appear to be, stories of actual ancient conflicts with the megafauna.

This story comes from the Iroquois as one of the oldest traditions of the tribe. Much of the story matches what we know of the Event, including falling stars, an explosion that knocked over forests, extinctions, and flooding down the Hudson River. In addition, the story includes a giant animal called a Horned Serpent, which, curiously enough, is a term that Native Americans frequently applied to ominous comets with their large heads and snakelike tails.

We obviously cannot view these stories as accurate history, but only as distant reflections of what happened. Even so, the Iroquois account is chillingly close to what seems to have happened 13,000 years ago.

‿

The Monster Mammoth and the Horned Serpent

Long ago, a small band of the Iroquois camped along the banks of a great river. They were going about their business when, without warning, a monstrous Mammoth attacked them. As they struggled to defend themselves, a gigantic Horned Serpent crawled out of Lake Ontario to assault them as well. These two animals feared that human beings were becoming too powerful, so they had decided to destroy them.

After a long, ferocious battle, the People surrounded Mammoth and hurled dozens of long spears deep into its huge, hairy body. At last, with one loud, shrill trumpet, the mighty Mammoth toppled over dead, and seeing that, the great serpent quickly retreated.

That night around roaring campfires, the tribe feasted on Mammoth's roasted meat, and it gave them strength after the battle. They knew they would need it in the morning, for they expected the Horned Serpent to assail them again.

Fig. 19.7. Members of the Iroquois Tribe. *Source: Library of Congress*

The next day, the Horned Serpent attacked with its foot-long fearsome teeth and claws, as well as its poisonous breath that could kill people at a distance. The warriors fell by the dozens as a deadly cloud spread quickly throughout the entire camp.

Finally, seeing their men dying before their eyes, the women burned the last of their sacred tobacco as an offering and pleaded with the Great Spirit to save the tribe. Moved by their chanting, the Great Spirit sent down the spirit Thunder to fight the Horned Serpent.

In a vicious battle, Thunder threw great bolts of lightning to strike the serpent. The sound shook the Earth, and the flashes were so bright that the People shielded their eyes, covered their ears, and hid in fear.

Gradually, Thunder's onslaught drove the serpent back into the deepest part of the lake, but he could not kill it. In one final attempt, Thunder hurled down the most powerful thunderbolt ever seen. The concussion was so great that the mountains shook and entire forests blew over. The stars broke loose from the sky, and some came falling down to Earth.

Fearing for the safety of the tribe, Thunder tried to catch the stars, but he could not reach them all. The falling stars hurtled right toward the Iroquois camp, hissing with a fiery glare. With a ferocious blast and scorching heat, a star smashed into the Earth near their camp, blasting earth and trees in all directions.

Another star fell right into the lake on top of the Horned Serpent, wounding it with a huge explosion of steam. The great serpent thrashed its tail in pain, and each whip of its tail sent gigantic waves coursing down the river valleys and surging over the hills in a series of colossal floods.

When a 100-foot-tall wall of water roared down the river valley toward them, the People rushed from their lodge, heading for high ground and leaving everything behind. The whole tribe ran for the hills as the huge flood raced down on them faster than an eagle can fly. Most did not make it. Nearly all of them disappeared under the roaring waters. Of the entire tribe, only six families climbed to safety.

Gradually, over time, these few remaining people multiplied, grew strong, and spread across the land. As each family grew large enough, it set out to form its own band. Later, after many seasons had passed and the memories had grown dim of the great battles to save humankind, the six families who survived became the Six Nations of the Iroquois.

RETOLD FROM E. JOHNSON, 1881

POINTING TO THE SPOT

By combining our own bay alignment research with that of Prouty, D. Johnson, Eyton, and Parkhurst, we have the compass headings of more than 1,000 large bays, and when we map the resulting lines, a clear and surprising picture emerges. Most of the lines meet in Wisconsin west of Lake Michigan. Other lines point farther north into Canada (fig. 19.8). So is it possible there were multiple craters, and therefore multiple pieces of the comet? The evidence suggests that this is the case.

Using Martian elliptical craters as the model, the bay lines add new support for our new Event theory, which holds that one or more large pieces of a comet *hit the ground in the Midwest or exploded above the surface there,* blasting out a cloud of ejecta that crashed onto the Atlantic coast to form the bays.

QUESTION: Does this mean that the comet really hit Wisconsin?

Fig. 19.8. Bay axis lines extend into the Midwest. The darker lines converge in Wisconsin near the Great Lakes, and the lighter lines converge farther north in Canada. *After Eyton and Parkhurst (1975)*

Actually, the comet did not land in Wisconsin, but to understand why it didn't, we need to factor in just one simple fact: the Earth rotates—and it rotates *very* fast. A point on the surface of the ground in Wisconsin would be moving from west to east at a remarkable 700 miles per hour, faster than a jet and about as fast as a bullet. That motion is great enough to create an illusion, distorting the flight path to make it seem that it started farther to the west.

To find the proper flight path, we took the known facts, made a few assumptions, and then calculated the correct direction and distance to the east. The result was surprising: the impact site was in or over Lake Michigan. Now, let's see if there is any evidence for a crater or an overhead explosion in that area.

QUESTION: Could it be that Lake Michigan contains the missing crater?

Now that you have seen so many Carolina Bays, look at the satellite image of Lake Michigan and notice the shape of the lower part of the lake's south end. It looks like two stacked ellipses (fig. 19.9).

Are these impact sites? Possibly, although this idea may seem hard to accept at first, since most people assume that Lake Michigan has been there forever. Surprisingly, scientists can find no evidence that Lake Michigan or

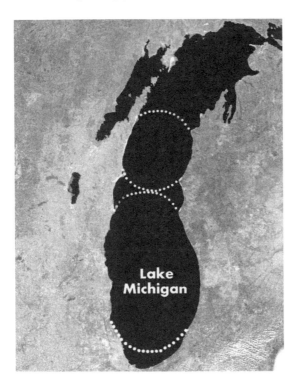

Fig. 19.9. The dotted lines mark two rough ellipses that look like Carolina Bays. In addition, the southern end of the lake has a much smoother, rounder outline than any of the other Great Lakes. Compare the south end with the rough, jagged upper shoreline of the lake. *Source: NASA*

any of the other Great Lakes existed prior to the end of the last Ice Age. Although many researchers assume that there were some earlier large lakes in the area, there are no scientific facts to support that.

In fact, the USGS and other organizations have drilled and extracted numerous cores from in and around Lake Michigan, and they have never found any sediment in the lake that predates the Event, although older layers exist on land all around the lake. The oldest sediment above bedrock found in the bottom of Lake Michigan is called the Haeger Till, a glacial formation that dates to around 14,000 to 16,000 years ago (Foster and Colman, 1991). All this means that Lake Michigan most likely formed quite recently in geologic terms, or if any prior lakes existed, *something* removed all evidence of them.

The age of the Haeger Till is within the 14,000- to 20,000-year range of dates by Riggs for the formation of bays. So could a comet, helped by the glaciers, really have created the distinctive shape of the lake? If it did, we need more evidence.

WHAT THE EVIDENCE SHOWS

- The bay lines point toward the Great Lakes and Canada as possible impact sites.
- Carolina Bay axes in the south tend to point more to the north.
- Carolina Bay axes in the north tend to point more to the west.
- Some bays are far larger than the modern cities over and around them, providing evidence for the awesome power of the Event.
- After correcting for Earth's rotation, the bay lines point to Lake Michigan.
- Lake Michigan's shape is somewhat similar to a very large Carolina Bay.
- The oldest sediments from Lake Michigan date to about the time of the Event.

Before we investigate Lake Michigan further, there is another mystery to solve. So far, we have found bays only along the Atlantic coast and not elsewhere. Why is that? If a comet actually landed in Lake Michigan, it should have sent a shower of ejected debris in many directions, not just to the southeast. It should have created bay craters in these other areas as well, so why don't we see them? Let's take a side trip to the Great Plains to find out. We'll return to Lake Michigan a little later.

20

BAYS ON THE HIGH PLAINS

QUESTION: Are there hidden bays elsewhere that the Event created?

Zanner (2001), in a conference presentation entitled "Nebraska's Carolina Bays," suggested that there are baylike features 1,200 miles west of the Carolinas out on the Great Plains. Called rainwater basins, the largest one is more than four miles long. There are many hundreds of these basins in Nebraska, and they look very much like Carolina Bays, since they are elliptical, shallow, and flat-bottomed, with rims that overlap other basins (fig. 20.1).

The one feature that is very dissimilar, however, is their orientation; they point to the northeast rather than the northwest. This actually strengthens the impact theory, however, since the axis direction of these basins typically points toward the northern part of Lake Michigan. This means that of all

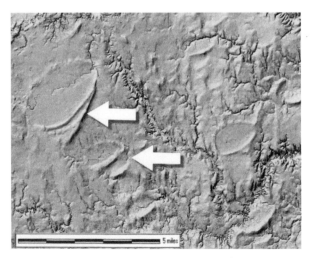

Fig. 20.1. DEM image of rainwater basins in Grand County, Nebraska, showing multiple overlapping features that look just like Carolina Bays.

214

the Carolina Bays or rainwater basins ever found, perhaps several million of them, every one of them points roughly toward Canada and the Great Lakes region.

AGE OF THE BASINS

One important fact is that there are no rainwater basins in the broad floodplain of the Platte River. If wind and water created them, they should be there too. This helps set a lower limit for the age of the basins, because it is widely accepted that the broad floodplain of the Platte formed near the end of the Ice Age, roughly 14,000 to 10,000 years ago. The rapidly melting ice sheet sent mighty floods surging down the Platte toward the Mississippi River, carving a visible channel through Nebraska. The present-day Platte follows only a very narrow channel down the center of the floodplain.

In addition, Zanner mentions that previous research has dated the underlying surface of the basins at about 31,000 calendar years ago. Since the basins formed after this surface formed, the rainwater basins are between 31,000 and 10,000 years old. Given that the basins are still visible, it is most likely that they date to the more recent part of this range of dates, meaning about 20,000 to 10,000 years ago. This range closely matches the estimate of Riggs for the Carolina Bays, and it includes our 13,000-year date for the Event.

We now have a date range for the basins, but, more important, we know that not one has formed since then. If the wind-water or solution theory created them prior to 10,000 years ago, it should be creating more of them today on the Platte River floodplain, and clearly this is not occurring. If, however, they formed from a single, very rare cosmic impact, we should not expect to see any more of them currently being formed.

In summary, we have seen that the impact theory explains the Carolina Bays as well as the rainwater basins in Nebraska. If so, the angle of impact points to the southwest. Could there be any more bay basins off in that direction?

HOLES ACROSS KANSAS

The next 3-D image shows the area around Wichita, Kansas, a city of nearly 350,000 people. Five large basins show up, the largest one about three miles long, and all show classic bay characteristics: elliptical shape; shallow, flat bottom; and raised rim (see fig. 20.2). Like those in Nebraska, they are oriented toward the northeast.

Three of the crater basins, ranging from about one to two miles in diameter, are located within the modern-day city limits. As you can imagine, if these same impacts happened today, it would be catastrophic for the entire city of Wichita.

Now, let's follow the trail of the basins and meteorites farther off to the southwest. There are more basins in Texas, which has some of the biggest crater basins of all.

SALTY SALINAS OF TEXAS

Just north of Odessa and Midland in the panhandle of Texas, there are hundreds of eroded basins, known locally as salinas because they often contain salty water. The salinas, which are similar to the bays and basins we have already seen, form a huge field that covers more than 10,000 square miles, the largest such bay field beyond the Atlantic coast (fig. 20.3).

Holliday and coworkers (1996) took core samples from beneath twelve dry lake beds in the area and found radiocarbon dates ranging from 16,000 to 20,000 years ago. He concluded that because these dates are from the underlying formation, the salinas formed more recently. This puts the creation of the salinas at about the same time as that of the Carolina Bays and

Fig. 20.2. This DEM image shows a cluster of baylike basins near Wichita, some of which are within the city limits.

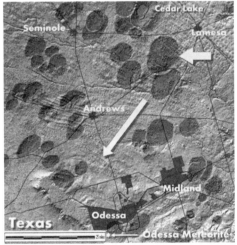

Fig. 20.3. In this group of salinas, the largest (short arrow below Lamesa) is about fifteen miles long, making it one of the largest baylike features ever found. As shown by the longer arrow, the long axes of the salinas align roughly with an impact flight path from the Great Lakes. Note that many of them overlap.

the rainwater basins. If that is the case, the huge elliptical basins across Texas may have been blown out of the soft earth of the plains by flying debris from the explosions to the northeast.

THE SURVIVORS: THE PAWNEE

The Clovis-era inhabitants of the area most likely would have seen impact-related megafloods. Do the tribes who lived there mention them? Here's a story from the Pawnee tribe, who roamed parts of Nebraska and Kansas. In it there are several common elements with the Event: heavy rains, continent-wide flooding, the extinction of the large animals, and the repopulation of Earth by a few people. Even though the story is told in an allegorical style, the similarity with the Event shines through.

⚜

Stuck in the Mud

In the beginning, before making People, Tirawa, the Creator, decided to create giant creatures to live in this world. Tirawa made them in many shapes, but in several ways, they were all alike—they were very big and very strong. Today, you might call them ugly, but Tirawa was happy with them.

After the giant creatures began to roam all over the Earth, they gradually forgot about the Creator. Being so big and powerful, they did whatever they wanted.

After a while, they began fighting with each other to see who was the greatest and most powerful among them. This led to many fierce struggles, and their constant fighting tore up the forests, dug up the prairies, and knocked down

Fig. 20.4. Two Pawnee. *Source: Library of Congress*

the mountains. Because they were so strong, there was great destruction.

Finally, Tirawa became angry with the giant animals. Stretching out a hand over the land, Tirawa called on the waters to rise. Rain began to fall, and water began to bubble up from deep in the Earth. Rivers overflowed their banks in a great flood that spread across the land. Everywhere there was water, but it was not deep, and the giant animals were tall, so at first they were not afraid. But then their feet began to stick in the mud, and they sank into the soft ground, first to their ankles, then to their knees, and then to their shoulders. They tried to escape, but the more they struggled, the faster they went down, until at last every one of them sank beneath the water into the mud.

When Tirawa saw that all of them were finished, the god waved a hand over the land, making hot breezes blow and causing the sun to dry out the Earth. Next, Tirawa fashioned a man and a woman, but Tirawa made them smaller, not giant-sized, so that they would not be so destructive. Tirawa also gave them some corn, showed them how to grow it as food, and sent them out into the land. They lived and multiplied, and today all Pawnees are descended from those two original people.

Now, when the Pawnee walk along the riverbanks, sometimes they find giant bones sticking out of the silt and mud. These are the bones from the animals that Tirawa drowned. They are there as a reminder not to forget the Creator.

RETOLD FROM GRINNELL, 1889

WHAT THE EVIDENCE SHOWS

- Some basins in Nebraska, Kansas, and Texas look just like Carolina Bays.
- Around Midland and Odessa, Texas, one field of basins covers more than 10,000 square miles.
- Like bays, the basins align with each other and point to the Great Lakes region.
- Like bays, no basins are forming today, suggesting that wind and water did not create them.
- The age of the basins is consistent with the date of the Event.

It seems that there are bays all across the Plains, suggesting that the Event reached deep into the U.S. Southwest, just as it did into the Southeast. In addition, there is another odd feature found on the High Plains that may connect to the Event.

21
STREAKS ACROSS
THE PLAINS

QUESTION: Is there any evidence for the debris that usually is ejected from an impact?

With the impact theory, one of the problems for the bays is locating the normal evidence we would expect to find from such an event. Whether meteorites explode in the atmosphere or hit the Earth, they blast small shattered pieces of the original body across the Earth near the site of entry. We often find those broken pieces in large elliptical clusters called strewn fields. There are many of these fields on Earth, so if we had an impact or aerial explosion 13,000 years ago, we might reasonably expect to find such a strewn field in North America.

To a certain extent, we have already found debris in a strewn field, in the form of magnetic particles and spherules, and, of course, we believe that ejecta created the bay craters, so they are a *type* of strewn field. There is another intriguing possibility, however, out on the Great Plains.

TRAIL OF METEORITES

Let's begin by looking for meteorites in Kansas. The Natural History Museum (NHM) in London maintains the widely recognized Meteorite Catalogue Database, which contains records for all verified meteorites found. They have two classifications for them, the first of which is *falls*, a term for recent meteorite events that people have physically observed. We can ignore that group, because falls all happened recently, but the other group is of interest.

Finds is the term, as you might guess, for meteorites that people have found on the ground but which fell sometime in the past. Searching the NHM database, we found 111 registered stony Kansas meteorite finds, which we plotted on a state map according to the location provided.

Next, we were curious about whether there was a pattern to the meteorites. Because seeing patterns can be subjective, we decided to "connect the dots" in a pattern analysis. That is, we drew lines from each meteorite dot to its closest two neighbors to form chains of dots. Since meteorites tend to cluster into strewn fields, we expected this method to reveal a pattern, if there was one. We knew there might not be any; because meteorites fall onto Kansas every year, it was possible that there was just a random pattern of overlapping strewn fields with meteorites spread somewhat evenly around the state. That was not the case, however.

The picture that emerges shows a distinct pattern with the meteorites clustering into long streaks. The most unusual result is that each of the three long lines points toward Lake Michigan (fig. 21.1). That result is beyond random chance.

But the fact that this pattern seems connected to the Event does not mean that all those meteorites came from that impact. Many fell both before and after the bays formed. Nevertheless, enough of them, perhaps 15 to 25 percent, fell at the same time that they made this pattern visible. Here, then, is another situation in which we can link baylike features and meteorites to a possible impact near the Great Lakes.

DATES FOR THE METEORITE CHAINS

There is one important question: Do many of these meteorites date to the time of the Event? There are several clues to the answer, the first of which is the age of the Kansas surface on which people found most of them. Earlier, we mentioned the research of Zanner (2001), who indicated that the rainwater basins of Nebraska and Kansas lie on top of a formation that dates to

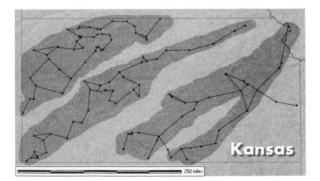

Fig. 21.1. A connect-the-dots map of Kansas, showing 111 meteorites that fell there in the past. The lines point toward Lake Michigan. *Data from the Natural History Museum, London*

the end of the last Ice Age, from about 10,000 to 30,000 years ago. So that fits. Most of the meteorites on our map probably fell less than 30,000 years ago, or time would have buried them.

Another clue comes from Jull (1993), who found that, in dry regions, stony meteorites have an average half-life of about 12,000 years, which means half of each will erode away to nothing in that time. Using his figures, it works out that for any meteorites that fell 36,000 years ago, nearly 90 percent of them would have disappeared. So we can be certain that most of the meteorites shown on our map fell less than 36,000 years ago, with many of them having fallen much earlier than that. But did many of them fall about 13,000 years ago? That would have to be the case for the pattern to line up with the Great Lakes.

To determine the likely age of the streak pattern, we turned to radiocarbon dating. Over the years, independent researchers have dated meteorites in published papers, and we found fourteen studies giving the dates for sixty-one meteorites found across the United States. We arranged the dates into 2,000-year groups and graphed the results.

We found an obvious and large peak at the exact time of the Event, which occurred around 13,000 calendar years ago (fig. 21.2). During the entire 30,000-year record, about 10 percent of the meteorites fell exactly at the time when the comet impacts occurred, and just a few thousand years earlier an equally large 10 percent peak occurred about when we think the second shock wave arrived. These levels are two or three times higher than expected, adding confirmation to other evidence that something extraordinary happened in North America at those times.

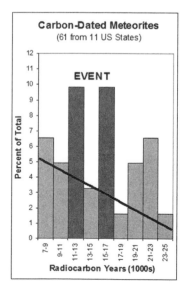

Fig. 21.2. For peaks in dated meteorites, about 10 percent fell 13,000–11,000 radiocarbon years ago and another 10 percent fell a few thousand years before that. These were the two last phases of the Event.

PATTERNS ON THE PLAINS

After finding that pattern in Kansas, we wondered whether it spanned the entire Great Plains. To answer that, we mapped the locations of 622 meteorite finds from the Natural History Museum's database for ten southwestern U.S. states. Together, they cover a square with 2,000 miles to a side, totaling 4 million square miles.

We found that the same streaks and clusters that crossed Kansas extended farther southwest into neighboring states. In addition, many more streaks appeared across most of the states in the lower Great Plains. As you can see from figure 21.3, all of them align with the Great Lakes.

Next, to see what the streaks looked like over the entire continent, we mapped the meteorites from the NHM for the United States, Canada, and Mexico, marking the locations of more than 1,200 individual meteorites. As before, they form a pattern of giant lines and clusters that point toward Lake Michigan. The longest series of lines extends nearly 2,400 miles southwest from Lake Michigan and into California almost to the Pacific Ocean, and a second long one extends through Texas almost all the way across Mexico (fig. 21.3).

The easternmost series of lines reaches from Lake Michigan through the Carolinas and Florida to the Atlantic Ocean, a distance of nearly 1,500 miles. The northernmost lines in Canada are less distinct, but they still cross nearly 1,000 miles of the Canadian provinces of Manitoba, Saskatchewan, and Alberta.

If the Event created these patterns, it blanketed an immense area with

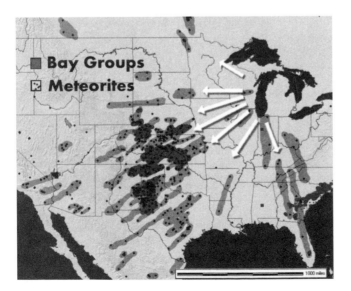

Fig. 21.3. This map shows about 1,200 North American meteorites. Long streaks extend radially away from the Great Lakes. The heaviest clusters of meteorites are over the heaviest concentration of Great Plains baylike basins, shown in darker gray. *Data from the Natural History Museum, London*

ejecta comprising meteorites, magnetic particles, spherules, and impactor fragments large enough to form the bays. North America makes up about 5 percent of the Earth's surface, and the Event appears to have covered it all with hot, high-velocity debris.

THE CAROLINA BAY METEORITE PROBLEM

The mounting evidence suggests that impact debris formed the rainwater basins and salinas on the Great Plains, but that presents a problem with the Carolina Bays. How can we explain the fact that almost no meteorites have been found alongside the bays? The simple answer is this: The Atlantic coast is much wetter than the Great Plains, causing stony meteorites, as well as iron ones, to erode faster. Whereas the half-life of stony meteorites in dry areas is about 12,000 years, in a wet environment like the Carolinas, the half-life may have been only 2,000 years. This means that after 13,000 years, which is the proposed date of bay formation, nearly 98 percent of the meteorites associated with the Carolina Bays would have disappeared. Originally, many of them probably fell on the East Coast, but if so, most are gone by now.

THE SURVIVORS: THE AZTEC AND TOLTEC IN MEXICO

In the state of Chihuahua, archaeologists found two large iron meteorites inside abandoned ruins. The Casas Grandes meteorite, found in 1867, weighed in at an amazing 3,400 pounds, and the Huizopa meteorite, found in 1907, weighed 286 pounds. The tribes must have considered them sacred, for they wrapped them in a ceremonial burial cloth. They also constructed special religious buildings around them where they had been discovered; certainly the larger one would have been very difficult to move. We cannot say for certain, but it is possible that these revered meteorites fell during the Event (Akridge, 1996).

So far, most of our tribal stories have come from the East Coast and from the Great Plains, but the following is a story from Mexico. Our map shows that a rain of meteorites fell across Mexico, just as recounted in this story, suggesting that the effects of the impact reached throughout the Western Hemisphere. This ancient Toltec-Aztec tale, which says that clouds obscured the Earth for twenty-five years, may originally have come from Event survivors.

Blood-Rain from the Sky

During the third age, Titlacahuan, the Creator, came to warn the People, saying, "This age is ending, and soon the Heavens will come crashing down." But many People no longer paid any attention to the gods. Others said, "We have plenty of time. Anyway, maybe it will not happen."

Before long, a heavy rain of flaming firestones and blood began to fall from the sky. It fell on houses and they burst into flames. It fell on fields and the crops were scorched. It fell on the forests and they were consumed. The People sought shelter, but their clothes burst into flames and they perished. Shaking caused by the falling firestones made some of the mountains explode, so that even more fire and rocks fell on the People. There was fire from above and fire from below; it came at them from all directions.

Most People did not know what to do. Some turned themselves into birds and flew above the storm. Others hid underground. But no matter what they did, most People perished.

At last, when it was over, thick, dark clouds covered the land for twenty-five years. The few survivors then began to create a new world, the one in which we now live. After the trouble was over, the Creator came back to tell the survivors that this current world will be destroyed by earthquakes one day. The People agreed among themselves never again to forget the Creator.

Even today, to remind us of those terrible days, People still find among

Fig. 21.4. Aztec statue. *Source: Library of Congress*

the river rocks and gravel the fire-starting stones [iron meteorites] that fell from the skies back then.

STORY FOUND IN A 1558 SPANISH MANUSCRIPT;
RETOLD FROM MARKMAN AND MARKMAN, 1992

WHAT THE EVIDENCE SHOWS

- Streaks of documented meteorites extend across North America.
- The streaks converge on the Great Lakes region.
- A significantly higher number of meteorite falls occurred around the time of the Event.
- The starburst pattern in North America is similar to impact craters on Mars.
- Tribal stories describe meteorite barrages such as might have happened during the Event.

Scientists learn most of what they know about meteorites by analyzing them with sophisticated instruments. Often when they do that, they find the signatures of supernovae and of other celestial events. Let's take a look at some of those telltale traces.

22

EARTH, THE RADIATION DETECTOR

QUESTION: It is hard to imagine how a supernova millions of miles away can create new elements on Earth, such as radiocarbon. How can that happen?

Even though cosmic rays from a supernova are extraordinarily small, they move so fast that they do significant damage when they hit other atoms, such as atmospheric nitrogen. In addition to producing a cascade of various particles when they smash into the atoms of the atmosphere and on the ground, cosmic rays produce radioactive isotopes, some of which have half-lives of thousands or even millions of years. The most important cosmogenic isotopes, the ones produced by cosmic rays, are carbon-14 (^{14}C), beryllium-10 (^{10}Be), aluminum-26 (^{26}Al), chlorine-36 (^{36}Cl), and calcium-41 (^{41}Ca). For our purposes, ^{14}C is the most important, since supernovae produce a lot of it when their radiation strikes our atmosphere. As we have seen, there was a sudden, sharp rise in radiocarbon about 41,000 years ago.

QUESTION: Maybe something besides the supernova or the impacts caused the rise in ^{14}C. Is that possible?

Let's look at what happens to the ^{14}C that cosmic rays produce in our atmosphere. The ^{14}C becomes radioactive carbon dioxide ($^{14}CO_2$), which joins the rest of global CO_2 by dispersing into soil and vegetation on land and by dissolving into the ocean. This movement of $^{14}CO_2$ is part of the larger global carbon cycle.

To understand the Event, which included a major increase in radiocarbon, we need to understand how radiocarbon levels can increase or decrease. First, we need to review one key point: to create the major increase that we see in the radiocarbon record, something had to add a lot of radiocarbon to the atmospheric storehouse. Scientists know that the atmosphere contains mostly nonradioactive CO_2—there are only 750 tons of radioactive $^{14}CO_2$, or less than 2 percent of about 41,000 tons total CO_2. The ^{14}C in the atmosphere is in equilibrium with the carbon in the land and the oceans, and cosmic rays currently produce about 5 tons of $^{14}CO_2$ per year in the atmosphere, which offsets the amount that decays in the air, on land, and in the sea. That 750-ton amount is the part of interest to us, since something doubled global ^{14}C to about 82,000 tons about 41,000 years ago. Radiocarbon in the air must have suddenly increased by a factor of more than 50 before dissolving into the ocean, still leaving 1,500 tons in the atmosphere to decay away.

Could the excess radiocarbon have come from the earth? A lot of regular carbon (^{12}C) exists in the ground, contained in coal, oil, and other forms of ancient carbon, but those sources contain no radiocarbon. This means that, if volcanic eruptions or the burning of fossil fuels released these ancient sources of carbon into the atmosphere, the fraction of radiocarbon in the CO_2 pool would become more diluted, so it would decrease, not increase, as we see at the Event. So this eliminates volcanoes and coal fires, for example, as causes of the radiocarbon increase that we see in the past.

The oceans store a lot of radiocarbon, so let's see how they might affect things. If the oceans were suddenly to release thousands of tons of CO_2 into the atmosphere due to sudden changes in ocean temperature or by stirring up ocean sediment, for example, the small proportion of $^{14}CO_2$ released from the ocean would produce no change in the atmospheric ratio of $^{14}CO_2$ since the ratio is virtually the same in the ocean and the air. On the other hand, if the global carbon pool were to suddenly decrease in size because of changes in ocean currents, the amount of ^{14}C in the atmosphere could double, but it would not occur overnight; it would take 10,000 years for that to happen.

So what else could suddenly produce another 41,000 tons of radiocarbon? Supernovae produce gamma rays, neutrinos, and cosmic rays, all of which can produce ^{14}C in the atmosphere. Calculations by Tommy Rauscher and colleagues showed that a type II supernova, which is twenty-five times the mass of our Sun, would produce total carbon enriched in ^{14}C by a factor of 10 million over terrestrial levels. Thus, a sudden burst of radiation, cosmic rays, or the influx of radiocarbon-rich debris from a supernova could dramatically increase the amount of $^{14}CO_2$ in the atmosphere, quickly doubling the entire global $^{14}CO_2$ pool. This scenario is the only one that clearly fits all the facts as we know them with regard to a major increase in radiocarbon levels.

QUESTION: Is there scientific proof that any supernova has increased Earth's radiocarbon levels?

Yes, there is scientific proof for an increase in Earth's radiocarbon levels due to supernovae. Nearly 1,000 years after supernova SN1006 exploded in 1006 CE, Paul Damon looked for such traces in dated annual tree rings. He chose the Big Stump Grove in Sequoia National Park, because some giant sequoias had been alive when the supernova occurred and were still around more than 1,000 years later. He also knew that tree rings can be accurately dated, because there is a tree ring grown for every year that the tree is alive. Damon's results are shown in figure 22.1.

He found that ^{14}C increased by an amount equivalent to adding 250 tons of ^{14}C to the environment instantly. As you may recall from chapter 21, the entire amount of radiocarbon in the atmosphere is only about 750 tons, so this event produced a one-third increase in atmospheric radiocarbon in only a few seconds, an increase equal to that from fifty years of normal cosmic radiation.

RADIOCARBON IN DEEP-SEA SEDIMENT

We have seen that historical supernovae leave their mark in the recent radiocarbon record. Figure 22.2 shows a much older prehistorical record of radiocarbon from Icelandic marine sediments. This chart uses the term *delta ^{14}C*, indicating the excess amount of ^{14}C existing at any time when compared to modern ^{14}C concentrations, which would read as delta ^{14}C = 0 percent. The chart shows three major increases in radiocarbon:

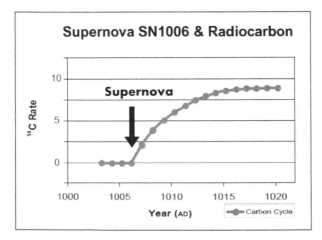

Fig. 22.1. The radiocarbon increase in tree rings in 1006 CE is attributed to SN1006. The event is spread over several years, as the ^{14}C mixed in the atmosphere and moved into the oceans. *Data from Damon et al. (1995)*

Fig. 22.2. Radiocarbon record from Icelandic marine sediments for the past 46,000 years. Sudden increases occurred around 41,000, 34,000, and 13,000 years ago. *Data from Voelker et al. (1998)*

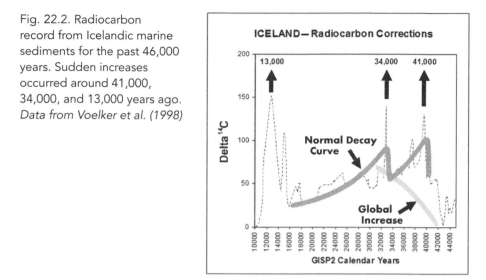

- *41,000 years ago:* At that time, global radiocarbon increased suddenly to delta ^{14}C = 150 percent, which equals a 250 percent increase in radiocarbon. This matches the first supernova radiation blast.
- *33,000 years ago:* The delta ^{14}C decreased up until that time, when another sharp increase occurred. The delta ^{14}C shot up by 175 percent. This matches the arrival of the supernova's shock wave.
- *13,000 to 16,000 years ago:* Radiocarbon again decreased to near-normal levels up until that time, when a third sharp increase occurred. This represents the arrival of the supernova debris cloud.

Such sharp increases in ^{14}C on Earth can be caused only by a sudden increase in the cosmic-ray rate or by direct deposition of radiocarbon from an impacting body. Between such events, global radiocarbon disappears with the expected ^{14}C half-life.

The thick gray curve in figure 22.2 indicates the expected decay of a sudden increase in radiocarbon with the ^{14}C half-life of 5,730 years. The thick, lighter gray curve indicates the rate that global radiocarbon would increase by doubling the cosmic-ray rate for an extended period. As the chart shows, that would result in a 70 percent increase in global delta ^{14}C over a period of 10,000 years.

This chart allows us to analyze the distance of a supernova. If a sudden increase in radiocarbon 41,000 years ago was due to a supernova, we could compare this with the ^{14}C increase in tree rings from SN1006 (see again fig. 22.1). SN1006 occurred 2,180 parsecs (7,100 light-years) from Earth and increased radiocarbon by 0.61 ± 0.16 percent. The supernova 41,000 years ago increased ^{14}C in Icelandic marine sediments suddenly by at least 150 percent. SN1006 data found that the increase occurred over a period of about ten years, while the Icelandic marine sediment samples data spanned periods of fifty years, so the instantaneous peak in radiocarbon was probably five times higher. If we scale the SN1006 radiocarbon increase to that of the 41,000-year-old event by the distance squared, we can estimate that the older event at 41,000 years ago occurred about 200 light-years from Earth (62 parsecs). This calculation matches very well with the estimated distance for the Geminga supernova. If the 34,000-year-old event results from the expanding supernova shock-wave shell traveling at 10,000 km/s, then the arrival 7,000 years after the original explosion is consistent with a distance of 230 light-years from Earth (72 parsecs), a figure that agrees quite well with the other estimate, given the large uncertainties.

QUESTION: So we can see that a supernova produces ^{14}C, but does it produce anything else that can be detected on Earth?

Throughout time, large quantities of other cosmogenic isotopes have been produced in supernovae, after which they were carried out into space by the expanding shock waves. Whenever Earth encountered such supernova debris, those isotopes rained down on the Earth and built up in sediment, tree rings, and the ice of glaciers. Scientists today can detect those ancient supernovae by studying the accumulation of cosmogenic isotopes in the geologic record, because there is no other known way that increases in certain of these isotopes can occur simultaneously on Earth.

For example, the radioactive isotope beryllium-10 (^{10}Be) is a key supernova-related element produced in the atmosphere by cosmic rays.

Figure 22.3 shows the record for ^{10}Be from a deep-sea drilling project in the Gulf of California near Guaymas Basin. McHargue and colleagues (1995) found distinct spikes in the ^{10}Be concentration at about 43,000 and 31,000 years ago, very close to where the ^{14}C spike occurred in the Icelandic data above. In their paper, McHargue and his coworkers argue that only a massive supernova close to Earth could account for the size of those spikes. The ^{10}Be peaks also closely matched sudden excursions in Earth's magnetic field at the same time. The researchers looked at all other explanations and ruled them out. Furthermore, they suggest that the double peaks appear to match the expected pattern of shock waves from a supernova. The accompanying changes in the Earth's magnetic field were also very sudden. (The magnetic field remains constant most of the time, disproving the claim by some that cosmogenic isotope-production variations reflect increased cosmic-ray rates from a changing magnetic field.)

Fig. 22.3. This Gulf of California ocean core chart shows [10]Be increases at 43,000 and 31,000 years, the times of the supernova blast and the first shock wave. Also, both correspond to two major magnetic excursions, times when the Earth's magnetic poles almost reversed polarity, but then returned to their original north–south alignment. *Data from McHargue et al. (1995)*

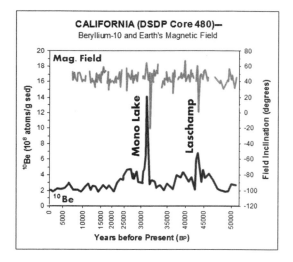

CALIFORNIA (DSDP Core 480)—
Beryllium-10 and Earth's Magnetic Field

The spike at 43,000 years is considerably weaker than the one at 31,000 years, whereas for the Icelandic marine-sediment data, the earlier radiocarbon peak was larger. To understand this, we need to go back to our scenario in which a supernova begins with a sudden gamma-ray flash and ends with a high-velocity particle and dust blast wave. It takes more energy to produce [10]Be than to make ^{14}C, so the lower energy from the initial gamma-ray flash would produce less [10]Be with a smaller spike, and that is what the record shows. The reverse shock wave and the debris wave carry in [10]Be that was formed in the explosion, so the spike is larger at 31,000. Both the [10]Be and ^{14}C data are consistent with the supernova scenario.

Figure 22.3 also shows the orientation of the Earth's magnetic field during the same time. The sharp fluctuations, called the Mono Lake and Laschamp excursions, occurred when the Earth's magnetic field temporarily reversed or fluctuated sharply. Each excursion tracks almost perfectly with the [10]Be data, suggesting a connection. McHargue and his team believe the supernova caused them, and the researchers discuss a means by which the incoming supernova-generated waves gave a strong shock to the Earth's field, causing it to do the flips that show up in the record.

McHargue and his coauthors also plotted the ratio of beryllium to aluminum (Be/Al) and compared it to other known sites, where, usually, Be and Al roughly increase or decrease together. As before, the researchers saw the same two peaks at 43,000 and 31,000 years, confirming that the Be came from outside Earth's system, such as from a supernova. In addition, the scientists found another anomaly at about 13,000 years ago, where they saw evidence for meltwater flooding, just as we saw in the Gulf of Mexico, the Atlantic, and the Arctic Ocean. Since the core sample the researchers studied was from the Gulf of California beyond the mouth of the Colorado River, it extends the reach of the massive meltwater pulses a long way from the ice sheets. It supports the idea that impacts suddenly melted the water and sent a huge surge racing down the Colorado into the Gulf of California. McHargue's team found no other surges besides that one.

LOOKING AT MORE ISOTOPES

The table in figure 22.4 shows additional stratigraphic data from around the world from sediments, stalagmites, tree rings, and ice cores which also show peaks in ^{14}C, ^{10}Be, or ^{36}Cl that may correspond to the three events. The dates do not match exactly due to many variables, but even so, the three-event pattern is clear. The ratios of Ice Age to modern global radiocarbon (R) were estimated from published data, and they indicate that the total cosmic-ray flux doubled about 40,000 years ago.

Peak values were comparable for all values, and the data are consistent with a cosmic-ray rate averaging two to three times the current rate from 40,000 to 13,000 years ago.

The most accurate ^{14}C data comes from the IntCal04 (Reimer et al.) data based on tree-ring measurements that are supplemented, for the oldest dates, with samples taken from coral reefs. These data have been carefully evaluated for consistency and are widely used for correcting radiocarbon dates. Figure 22.5 shows the excess radiocarbon for the past 16,000 years.

Site	Sample	Isotope	First Event Age	First Event Δ¹⁴C%	Second Event Age	Second Event Δ¹⁴C%	Third Event Age	Third Event Δ¹⁴C%	R[†]
Iceland	sediment	^{14}C	41,000	150	33,000	170	13,000	200	≈2.5[‡]
Bahamas	stalagmite	^{14}C	45,000	170	33,000	130	—		—
Dead Sea	sediment	^{14}C	50,000	170	28,000	60	—		—
Japan	sediment	^{14}C	31,000	80	23,000	60	—		—
South Africa	stalagmite	^{14}C	39,000	50*	35,000	100*	18,000	50*	—
IntCal98	tree ring	^{14}C	—		—		16,000	40	—
				Δ%[#]		Δ%[#]		Δ%[#]	
Gulf of California	sediment	^{10}Be	43,000	300	32,000	500	—		1.8
Vostok Antarctica	ice core	^{10}Be	60,000	150	35,000	500	—		2.3
Taylor Dome	ice core	^{10}Be	41,000	150	37,000	200	15,000	600	3
GRIP Greenland	ice core	^{10}Be	40,000	300	—		—		1.5
GISP2 Greenland	ice core	^{10}Be	40,000	200	—		13,000	300	2.4
GRIP Greenland	ice core	^{36}Cl	38,000	200	32,000	300	—		≈2

[†]Ratio of the average concentrations (45,000-15,000 BP)/(<10,000 BP).
[‡] Calculated from published data and corrected for ^{14}C decay.
*Few data points were measured for these data.
[#] Ratio of peak to normal isotope concentration.

Fig. 22.4. These peaks in ^{14}C, ^{10}Be, and ^{36}Cl indicate that three cosmic events have affected the Earth during the past 50,000 years. *Data from multiple sources*

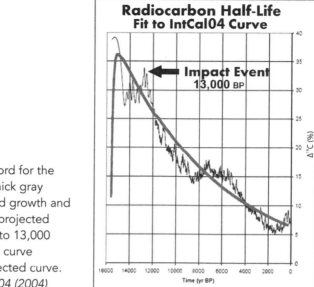

Fig. 22.5. Carbon-14 record for the past 16,000 years. The thick gray curve shows the expected growth and decay of ^{14}C following a projected supernova event 16,000 to 13,000 years ago. The actual ^{14}C curve closely matches the projected curve. *Data from Reimer, IntCal04 (2004)*

Radiocarbon increased between 13,000 and 16,000 years ago by about 35 percent and then declined continuously until modern times. The thick gray curve shows the expected rate of disappearance of the excess radiocarbon, assuming it decayed with the ^{14}C half-life of 5,730 years. After 13,000 years ago, radiocarbon production, and hence cosmic-ray rates, returned to normal, indicating that the effect of the supernova shock wave had passed. Small oscillations in the recent radiocarbon record probably result from variations in the Earth's magnetic field, which protects us from solar cosmic rays.

ISOTOPES IN THE ICE CORES

Recent independent ice-core research has confirmed the presence of increased cosmic radiation about 41,000 years ago, just as we would expect from a supernova. Nishiizumi and colleagues (2005) studied the GISP2 ice core from Greenland and measured the levels of ^{10}Be, ^{36}Cl, and ^{26}Al, three radioactive isotopes that come from supernovae. They found that the three isotopes rose in unison around 41,000 years ago to levels 300 percent of the values before and after that time. The researchers conclude that such peaks were "due to higher cosmic ray intensity at the peak period," and they suggest that the cause may have been a supernova.

REVIEW OF ISOTOPES

The cosmogenic isotope data are all consistent with the following:

First, a supernova exploded about 200 light-years (60 parsecs) from Earth 41,000 years ago, bathing Earth in cosmic radiation and increasing global radiocarbon by 150 percent, along with major increases in ^{10}Be, ^{36}Cl, and ^{26}Al.

Second, about 7,000 years later, another event increased global radiocarbon substantially again. The shell of debris from a supernova initially travels at about 6,000 miles per second (10,000 km/s), so assuming that ejecta from the supernova reached us at that speed in 7,000 years, the supernova would have been 230 light-years (72 parsecs) from Earth, which agrees very well with the distance above. The second event most likely involved the arrival of high-velocity, isotope-rich material ejected from the supernova.

Third, 20,000 years after the second event, at the end of the Ice Age, the last major event occurred, bringing yet more radiocarbon to Earth, most likely from the supernova debris wave passing the Earth. The arrival time of this event is consistent with the shock speed of the supernova remnant, which we would expect to have been moving faster than 2,000,000 miles per hour (1,000 km/s). All three events from the supernova correspond very well with the evidence in the geological record on Earth.

THE SURVIVORS: THE NAVAJO

The third event in the supernova scenario includes impacts on Earth by supernova debris, comets, and asteroids. The following story from the Navajo describes many events that are consistent with an impact to the north of the tribe in Canada and the northern United States: the explosion, the storms of black sand and ash, the heavy snowstorm from the north, the flooding, and the extinctions.

☀

The End of the Third World

During the last world before this one, men, women, and animals fought constantly. Begochiddy (son of the Creator), called Golden Child of the Sun, warned the People, "If you do not stop all this fighting, a great and terrible flood will come upon you that will destroy the world."

For a while, the People listened and tried not to fight, but then Coyote,

Fig. 22.6. A Navajo. *Source: Library of Congress*

who loved to cause trouble, could no longer restrain himself. He stole a child from a river monster, and war broke out again.

Then one day in the midst of a big battle, the People were startled by a great explosion that came from all directions. Suddenly a fierce storm of black sand and ash blew in from the east. Very soon, a storm of harsh yellow sand blew in from the west. Not long after that, a storm of stinging ice and blinding snow that looked like white sand came down upon them from the north. Then, without warning, a torrent of water poured out of the Earth and began to rise, swirling all around them higher and higher. Most splashed around helplessly until they sank under the waves. Only a few people and animals managed to climb into a giant hollow reed that floated toward them on the rising water. Before long, nearly every living thing in the world had drowned.

Seeing that it was over, Begochiddy waved a hand over the water, and it flowed back down into the Earth. Then Begochiddy called on the giant dust devils to spin out across the wet Earth and dry it out. When they finished, the Fourth World began.

RETOLD FROM BRUCHAC, 1991

WHAT THE EVIDENCE SHOWS

- Many radioactive isotopes support a three-phase event at 41,000 years ago, 34,000 years ago, and 13,000 years ago.
- The three phases were (1) radiation, (2) an initial shock wave, and (3) a debris wave.
- Gulf of California cores show a ^{10}Be peak at 41,000 and 33,000 years ago.
- The gulf cores also record a major pulse of meltwater around 13,000 years ago.
- The same core indicates that the supernova caused the Earth's magnetic field to flip briefly.
- In ice cores, the supernova isotopes ^{10}Be, ^{36}Cl, and ^{26}Al all peak at 41,000 years ago.
- The timing of all three events is consistent with what we know about supernova remnants.

Next, let's look at how supernova radiation has affected your genes and, oddly enough, your blood type.

23

SUPERNOVA 41,000 BP

QUESTION: When the supernova went off, what kind of damage would the radiation have caused?

As we have demonstrated, cosmogenic isotope data suggest that 41,000 years ago a supernova exploded about 200 light-years (60 parsecs) from Earth. Initially, a flash of light lasting a few seconds would have dazzled observers, even with their eyes closed, on the side of Earth facing the super-nova. The experience would have been much like that of a Hiroshima sur-vivor, except that no blast wave followed. Instead, the sky burst into an array of color as the enormous wave of cosmic rays lit up the atmosphere. A giant new star, larger than the moon, appeared in the sky, visible night and day, and became increasingly brighter for twenty-one days as the cloud of dust ejected from the supernova expanded and thinned. The light from the supernova, powered by X-rays and gamma rays from the decay of ^{56}Ni produced by the explosion, waned slowly for many months as ^{56}Ni decayed with a half-life of 77 days. The remnant of the supernova remained visible in the night sky for about a decade, growing dimmer as it faded from view.

The light from the initial supernova explosion included a massive burst of gamma rays, high-energy photons that produced nuclear reactions in the atmosphere, creating ^{14}C, ^{10}Be, and other cosmogenic isotopes. At sea level, living things were exposed to 300 rems, a measure of radiation exposure that is seven times the worst exposure of people living near Chernobyl. It was a lethal dose for some plants, animals, and people. All living things underwent radiation stress and injury, often leading to DNA damage, decreased fertility, mutations, and death. The ozone layer suffered wide-spread damage, exposing everything to dangerous ultraviolet radiation for many years. Nudged from their orbits, comets and asteroids rained down on the solar system, causing great destruction for years. Before long, hypervelocity particles from the explosion reached Earth, peppering our atmosphere with tiny projectiles. We saw this in the impact particles in the mammoth tusks. Cosmic rays increased cloud formation, causing climate changes. We found such a change in the sudden drop in ocean temperatures that took us to the coldest climate in 150,000 years.

About this time, major extinctions of large animals, the megafauna, occurred in Australia, Southeast Asia, and possibly Africa. These included every mammal in Australia larger than about 200 pounds (>100 kg), including the giant wombat, the marsupial lion, and many others. This occurred shortly after the arrival of humans in Australia, and some scientists blame their hunting practices and early use of fire to clear the land. Other researchers, such as Price (2005), believe that climate change did in the animals; they found that forty-four species disappeared, including land snails, frogs, lizards, small mammals, giant wombats, and kangaroos. Their theory is more plausible than overhunting, since it is hard to imagine primitive people killing and eating enough of all those creatures to wipe them out. A well-placed supernova could have achieved this same effect overnight, however, directly through irradiation, by suddenly altering the climate, and by setting the fires that destroyed the animals.

The era around 40,000 years ago was a period of major change in the evolution of humankind. Neanderthals began to decline, and Cro-Magnon people mysteriously evolved into modern humans. They expanded into Australia, and very possibly into the Americas. Around the same time, people may have domesticated the first dogs from wolves in East Asia. In addition, a mutation in human brain size appeared at that time that coincided with the emergence of traits such as art and music, religious practices, and sophisticated toolmaking techniques.

QUESTION: While it is easy to see how radiation could be lethal, it is hard to imagine how a distant supernova could affect our way of life otherwise. How is that possible?

It happens indirectly, through mutation. Major changes in species, including humans, occur through mutations, and the high cosmic-ray rate from the supernova would have accelerated the mutation rate dramatically. One example that appears to bear this out is the evolution of blood types in modern humans. Our early ancestors had only type O blood. Around the time of the supernova, mutations most likely occurred, creating blood types A and B. Types A and B blood are from dominant genes, so they spread through the population and became more common.

TYPE B BLOOD FROM ASIA OR AFRICA

DNA evidence suggests that B type blood probably originated in Central Asia or Africa, where the percentage is uniformly highest. Because the percentage is still very low in Australia and the Americas, it seems unlikely that it originated in either of those two places. Some geneticists conclude that type B is the youngest blood type, which appeared no earlier than 15,000 years ago and later than 45,000 years ago, and if so, this distribution seems inconsistent with early Americans originating in Asia and traveling across the Bering land bridge. If they had done so, there would be a lot more type B in the Americas. Other lines of genetic evidence support an Asian connection or a Polynesian origin, so the true homeland remains unclear. No matter where these blood types came from, however, the supernova could have produced them.

TYPE A BLOOD FROM ASIA OR AFRICA

For type A blood, the picture is more complicated, with apparent origins in Europe, Canada, and Australia. Again, there is little evidence that type A spread from Asia to the Americas. Instead, paradoxically, it appears to have arrived in the Americas from Europe long before Columbus did.

Is it possible that the Indians came from Europe? That idea seems far-fetched according to traditional views, and yet, according to Dennis Stanford, of the Smithsonian, and Bruce Bradley (2000–2002), there is intriguing evidence connecting Clovis flint-knapping technology to the Solutrean flint technology in Spain at the end of the Ice Age. In addition, Clovis points are very unlike flint points from Asia, their supposed land of origin. Since blood types show a connection with Europe, perhaps there is one. Rather than Asians, maybe the Solutreans really discovered the New World—or perhaps others did, because, remarkably, recent studies of early South American skulls suggest aboriginal or African origins.

TYPE O BLOOD FROM THE AMERICAS

Although type O blood is common everywhere, it is nearly universal among natives of South and Central America, and much more common in North America than in Asia or Europe. If people populated the Americas from Asia at the end of the Ice Age after types A and B arose, those people neglected to bring their normal distribution of blood types with them.

Another blood-typing system has been used to demonstrate the Asian origin of Native Americans. Called Diego, it evolved recently as a mutation, and all Africans, Europeans, East Indians, Australian Aborigines, and Polynesians are Diego-negative. East Asians and Native Americans are the only people that are Diego-positive. But Diego-positive is more common among Native Americans than among East Asians, raising the question of who got these genes first. From blood types alone, a case can be made that the oldest indigenous people are the Native Americans!

THE CAUCASIAN–ASIAN GENETIC SPLIT

The supernova may have a link to at least two of the races, Asian and Caucasian, as suggested by research from Nei (1982) and Gong and associates (2002). They presented genetic evidence showing that the two races split off from each other about 41,000 years ago, meaning that some major mutation occurred at that time. In addition, around the time of that mutation, these researchers found evidence that every Paleo-Indian (Asian race) that ever lived evolved from just one small group of individuals, who were the survivors of a major population decline, called a bottleneck. Many geneticists agree that all humanity endured a serious population decline sometime between 70,000 and 30,000 years ago, which some scientists attribute to an enormous eruption of Toba, a supervolcano in Southeast Asia. The unknown catastrophe decimated the human race, which dropped from perhaps a million to fewer than several thousand individuals. Scientists believe that every person on Earth is a descendant of one of just a small group of human couples living at that time.

We propose another possibility for that cataclysmic bottleneck—radiation from the supernova. The radiation levels were deadly enough to cause widespread casualties among early humans, and, in addition, intense radiation is capable of producing the major mutations that account for the skin color we see in the Asian and Caucasian races.

Altogether, the genetic and blood-type evidence indicates that many more mutations occurred in Asia and Europe around 41,000 years ago. Supernova radiation blasts last for only a few seconds, so the greatest radiation would fall on that portion of the planet facing the supernova at the time of the initial explosion. Because of the genetic evidence, we believe the supernova focused mostly on Southeast Asia and Australia. The genetic timing could be coincidental, and it will take more research to find out, since the exact dating of genetic events is still in question, but, once again, here is a possible link to the supernova 41,000 years ago.

HUMANS WITH LARGE BRAINS

Scientists have discovered that several genes regulate human brain size, and they know that when these genes malfunction, babies are born handicapped with severely small brains, a condition called microcephaly. In recent research, Evans and associates (2005), who studied ethnically diverse populations, found evidence that one of the key genes, called microcephalin, mutated about 37,000 years ago to allow people to be born with larger brains. Now, brain size does not equate with intelligence; many small-brained animals and people are very smart. But this mutation must have allowed some particular advantage to those who had it, or we would not find it in nearly everyone on the planet today. Evans and his team point out that, at the time of the mutation, art, music, and advanced toolmaking suddenly appeared to flourish, suggesting that there was a direct connection. Did that enormous flash in the sky directly lead to cave paintings and Vincent van Gogh's *Starry Night*? Did it bring about bone-flute music and Beethoven's *Ninth*? If so, it began with a dazzling light show in the sky 41,000 years ago.

ANIMAL EXTINCTIONS FROM THE SUPERNOVA

At about the same time, mass extinctions of large animals occurred in Australia and East Asia. Modern humans' domestication of dogs seems to have occurred then in East Asia. Beetle evidence found in eastern England suggests that, from 43,000 to 41,000 years ago in Europe and Asia, the weather suddenly became temporarily much warmer, before returning to much colder and drier conditions. Similar climate changes are coincident with nearby historical supernovae.

All this evidence supports the possibility that the supernova 41,000 years ago occurred mainly over the skies of Asia and Australia, but also partly over Europe and Africa. Large populations of humans and animals most likely were killed or sterilized by the radiation exposure. Survivors' genes developed mutations that spread through much of the world, but failed to reach the most distant populations in South America, which was on the opposite side of the planet at the time of the supernova explosion.

UNCOVERING THE "GENETIC ADAM"

Geneticists are able to look at genes and, through a complex set of markers, estimate the approximate ages of various groups of people. They also can do this by gender, because men and women have somewhat different genes. Throughout the last several hundred thousand years, the genetic makeup of men and women has changed, and according to the frequency of the mutations throughout the world's population, we can estimate the age of those changes.

According to S. Wells (2004), one male genetic mutation, called M89, arose in Iraq about 50,000 to 40,000 years ago. The ancestors of M89 Man moved into Asia and lived along the edge of the great Asian ice sheet, hunting mammoths and other megafauna. After that, the man's descendants expanded to include nearly everyone alive today in Europe and the Middle East, along with most Asians and Native Americans, a fact that argues for a migration from Asia to the Americas.

Another earlier key mutation exists in nearly every male on the planet; scientists believe it first occurred in a person whom they call Genetic Adam. Using computer projections, geneticists estimate that Genetic Adam lived between 100,000 and 40,000 years ago. Some scientists believe the date was closer to 100,000 years ago, but if it was nearer to 40,000, then Genetic Adam was alive when the supernova hit.

The above dates are only approximate, because mutation rates can only be estimated, but there is one fact involving the mutation rate that, for all these mutations, favors the younger dates nearer to 41,000 years. The problem is that geneticists assume a constant rate of mutation, but they do not factor in the radiation from the 41,000-year-old supernova. That potent burst of radiation greatly speeded up the mutation rate, making all genetic-age estimates too great. Thus, if geneticists estimate that the population bottleneck was 70,000 years ago, it is more likely that the true date is closer to 41,000. If so, then Genetic Adam and the decimation of humans, known as the bottleneck, may have occurred exactly when the supernova dazzled our ancestors in the skies over Asia and Australia.

Calm returned after the initial flash of the supernova explosion and the eventual disappearance of the bright new star in the sky, but the world had changed. Even though the star was gone, the effects lingered; the event profoundly altered both humans and animals. That brilliant burst of radiation silently rearranged humankind's DNA and unleashed a burst of brilliant cultural creativity that progressed from launching innovative new mammoth spears to launching innovative new rockets to the moon.

THE SURVIVORS: DOGS OF THE OJIBWA TRIBE

Since humans and dogs have befriended each other for so long, there are many cultural stories about their relationships. In this one from a Native American tribe, the Ojibwa, we have an ancient story about how the two got together. The story's setting is during a time of natural disasters, just as would have occurred from the supernova radiation and debris. This fits with the domestication of dogs, which scientists believe took place sometime between 41,000 and 13,000 years ago.

✍

The First Dogs Make a Hard Choice

In ancient times, people and animals lived in harmony. All was well until one day the Earth began to shake violently, knocking the first man and woman to the ground and toppling tall trees in the forest.

As the couple crouched on the ground in fear, with a grinding rumble the Earth in front of them split apart in a great crack that cut across the planet from horizon to horizon. As it opened, hot wind blew dust and sand into their faces.

The crack grew ever wider, and the first man pointed to the far edge of the growing canyon. "Look! There!" he shouted to his mate. Horses, cattle, donkeys, deer, elephants, dogs, and all the other animals were trapped on the other side. The couple was alone on their side of the great crack.

As the first man and woman watched, the creatures turned one by one and fled into the forest to escape the growing rift. Before long, all had disappeared, except for the first two dogs, a male and a female. Fearful of the mighty canyon, the two dogs first stared at the pit, then looked over at the couple, and then glanced back at the pit. They wanted to flee like the other animals, but their loyalty to the man and woman made them hesitate.

Finally, the dogs ran toward the chasm and leapt as high and as far as they possibly could. The female landed on the other side, but the male's leap was too short, and he hung by his front legs, slipping from the edge. Racing forward, the first man grabbed the dog by the neck and pulled him to safety.

After that, bound by their mutual loyalty, people and dogs have prospered together. Many times, people have saved the lives of dogs, but just as many times, dogs have saved the lives of men and women. Dogs made a choice long ago to stand beside humans, and it has worked out well for both.

RETOLD FROM THURSTON, 1996

WHAT THE EVIDENCE SHOWS

Scientists suggest the following occurred about 45,000 to 30,000 years ago, during supernova time:

- Cro-Magnon humans appeared suddenly. In nearly all respects, they were like modern people.
- Neanderthal humans, the so-called cavemen, began to disappear or assimilate.
- Our ancestors domesticated dogs from gray wolves, very possibly in Southeast Asia.
- Art suddenly emerged as a sign of human creativity, and we find it on many cave walls.
- Some scientists believe sophisticated speech arose at that time of the supernova.
- The human brain became larger and capable of much more complex thinking.
- Mass extinctions occurred, primarily in Australia, which faced the supernova radiation.

Now, let's see what happened when the supernova debris wave collided with Earth and the rest of the solar system.

24

SOLAR SYSTEM IMPACT

QUESTION: If the supernova blast hit the Earth, did it affect the sun, moon, and other planets?

If the supernova happened the way we think, we should see evidence of this elsewhere in the solar system. Fortunately, the Apollo astronauts collected numerous rocks and sediments from the moon, which can be analyzed to search for clues about these events. Herbert Zook, a NASA scientist at the Johnson Space Center in Houston, found evidence, in lunar rocks, of a recent increase in cosmic rays. In figure 24.1, the density of 10-micron-deep cosmic-ray tracks is plotted against the age of lunar rocks.

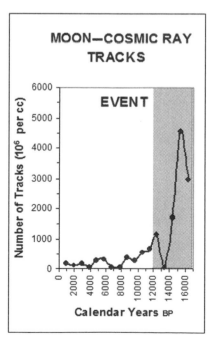

Fig. 24.1. Lunar cosmic-ray impact rate over the past 16,000 years. It was about twenty times higher during the Event. *Data from Zook (1980)*

245

Many more cosmic-ray tracks are found in the microcraters of older rocks, suggesting that cosmic-ray rates were higher in the past, especially around 16,000 to 12,000 years ago, during the arrival of the debris wave from the supernova. Zook believed that the tracks came from massive solar flares that erupted from the sun, although similar tracks would also be produced by the supernova.

So is one or the other the real scenario? It is possible that both scenarios are true, because the debris wave of the supernova could have seriously disrupted the sun, causing a prolonged period of flare eruptions. This is consistent with research from Lal and coauthors on cosmic-ray effects on the GISP2 ice core. These researchers found evidence for a sudden large increase in solar activity. Those enormous bursts of solar particles would have caused major problems for life on Earth.

The table in figure 24.2 shows the track and crater data from moon rocks in another way. Here, modern cratering rates, as determined by Apollo and Skylab missions, are used to determine a "crater age" for each rock. Track rates from glass on the Surveyor 3 satellite are used to determine a cosmic-ray "track age" for the exposed surface of the rocks. A third date is determined from the amount of the cosmogenic isotope ^{26}Al on the surface of the rock compared to the modern production rate corrected for erosion. Aluminum-26 is produced by cosmic rays, so the age determined by this method should be comparable to the "track age," as it is. Both ages appear much older than the "crater age," indicating that there were many more cosmic rays bombarding these rocks in the past than in the present.

Also, the cosmogenic isotope ^{59}Ni was also found in higher-than-expected concentrations near the surface of these rocks, confirming a high cosmic-ray rate in the past. Finally, Zook determined that the ^{14}C concentration in lunar surface soils was three times the expected amount. Based on all the evidence, he concluded that a significant increase in cosmic rays

Lunar Rock	1. Crater Age	2. Track Age	3. ^{26}Alum. Age	Ratio of #2/#1	More Past Radiation
12054	26,500	175,000	~150,000	6.6	Yes
15205	15,300	80,000	~100,000	5.2	Yes
60015	20,000	150,000	---	7.5	Yes

Fig. 24.2. Comparison of ages from cratering, cosmic rays, and ^{26}Al rates in lunar rocks. All show more cosmic radiation in the past. *Data from NASA*

occurred during the past 20,000 years, although he was unsure of the exact date and agreed that it could extend further back in time. We think all the newer evidence extends the date to 41,000 years ago.

Later, with improved technology, Tim Jull of the University of Arizona and colleagues more thoroughly analyzed the cosmogenic isotopes in lunar rocks. They reported, "Over the last 20 ka [20,000 years], there cannot have been more than one event $>5 \times 10^{13}$ protons/cm^2." To put this in perspective, the modern cosmic-ray rate is about 1 proton per cm^2 per second, so such an event would produce 170,000 years' worth of cosmic rays! If we extend Jull's limit to 40,000 years, then about 15,000 years' excess of cosmic rays would have been required, enough to more than double global radiocarbon. Notably, Jull's team worded its conclusions negatively, perhaps to mute the earlier conclusions of Zook. It has been unpopular in much of the research community to consider seriously the possibility that a recent increase in the cosmic-ray rate could have occurred, because it is a phenomenon that can be explained only by the explosion of a nearby supernova.

ANCIENT LUNAR SPHERULES

Timothy Culler and Richard Muller of the University of California, Berkeley, have analyzed lunar spherules using the potassium–argon dating technique. These spherules were produced by meteor impacts on the moon that scattered the spherules long distances over the entire surface, and the age of

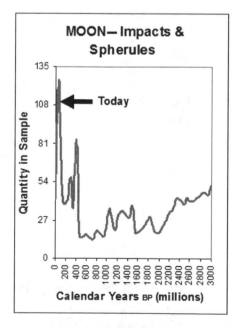

Fig. 24.3. The rate of production of lunar spherules over the past 4.5 billion years. *Data from Culler (2000), in figure 4*

each spherule dates the impact that formed it. Figure 24.3 shows the number of lunar spherules grouped by age. Impact rates were fairly low 3 billion years ago and declined markedly until about 400 million years ago, when rates climbed abruptly. This record of increased impacts on the moon corresponds closely with the major extinctions on Earth 440 million years ago.

The most dramatic increase corresponds to the time of the dinosaur extinction 65 million years ago, and it continues to this day at a higher level than for most of the last 3 billion years. Clearly, the moon has been under a major meteor and comet assault in recent times, and this means that Earth has suffered from the same fate. According to Muller's research, we are still at one of the highest levels of bombardment that the Earth has seen in the last 3 billion years. There is absolutely no evidence that it is over; instead, all the evidence suggests that it is continuing.

RADIOACTIVE HOT SPOTS ON THE MOON

Many of the Clovis-era magnetic grains are abnormally high in radioactive thorium, uranium, and potassium-40, so we set out to determine the significance of this as it pertained to our theory. Our first stop in the search for answers took us back to the moon.

NASA orbited the Clementine satellite in 1994 and the Lunar Prospector satellite went around the moon in 1998, both carrying gamma-ray detectors capable of measuring levels of the radioactive isotopes thorium-232 and potassium-40 on the moon's surface. Figure 24.4 shows the distribution of these elements only on the near side of the moon; the far side has very little of either. Notice that distribution of the darker "hot spots" for both isotopes is nearly identical, and both are only in the northern hemisphere, in a rough ring around a large lakelike crater called Mare Imbrium.

Lunar rocks from the regions of high thorium and potassium concentrations are called KREEP rocks, short for potassium (K), rare earth elements (REE), and phosphorus (P). These elements are usually not found together. We have KREEP-type rocks at many of our Clovis-era sites, and here they are on the moon as well. They will show up again in our story.

The position of these hot spots cannot be accidental. That "bull's-eye" on the moon is the only region with high concentrations of radioactive potassium and thorium. Some researchers suggest that an unusual geological process placed these elements nowhere else except on this small part of the moon, where a subsequent impact threw them out across the moonscape. That may be true, but we wonder if these spots were "painted" on the northern hemisphere of the moon instead, during a single barrage of supernova or cometary debris. We cannot be certain at this point, since there is insufficient evidence, but this seems to be a crucial piece of information. Keep this unique northern hemisphere hot spot in mind, since we will see it appear three more times in our solar system.

RADIOACTIVE MARS

Radioactive potassium and thorium analyses have also been done on Mars with the gamma-ray spectrometer on the Odyssey satellite launched in 2001. Figure 24.5 shows Martian potassium-40, which, along with thorium, appears in the highest quantities in the northern hemisphere, just as on the moon.

This could have happened in two ways on Mars. First, the same unusual geological process could have occurred on Mars as on the moon. No one currently makes that claim, however, because it is clear that Mars, being a much larger body, did not cool the same way that the moon did, so that theory does not seem to fit. The only alternative, which is the most plausible one, is that some type of radioactive impactor crashed into the northern hemisphere of Mars, just as on the moon.

It happened on Venus, too, although the data are sparse. Earlier studies of Venus with the Soviet Venera missions that landed a gamma-ray spectrometer on the surface also found enrichments in ^{232}Th at midlatitudes of the planet's northern hemisphere.

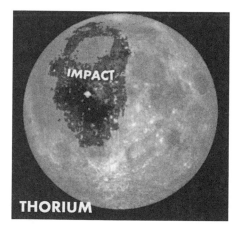

Fig. 24.4. Map of the only large areas of thorium on the moon; radioactive potassium has an identical pattern. Both are in the moon's northern hemisphere. *Adapted from NASA/JPL*

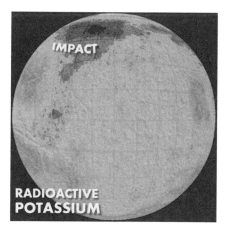

Fig. 24.5. Odyssey satellite map of radioactive potassium in the midlatitude regions of Mars. Thorium left the same pattern. *Adapted from NASA/JPL/University of Arizona*

HOT SPOTS ON PLANET EARTH

As you might guess, we think the same thing happened on Earth. Although data are not readily available to us outside North America, we suspected that radioactive thorium and potassium are most highly concentrated in the Northern Hemisphere of the Earth, so we looked first in Canada, which, according to the flight lines of the Carolina Bays, may be one of the main impact sites. In Canada and the United States, detailed measurements of the geographical distribution of ^{232}Th and ^{40}K have been performed with gamma-ray spectrometers flown on airplanes, rather than with satellites, as on the moon. Not all areas of Canada have been completed, so the maps do not yet cover the entire country. But even so, as we can see in figure 24.6, thorium forms an elliptical pattern around Hudson Bay, which, incidentally, is not a Carolina Bay, but rather a true ocean bay. It is remarkable that potassium, thorium, and uranium form nearly identical patterns yet have very different chemistries, and move around much differently in the environment. For example, thorium is less soluble than uranium or potassium, and therefore the two should be distributed differently after millions of years. Because they so closely overlap each other, all three are probably recent arrivals.

As to the deposits that cover the Rocky Mountains, most scientists think they were part of the natural mountain-building processes. They may be correct, but we found that the deposits do not line up very well with the

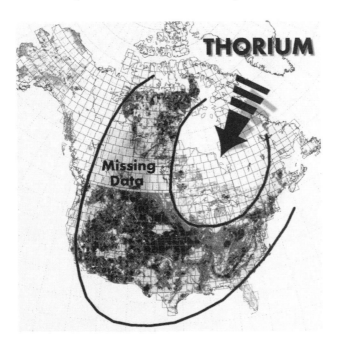

Fig. 24.6. Thorium appears to form a ring around Hudson Bay in Canada and extends into the southwestern United States. Potassium and uranium form nearly identical patterns. Even though there is missing data for some areas, the pattern is still evident. *Data adapted from the USGS and Natural Resources Canada*

mountains; in fact, some of the most uniform deposits are spread across the Great Plains of Nebraska and Kansas, far from any mountains. As before, we suggest that a large dustball comet exploded to the north, dusting the area with the radioactive elements and with KREEP.

Even though the visual evidence is striking, we have no proof that our theory is true—it is only speculation at this point. If it is true, then at least two enormous cometary bodies exploded over the heartland of the United States and Canada, blasting high-speed particles and tornado-force, super-heated air across the continent. If this did occur at the end of the Ice Age, it is little wonder that the mammoths disappeared, along with about 40 million other animals. At the same time, other, similar dustballs may have slammed into the moon, Mars, Venus, and the sun, creating a barrage unlike anything else that we see in the geologic record.

A LUNAR KREEP ROCK IN THE DESERT

Recently an unusual meteorite, named Sayh al Uhaymir 169, or SaU 169, was found in the Oman desert in the Middle East. Its composition is different from most other meteorites, in that it is very similar to lunar KREEP, so scientists concluded that SaU 169 originated on the moon from the bull's-eye area that you saw in the lunar image above in figure 24.4. The meteorite is extremely rich in potassium, rare earth elements, phosphorus, thorium, uranium, and potassium oxide, just as are the magnetic grains we have found at Clovis-era sites in the United States. In addition, based on the ^{14}C and ^{10}Be concentrations in SaU 169, the meteor arrived on Earth 13,000 to 9,500 calendar years ago, at a time consistent with the age of the particle layer.

Perhaps a large cosmic object struck the moon, precisely in the KREEP-y part of Mare Imbrium, and knocked off tons of debris, which then crashed into Earth, spreading KREEP across North America. That is one of two possible ways that it got here. In the second way, the impactors 13,000 years ago may have been KREEP-rich bodies themselves. We don't know if either theory is right, but it is intriguing that this KREEP meteorite from Oman fell during the Clovis period and that its chemistry so closely matches the KREEP-like material that fell across North America at the same time.

WHAT DOES IT ALL MEAN?

Even though the universe is full of unlikely events, we think it improbable that all the phenomena we have described are isolated and unrelated. One of the best explanations is that the supernova shock waves passed through our part if the galaxy, producing a widespread bombardment of our entire solar system. The abundant potassium and thorium located almost exclusively in the northern hemispheres of the Earth, the moon, Mars, and possibly Venus suggests that these elements were deposited by massive impact events that came from one particular direction, from the north as we view it on Earth.

THE SURVIVORS: THE TOBA AND PILAGÁ TRIBES

Even though tribes in South America were far away from the impacts in Canada, Michigan, and Utah, nevertheless, stories about impacts show up in their cultures. Could these impacts have been more widespread than we have investigated so far? Or is this just a story handed down from eyewitnesses who moved to South America later? We believe that the impacts reached at least into the northern part of South America, as described in the following story from the Toba and Pilagá tribes, who live in Bolivia, Paraguay, and Argentina.

If an icy comet hit the Earth, we would expect to see both burning rocks and giant icy hailstones falling from the sky, as recounted in this story. Also, the impact would have blown a lot of dust and soot into the atmosphere, which would have produced extended darkness.

༻

Long Night of Fire

One night, without warning, huge jaguars with giant yellow eyes attacked the moon. The battle went on for hours, and as the People watched in horror, the monster-cats tore great pieces out of the moon, turning it red with blood.

Fig. 24.7. Jaguars devour the moon. *Composite from NASA and the Library of Congress*

Portions of the moon fell away, leaving ghostly white trails across the sky.

The battle went on throughout the night until at last the sun came up and the jaguars went away, but the trouble was not over. Suddenly, giant fireballs from out of the disk of the sun hurtled to Earth. Thousands of burning stones and huge lumps of icy hail fell at the same time. They blasted enormous clearings among the trees and set fire to the jungle. Roaring fires leapt up all around the People, until the fire nearly surrounded them.

To save themselves, some People floated up on the wind like sparks from a fire. They went up and up into the sky until they became stars. They are still shining in the night sky even today. Others slipped into the cool water of the rivers and turned themselves into crocodiles so they could stay underwater to escape the fierce heat.

After the barrage finally stopped and the fires burned out, a black cloud enveloped the sun and a deep darkness fell over the land. It was many moons before the darkness disappeared into the north and life returned to normal.

RETOLD FROM WILBERT, 1975

WHAT THE EVIDENCE SHOWS

- Cosmic-ray tracks on the moon point to a heavy bombardment 16,000 to 12,000 years ago.
- Lunar spherules indicate we still are in a highly dangerous period of heavy impacts.
- Radioactivity and KREEP form an odd bull's-eye pattern in the moon's northern hemisphere.
- Radioactive thorium and potassium form large blotches in the northern hemisphere of Mars.
- Thorium, potassium, and uranium form large rings around Hudson Bay and Utah.
- The marks suggest that dustball-comets caused impacts throughout our solar system.
- They quite likely hit every large planet or moon in our system, as well as the sun.
- The comets seem to have been rich in radioactivity and KREEP.
- The evidence suggests that they came roughly from the north, as viewed from Earth.

Now let's take a closer look at the KREEP-y stuff that shows up at the Clovis-era sites.

25

CHEMISTRY OF
THE COMET

QUESTION: Okay, so the distribution of thorium, potassium, and uranium on Earth matches that on the moon. How can we be sure whether the Clovis particles came either from there or from the supernova?

If the particle layer was deposited by an impact at the end of the Clovis era, the chemical analysis can tell us a lot about the origin of the event. Did the material we find come from the impacting body, or was it sediment from the impact site that was thrown across the continent? Was the impact from a comet, meteor, or something else? We chose some powerful analytical methods to answer these questions.

We sent samples from Clovis-era sites to Budapest for PGAA analyses. PGAA is a nuclear testing method for determining various chemical elements that are in a sediment sample. It is similar to the traditional method, called neutron activation analysis (NAA), in which samples are irradiated inside a reactor, producing radioisotopes whose decays are then measured. NAA is often very sensitive to low concentrations of some elements, and PGAA can analyze all elements, although with less sensitivity. Our NAA analysis was conducted by Becquerel Laboratories on the McMaster University campus in Hamilton, Ontario.

We analyzed magnetic particles from several Clovis-era sites. These results are shown in table B.1 in appendix B. The mineral composition at of each set of magnetic particles can be inferred from the data in table 1. Particles from Gainey are composed by weight mainly of quartz (SiO_2), feldspar ($K,NaAlSi_3O_8$), magnetite (Fe_3O_4), and various additional compounds rich in calcium, magnesium, and aluminum. These minerals are common on Earth, and the composition is similar in many respects to the average composition of the continental crust, as reported by Holland and Lampert. Terrestrial concentrations of titanium (Ti) and manganese (Mn) and many trace elements, including thorium and uranium, are all considerably lower than the concentrations found in the Clovis layers. It appears that the quartz, feldspar, and magnetite were ejected from the crater of an impact event, but could the Ti, Mn, and other elements have come from an impact? Let's look at the evidence.

A RETURN TO THE LUNAR HOT SPOTS

Figure 25.1 shows titanium concentrations on the moon measured using reflected light by the Clementine lunar orbiter, launched in 1994. These concentrations are especially high in the Mare Imbrium region, which also has high levels of iron, along with radioactive ^{40}K and ^{232}Th. Notably, there are no significant enrichments of iron or titanium on the dark side of the moon. As with the radioactive elements, they exist only on the near side's northern hemisphere.

Concentrations of titanium at the darker spots on the map exceed 7.5 percent, which is far above the normal terrestrial abundances of 0.66 percent. The highest titanium concentrations are more than about 15 percent, and they correspond to high iron concentrations, up to about 19 percent. Similar concentrations are seen in the magnetic particles at Murray Springs and Topper.

Fig. 25.1. Distribution of titanium on the surface of the moon. It is roughly in the same areas as iron, thorium, and potassium, with almost none on the far side. *Adapted from NASA*

HOT SPOTS ON EARTH

When we return to Earth to look for a titanium signature, it shows up in the southwestern United States in exactly the same place that we found the pattern with thorium and potassium (in chapter 24, figure 24.6). In fact, the same ringlike pattern exists not just for one or two elements; all the following form a similar ring around the Southwest: cobalt, cerium, cesium, hafnium, iron, lanthanum, lithium, magnesium, manganese, nickel, scandium, strontium, vanadium, ytterbium, zinc, and zirconium.

Many on the list above are rare earth elements, and nearly all are found at high levels in the Clovis-era sediment. In fact, the lunar values are all comparable to what we find in the Topper magnetic particles, which seems to be the purest sample from the Event that we have found. At that site, the magnetic particles are composed almost entirely of various compounds of iron, titanium, and manganese. That is very unusual for terrestrial sources, but very common on the titanium-and-iron-rich lunar hot spots. Similar relative concentrations of Mn/Ti are observed at Gainey, Topper, and Murray Springs.

Even though diluted by terrestrial elements, we can see through the "clutter" by using titanium as a benchmark. That is clear in figure 25.2, where we compare titanium to the rare earth elements (the "REE" in KREEP), along with thorium and uranium, at the Murray Springs, Topper, and Gainey sites, and to those on the moon. The ratio plots from the four sites overlap nearly perfectly, providing clear evidence that the material is very similar. The close overlap is unusual when you consider that the Arizona site is about 1,650 miles away from the Michigan site, which is about 640 miles from the South

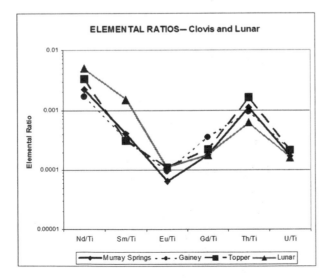

Fig. 25.2. The ratios of various rare earth elements, plus thorium and uranium, from Topper, Murray Springs, Gainey, and the moon, as compared to titanium.

Carolina site, and all have very different geologic settings and climates. It is even more compelling when you consider that the moon is 240,000 miles away from the other sites. Because of the similarity over such wide distances, we think that these minerals are related to the impacting body, either coming in with the body or being ejected by the impact.

Figure 25.3 compares trace-element concentrations in the magnetic particles at Murray Springs with the site's adjacent sediments. Iron-rich black sand from the surface of the site contains little titanium, whereas the nearby black mat and adjacent sediments have trace-element concentrations similar to the average terrestrial values. The magnetic particles at Murray Springs, however, are nearly ten times richer in trace elements than the local sediments.

The comparison of the surface magnetic grains at Murray Springs and the Clovis-era magnetic grains is particularly important; both are magnetic and visually indistinguishable. One would naturally assume that all magnetic grains at any site should be nearly identical, because they should have come from the same local source rocks. Instead, we find that they are radically different, with the Clovis-era magnetic grains containing nearly 1,600 percent more KREEP. Such differences show conclusively that the two types of grains, one on the surface of the wash (named "black sand" on the chart) and one in the Clovis layer, did not come from the same place. We think that the surface grains came from the Murray Springs area and that the Clovis grains came from beyond the local area.

If the Clovis grains are not local, where did they originate? As you can see by the KREEP bar on the chart, the Murray Springs magnetic grains

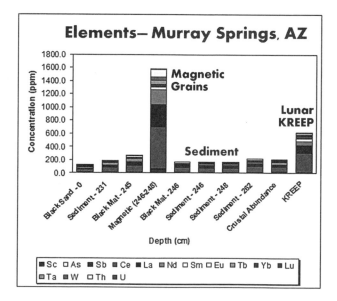

Fig. 25.3. Comparison of the magnetic particles at Murray Springs with the site's sediment profile of trace-element concentrations.

contain nearly twice as much of these KREEP elements as the actual lunar KREEP does. In other words, the Murray Springs grains are even KREEPier than KREEP, suggesting that both came from the same source. This is an important point, one worth repeating: the Murray Springs magnetic grains look much more like material from 239,000 miles away on the moon than like the magnetic grains twenty feet away on the surface of the Earth.

KREEP AT OTHER CLOVIS-ERA SITES

Similar comparisons of magnetic particles with adjacent sediments for other sites are shown in figure 25.4 (Topper, South Carolina) and figure 25.5 (Blackwater Draw, New Mexico). In all cases, the KREEP rare earth elements, along with thorium and uranium, are substantially higher in the particles than in the adjacent sediment. This is strong evidence that the magnetic particles include a significant component that comes from a common extraterrestrial source.

Now let's compare some of our Clovis-era sites with the lunar material. The Apollo 12 astronauts brought back KREEP-y sediment samples that were analyzed by Korotev and associates (2000) at Washington University, St. Louis. Their results are compared with the Gainey and Murray Springs

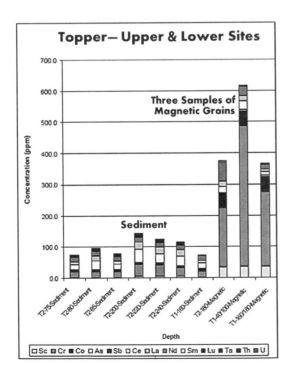

Fig. 25.4. Comparison of KREEP in magnetic grains to that in sediment at the Topper Clovis site in South Carolina. KREEP is substantially higher in the magnetic grains.

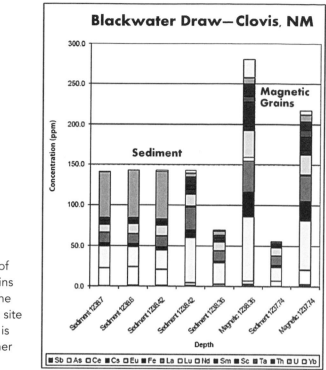

Fig. 25.5. Comparison of KREEP in magnetic grains to that in sediment at the Blackwater Draw Clovis site in New Mexico. KREEP is about 200 percent higher in the magnetic grains.

particles in figure 25.6. Once again, the similarities are striking. There is almost no difference in the ratios of rare earth elements, thorium, and uranium in the three samples, one from the moon, one from Michigan, and one from Arizona.

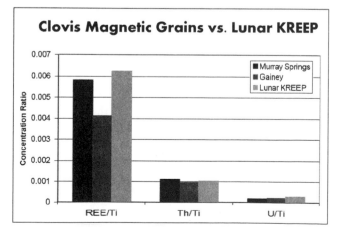

Fig. 25.6. Ratios of titanium to rare earth elements (REE), thorium, and uranium. Sediment from the Gainey and Murray Springs Clovis sites are compared to sediment from the moon's Mare Imbrium that was collected by the Apollo 12 astronauts.

POTASSIUM-40 ANOMALIES, THE "K" IN KREEP

We have shown that the magnetic-particle layer is rich in rare earth elements, thorium, and uranium and is quite similar to KREEP from the moon. So far, however, we have not looked at the evidence for enrichment of potassium (K), so let's see how it fits into the picture.

Potassium-40 is a naturally radioactive isotope with a 1.3-billion-year half-life that represents a tiny fraction of all potassium on Earth (about 0.0117 percent). This amount is very uniform throughout the solar system, except for meteorites, comets, or when a supernova is involved. Supernova shock waves produce lots of fresh ^{40}K that shoots out into space, and considerable ^{40}K is produced in iron meteorites by cosmic rays. If the debris of a supernova shock wave passed by the Earth 16,000 to 13,000 years ago, or we were hit by a comet or meteorite enriched by supernova-created ^{40}K, we should find excess amounts of the radioisotope in the sediment. If neither of those happened, we should find normal terrestrial levels of ^{40}K that are the same in the Clovis-era layers as they are in the layers above and below it. This is a key test of our theory.

In our research, we have measured potassium in various samples using two methods—by measuring ^{39}K, the most common isotope of potassium, with PGAA and by gamma-ray counting of the ^{40}K isotope of potassium. For the latter, Al Smith, a pioneer in this method, performed our initial measurements of ^{40}K at the Low-Background Counting Facility at Lawrence Berkeley National Laboratory. We analyzed potassium in chert artifacts from Clovis-era sites with both methods and found that the results could only be explained if the ^{40}K was enriched by 50 to 100 percent relative to ^{39}K. Analysis of modern chert and sediment found normal ^{40}K abundance, as expected. This means that the pieces of Clovis-era chert were on top of the ground 13,000 years ago, when the shock wave came through, and they were bombarded with material containing ^{40}K and with the microscopic iron grains found by Bill Topping.

Becquerel Laboratories also analyzed additional sediments from the particle layer for total potassium and for ^{40}K. These results are summarized in figure 25.7, which includes samples from across North America, including Alberta, Arizona, Michigan, and the Carolinas. They suggest that ^{40}K is enriched in the KREEP material deposited 13,000 years ago, sometimes at 350 percent of the expected amount. This can be true only if the impacting body did not come from our solar system. Enrichment of ^{40}K is also consistent with impact from a young cosmic body of some kind. The body itself may have been produced in the recent supernova, or, as the comet passed through

Fig. 25.7. Potassium-40 abundance in Clovis-era chert and sediment versus normal chert and sediment. All samples date to 13,000 BP, except for the Topper sample, which appears to date to the shock wave at 32,000 years ago. All Clovis samples show enriched ^{40}K.

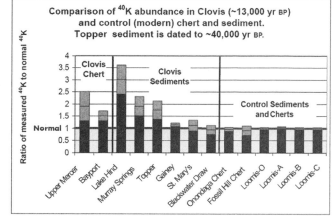

space, it may have swept up the supernova dust and debris. Either way, the impact body appears to have contained substantial amounts of ^{40}K.

NASA used similar methods to analyze the potassium in the lunar KREEP. The Lunar Prospector satellite measured gamma rays from ^{40}K as it flew over the moon; as you may recall, the map of radioactive ^{40}K shows the ringlike feature in the northern hemisphere. The satellite recorded only radioactive ^{40}K values, however, and could not measure the total potassium, both radioactive and nonradioactive. To find the total potassium in lunar rock collected in the Apollo missions, the samples were analyzed on Earth by conventional methods, but scientists did not measure the ^{40}K, only the total potassium.

Scientists used both measurements, the one for ^{40}K and the one for total potassium, to check the ratio, which should have been the same as the ratio throughout the solar system. To their surprise, they found that the ^{40}K was too high by 50 to 100 percent. It was a lot more than they expected. Puzzled by the discrepancy, NASA researchers assumed that the satellite had been calibrated incorrectly, and they ignored the difference. The only problem with that is that there was no indication the satellite was wrong. We believe the measurements were correct and that they confirm our measurements of the ^{40}K-enriched KREEP-like particles on Earth. We found high levels on parts of the Earth, and NASA found high levels on parts of the moon.

WHAT HAPPENED?

The distinct KREEP bull's-eye in Mare Imbrium on the moon has intrigued scientists from the first moment they recognized it. The usual explanation is that it is due to the slow cooling of many layers of rock billions of years ago, which allowed the KREEP to rise to the surface. Then, long afterward, large impacts moved the KREEP around, including a recent impact that launched this material to Earth.

Our discovery of a KREEP-rich particle layer across North America brings the above interpretation into question; in fact, there are many problems with that theory. If the conventional explanation is true, the main puzzle is how all the KREEP ended up at the same spot on the moon. If KREEP is so old, why is the ^{40}K isotopic abundance so anomalous? If KREEP formed that way, why do we see the same pattern on Mars and the Earth, when no geologist thinks those two bodies formed the same way that the moon did? All these questions point out problems with the standard theory.

A NEW THEORY

We propose a much different scenario in which numerous ultralow-density, KREEP-rich impactors collided with the Earth and moon, leaving KREEP behind in both places. It is a theory that easily explains all the facts that show up on both worlds. The impacting bodies were young, possibly produced by the supernova 41,000 years ago. Maybe a wall of expanding supernova material, in which dense bodies had begun to form, passed through the solar system. Such bodies could form if interstellar grains and dust formed nuclei with weak gravity that could pull together more dust to form larger bodies, in much the same way that scientists explain the formation of suns, planets, and comets.

Or the larger bodies may simply have been dustball comets that were expelled with the supernova remnant. We know that supernovae can eject neutron stars and pulsars, and we find small supernova-formed grains stuck in meteorites; but until now, no one has proposed that they can eject smaller bodies that vary in size between pulsars and dust grains. Logic says this should be possible, but we cannot say with certainty; it is just a working hypothesis. Nevertheless, little is actually known about supernovae, so this scenario might be the right one. It explains the evidence that we find throughout our solar system.

In spite of the uncertainty, the evidence tells us that, no matter where the KREEP came from, something arrived here 13,000 years ago. Whatever it was, it passed through the solar system in a giant wave that was billions of miles wide and millions of miles thick. After it moved on, the cosmic chemistry quickly returned to normal, but life on Earth did not. In the wake of the shock wave, millions of animals perished, and bright, radioactive scars still glow on every celestial body that was in the way.

THE SURVIVORS: INDIA

In this section, we looked at evidence for destruction on the planets. In chapter 24, we considered the possibility of massive solar flares. In the following story, which tells of King Manu and comes from India, many elements fit the supernova theory: a worldwide drought occurs; fires (solar flares) come from the sun; fires (comet impacts) destroy Earth and all the planets; then, massive waves (tsunamis) and flooding begin; and only a few people and animals survive.

⚹

Manu and the Magic Carp

King Manu's devotion was pleasing to the god Brahma, so Brahma decided to give Manu the gift of a very special fish, a magical carp. As Manu was praying one day, suddenly the wondrous fish fell from the sky into his hands. Realizing it was a gift from heaven, Manu carefully tended it as it grew larger and larger. After living in one ever-larger fishbowl after another, the magic carp finally became so huge that Manu had to release it into the ocean, where, before Manu's eyes, the fish grew to fill the entire ocean.

With that, King Manu laughed aloud, realizing that none other than Brahma could do such a thing. The Brahma-fish then spoke in a deep voice, saying, "The end of the age is rapidly approaching, Manu, and soon the world will be destroyed. You must prepare." The Brahma-fish told Manu that a worldwide drought would begin that same day, causing crops to wither and die. Soon, the great fish warned, scorching fires would fall from the sun and explode from within the Earth. They would be so fierce that Earth would be consumed along with the air itself, the clouds, and even the moon and planets that whirled through the sky. All would be scorched by the relentless flames, including the gods themselves.

"Soon after the great fire," Brahma told Manu, "the world will be cleansed by great floods. Seven clouds will rise as steam from the fire, and they will inundate the Three Continents, which will slip beneath the relentless waves." Brahma instructed Manu to build a mighty seagoing vessel and to take upon it a few chosen people and to make space for the seeds of many plants and creatures, so that they would survive the cataclysm.

Finally, the Brahma-fish told Manu, "The floods will be too great for you alone. Before they begin, weave a rope as thick as your waist. Then, when the waves and winds toss you about, look for me. As I swim past, toss the rope over the horn on my head, and I will pull you to safety. Do you understand, Manu?" Manu nodded and hurried off to start building the vessel.

Before long, Brahma's grim prophecy unfolded exactly as predicted. The great fish swam to Manu's boat and towed him to safety. Only the people and animal and plant seeds on Manu's boat survived the universal fire and the deluge. These survivors were entrusted to rebuild the New World.

RETOLD FROM FRAZER, 1919

WHAT THE EVIDENCE SHOWS

- The bull's-eye crater on the moon is dusted with iron and the rare element titanium.
- The hot spot in the American Southwest is high in iron, titanium, and twenty other rare elements.
- Titanium and KREEP ratios are nearly identical for all the Clovis-era sites and the moon.
- Potassium-40, a supernova isotope, is common in the bull's-eye on the moon.
- Potassium-40 is common in Clovis-era sediment and chert, but not before or after Clovis times.
- We think that either the comets kicked KREEP, thorium, and uranium from the moon to the Earth or the comets themselves contained the KREEP and brought it to all the planets.
- We suggest that the comets came directly from the supernova.

Now that we have seen some of the effects on chemistry of the impacts, let's try to find the impact sites.

COSMIC NEST. This dust shell in the Unicorn constellation is illuminated by the explosion of a red supergiant star, the type that accounts for the most violent supernovae.

Source: NASA, European Space Agency, H. E. Bond (Space Telescope Science Institute)

LASER SHOW. The star Eta Carinae is about to explode as a supernova. This image depicts the highly unstable star emitting blasts of laser light, in one of the rarest light shows ever discovered.

Source: J. Gitlin (Space Telescope Science Institute), NASA

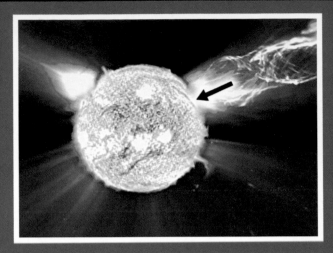

BOILING SUN. In June 1998, two sun-grazing comets passed behind and crashed into the sun (arrow). Almost immediately, an enormous solar flare erupted from the general area, suggesting that the impacts caused the explosion.

Artist's composite from NASA and Solar and Heliospheric Observatory images

BEDEVILED MOON. This mysterious red-and-yellow blot represents the only area on the moon with high levels of radioactive potassium and thorium. Did a comet or asteroid impact spread radioactivity across the moon not long ago?

Source: NASA

KAMIKAZE COMET. This artist's impression shows a giant comet colliding with the sun and triggering a massive solar flare. Both comet impacts and flares like this would be catastrophic for Earth. Pitted, melted rocks on the moon suggest that it was hit by both events thousands of years ago.

Artist's composite from NASA and Solar and Heliospheric Observatory images

COOKED FROM SPACE. Coming from a supernova that is light-years away, gamma radiation destroys half the Earth's ozone in ten seconds. No longer shielded from harsh cosmic radiation, widespread extinctions follow.

Source: Artist's conception by NASA

TROUBLE ON THE WAY.
Giant asteroids or comets
are suspected of causing
several of Earth's largest
known extinctions, including the disappearance of
the dinosaurs. Did one
also hit us at the dawn of
civilization?
Artist's composite from several
NASA photos

VANISHED MAMMOTHS. Some researchers propose that the Carolina Bays formed
when millions of Ice Age mammoths and
other animals became extinct.

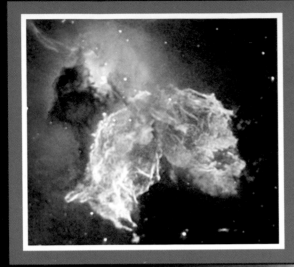

BURSTING THE BALLOON. Supernova N 63 exploded in space, but it did not produce a uniform shell. Instead, the detonation blew out the side of the gas bubble like a popped balloon.

Sources: NASA, European Space Agency, Higher Education Information Center, Hubble Heritage Team (Space Telescope Science Institute/ Association of Universities for Research in Astronomy)

GIANT BUG. This unstable, dying star in the Bug nebula is ejecting star-stuff into space in the shape of huge insectlike wings. The expelled material contains Earthlike hydrocarbons, water, and iron.

Sources: A. Zijlstra (University of Manchester Institute of Science and Technology) et al., European Space Agency, NASA

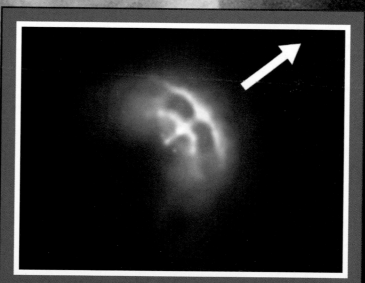

BOW AND ARROW. This Vela pulsar is the X-ray–emitting remnant star of a supernova explosion. After compressing to about twelve miles in diameter, it spins ten times each second and travels rapidly in the direction of the arrow.

Sources: G. Garmire et al. (Pennsylvania State University), NASA

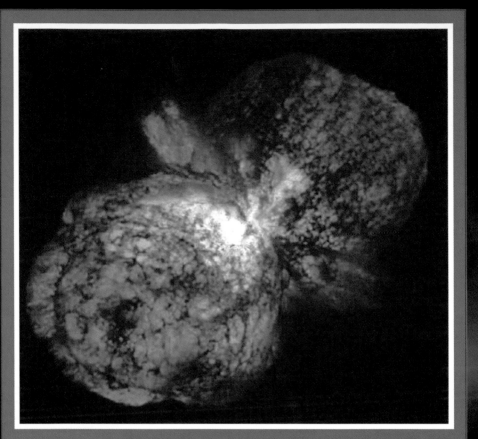

DUMBBELL. Eta Carinae also pumps vast clouds of hot gas into space in a dumbbell shape. No one knows when, but eventually the star will explode catastrophically as a supernova.

Sources: J. Morse (Arizona State University) and K. Davidson (University of Minnesota) et al., Wide Field Planetary Camera 2, Hubble Space Telescope, NASA

CHRISTMAS ORNAMENT. This ball-like cloud is from a supernova that exploded about 10,000 years ago. The expanding shell we see today consists of lumpy blobs of stellar material surrounded by thin gas.

Sources: S. Park, D. Burrows (Pennsylvania State University) et al., Chandra Observatory, NASA

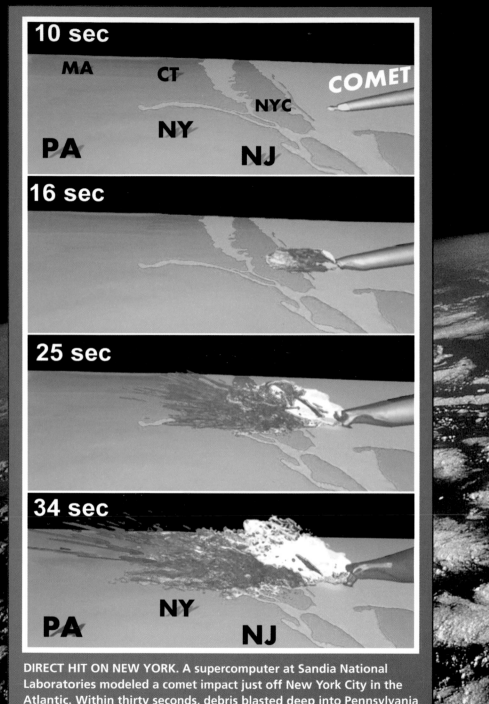

DIRECT HIT ON NEW YORK. A supercomputer at Sandia National Laboratories modeled a comet impact just off New York City in the Atlantic. Within thirty seconds, debris blasted deep into Pennsylvania and leveled most of New England. The aftereffects included a weeks-long global snowstorm and a years-long "impact winter."

Photos courtesy of D. Crawford and A. Breckinridge, Sandia National Lab, www.cs.sandia.gov/projects/comet.html

26

CRATER BELOW LAKE MICHIGAN

QUESTION: How can Lake Michigan be an impact site? Isn't it just a huge glacial lake?

Lake Michigan may indeed be a huge glacial lake. If a comet landed in or exploded above the lake, however, the impact or explosion might have rearranged bedrock and the lake sediment, creating shallow craters. In fact, there is some evidence suggesting that the deepest parts of the lake conceal possible shallow craters.

POSSIBLE BLAST MARKS AT THE BOTTOM OF LAKE MICHIGAN

First, let's look at the upper part of Lake Michigan. The lake bed contains several large depressions, or sub-basins, and the main one is called the Chippewa Basin. It is below sea level and about 925 feet deep, making it the largest and deepest in Lake Michigan. Of all the parts of Lake Michigan, the basin looks most like a traditional impact crater. But natural processes, other than impacts, create basins like this too, so let's look at the evidence to see whether this idea is plausible.

Unlike the other sub-basins in Lake Michigan, Chippewa Basin has a rough, irregular bottom (fig. 26.1), which some scientists attribute to the motion of the ice sheets traveling down the lake from the north. This explanation does not fit completely, however; as you can see in figure 26.1, cracks and ridges appear to radiate from the center. This is not what we would expect to see from ice sheets always moving in the same direction along the basin, which would cause the cracks and ridges to be parallel. But

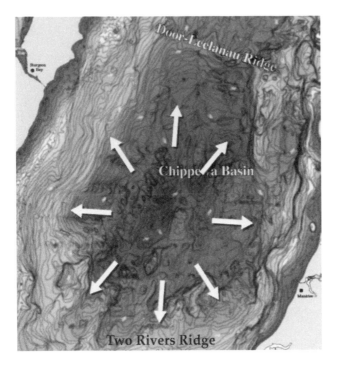

Fig. 26.1. Faint cracks and ridges appear to radiate out from Chippewa Basin. This is consistent with a crater. *Source: NOAA/ NGDC*

it is exactly what we would expect to see from an impact. In addition, the center of the basin shows a low-relief but distinct peak, which is typically seen in complex impact craters.

SCIENTIFIC SURVEYS FIND POSSIBLE IMPACT CRATER

USGS scientists spearheaded an extensive series of mapping and other surveys of Lake Michigan in the 1980s that included seismic soundings in which controlled explosions sent powerful shock waves far down to bounce back off the many layers of sediment and bedrock. The information the scientists recorded went into a computer to create simulated cross-sections of Lake Michigan, which look similar to medical X-ray images. Colman and Foster published quite a few of the seismic profiles beginning in 1990, and some of them show craterlike sub-basins.

One Lake Michigan seismic profile provides dramatic evidence for an impact when compared to the profile of what is widely accepted as a fifty-mile-wide crater under Chesapeake Bay. Both seismic readings show terrace faulting—that is, a stair-step pattern formed when large slabs of rock crack and slide downward after impact. This faulting is a classic signature of impact craters.

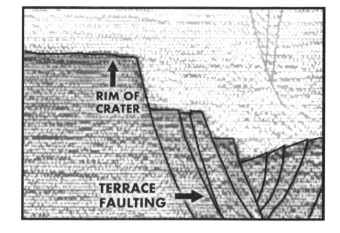

Fig. 26.2. This contrast-enhanced seismic image shows terrace faulting in Chesapeake Bay. (For a better comparison, the image has been reversed left to right.) *Source: USGS (1998)*

We start with the Chesapeake Bay crater's terraces, shown in a seismic profile from USGS (fig. 26.2). The largest fault block that slumped down into the crater is about 600 feet wide at the top, or about two football fields wide.

The next image is from the Chippewa Basin in Lake Michigan (fig. 26.3), showing nearly identical terrace faulting along the southeast margin of the basin. The largest block to the left is about a half-mile wide at the top, or about three times wider than the Chesapeake Bay blocks.

THE DIFFERENCES

In spite of the similarity between Lake Michigan and Chesapeake Bay, there are many differences, such as the fact that the Chippewa Basin is too shallow for a classic crater and the lack of typical meteorite-impact evidence.

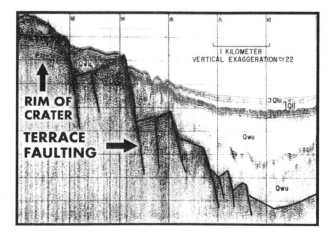

Fig. 26.3. In the Chippewa Basin, this contrast-enhanced seismic profile shows very similar terrace faulting. *Source: Colman and Foster, USGS (1990)*

Usually, around a true impact, we find rocks that are shocked, fractured, and deformed. But we don't think this was a meteorite impact; we think it was a comet impact or bolide explosion. Neither of those events has been identified clearly before, and if this was the first one to be found, we would expect the signature to be similar in some respects but different in others.

ANOTHER GREAT LAKES CRATER

The crater in Lake Michigan is debatable, but in the northeastern part of Lake Superior, there is what is almost certainly an impact crater at Charity Shoal, as shown in figure 26.4. Holcombe and colleagues (2001) investigated that crater but were unable to determine a date for the structure. One of their leading theories is that it dates from the end of the Ice Age, or otherwise the ice sheet would have ground away the half-mile-wide crater. As it is, the crater has a long, drumlinlike tail on the downstream side, indicating that the ice sheet moved across it after it formed. Because of this, they suggest that the crater formed at or just before the ice began its full retreat—in other words, at around 16,000 to 13,000 years ago, exactly at the time the other comets fell. It is possible that in this image, we are looking at the actual crater of one of the smaller comets from the Event.

WHAT THE EVIDENCE SHOWS

- Lake Michigan contains craterlike basins.
- Seismic profiles of Chippewa Basin match the Chesapeake Bay crater profile.
- Seismic profiles reveal terrace faulting, just as in the Chesapeake crater.
- Chippewa Basin shows apparent radial fracturing, as would happen in an impact.
- Other classic signs of a meteorite impact are missing, but we think this was a comet.
- A possible Event-related crater exists at Charity Shoal in Lake Ontario.

We believe that Chippewa Basin offers the best evidence for an impact in Lake Michigan. If so, then this Lake Michigan sub-basin, with a diameter of sixty-five miles, would be the fourth-largest crater ever found on the planet. In addition, a small but significant crater exists in Lake Ontario. Our search for craters does not stop here, however.

Fig. 26.4. A half-mile-wide crater is located in shallow water of Lake Ontario. *Source: NOAA/NGDC*

Do you recall the flight lines leading from the Carolina Bays into Canada? In a few more pages, you will see images of other potential megacraters. If they are accepted as impact sites, all of them will move to the top of the list of the world's largest impact craters. We will have more about these other craters in a bit.

First, there are some surprising new implications for our Lake Michigan story. If an icy comet landed in Lake Michigan, research shows that it might have behaved differently from how a stony meteor would have. Let's look at some of those differences and see how they match the evidence.

27

MOMENT OF IMPACT

QUESTION: If a comet hit Lake Michigan, what would the impact have looked like?

Although the potential crater in Lake Michigan is underwater today, when Lake Michigan formed it looked very different from how it does today. Some 20,000 to 13,000 years ago, the continental ice sheet entirely covered parts of the northern United States, and Lake Michigan was as much as a mile beneath the glacier. If a piece of comet landed there, it did not land on water or solid ground—it smashed into thick ice. This realization adds a surprising new dimension to our impact theory.

ICE SURPRISE

At the instant of impact, the comet's explosion would have created several very noticeable effects:

- It would have blown a massive hole through the ice, forming a crater that extended into the lake bottom.
- The downward force of the explosion would have sent high-pressure, high-velocity jets of meltwater rushing along the ground underneath the ice sheet.
- The flood surge would have carried with it mud, sand, gravel, and pulverized bedrock.
- The blast would have vaporized thousands of cubic miles of ice instantly and would have thrown huge chunks of ice into the air.

FLYING ICE

We will examine the massive meltwater surges later, but first let's look at the flying icebergs. The curved floor of the northern basin suggests that

270

the original ice-walled crater may have been as much as sixty-five miles in diameter. If so, that means that the impact blew out of the crater more than 3,300 square miles of ice that was about a mile deep. The size of the hole was nearly as large as present-day Connecticut or Rhode Island.

If the explosion had distributed the crushed ice evenly across the entire continental United States, the layer would have been about four to five feet deep, coming up to about the eye level of any standing adult. But not all the flying ice remained frozen; the heat of the explosion vaporized a lot of it. Possibly half to three-quarters quickly turned to steam, while the impact ejected the remainder into the air as lumps of fractured ice and icy water. These chunks would not have looked like normal water or ice—they were "burning"—that is, quickly turning into steam. The result probably looked like a barrage of sizzling, steaming snowballs, except with very important differences: they may have been miles in diameter and traveling at thousands of miles per hour. The icy debris arced out at a low angle in many directions, including toward the East Coast and southwest over the prairies. Within minutes, the chunks of sizzling ice struck the ground, exploding on contact to form the Carolina Bays and High Plains basins. They also probably exploded on a lot of landscape in between, but none of those craters has survived to the present day.

It may be hard to accept that flying chunks of ice could travel very far through the atmosphere, but scientists accept that icy pieces of comets can do it. If you watched images of the comet Shoemaker-Levy slamming into Jupiter, you know that it can happen. Scientists assume those icy chunks penetrated far into Jupiter's atmosphere. The same would be true on Earth.

STONES FROM HEAVEN: SOLVING THE RIDDLE OF THE BAYS

Now we are close to finding the last answers to the enigma of the shallow, elliptical shape of the Carolina Bays, rainwater basins, and salinas. To do so, we have to take a closer look at falling comets.

There was a time not long ago when most scientists believed that asteroids and comets never fell to Earth; in fact, at least one U.S. president thought so, Thomas Jefferson. In France in 1790, about 300 witnesses saw an impressive, fiery fall of meteorite fragments, after which they collected a large number of samples. In spite of the eyewitnesses and hard evidence, most scientists of the time scoffed at the idea and pronounced the whole thing a hoax, since such a thing as sky-rocks was "physically impossible." Even though two Yale scientists went against tradition and supported the facts as being true, President Jefferson supposedly remarked, "It is easier to believe that two Yankee professors would lie than that stones would fall from heaven" (Williams, 1996).

Eventually, scientists came to accept the idea of meteorite and comet impacts on the Earth and moon, but not during recent eras. For a long time, most felt that the large bombardment craters that we see on the moon and Mars occurred millions of years ago.

All that changed in 1994, when millions of people watched NASA images in awe as an icy chain of about twenty "ice-bullet" fragments of Comet Shoemaker-Levy collided with Jupiter, creating dark impact spots in Jupiter's upper atmosphere (fig. 27.1). Measuring two to three times the size of Earth, the largest debris clouds were created by comet pieces about 1.2 miles in diameter, bigger than most Earth cities. The comet could have hit our planet just as easily, and many scientists acknowledged for the first time that the threat of such giant impacts exists today. Impacts on Earth most likely would create similar dark clouds, producing a temporary "impact winter."

UNDER THE GUN

Modern space exploration has revealed the danger: we live in a dusty, crowded solar system, a "shooting gallery" of objects the size of tiny dust particles to huge comets that routinely cross paths with the Earth as it makes its way through the clouds of debris. We see this firsthand each fall in the Northern Hemisphere and each summer in the Southern Hemisphere when colorful meteor showers, such as the Leonids, Perseids, and Taurids, dazzle viewers with nighttime fireworks displays. Only recently, astronomers Clube and Napier realized that these dazzling displays are leftovers from

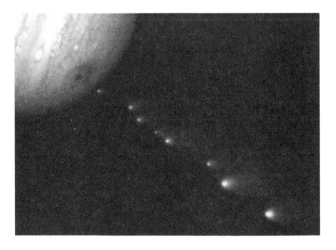

Fig. 27.1. A chain of fragments of Shoemaker-Levy heading for Jupiter. *Adapted from NASA images*

the breakup of a massive megacomet that occurred about 20,000 years ago. Clube and Napier calculated that after the megacomet broke up, the clouds of debris headed our way at regular intervals, a span that includes all the evidence we have for the Event. Even more disturbing, the two scientists believe that the threat is not over—the debris is coming back again. They think we are just now entering the most dangerous period during which it might return, but more about that later.

GIANT SNOWBALLS

Scientists who studied Shoemaker-Levy and other comets found that they are like huge, fluffy dustballs, made up of loose ice grains with a lot of open space between them, sometimes surrounded by nearly black carbon shells. It is not clear yet what is in the core, but it seems likely that comets vary from having stone or iron cores to having icy centers. After many circuits around the sun, the ice can vaporize, leaving behind the heat-resistant carbon compounds. Some very old comets may consist of only an immense black glassy carbon core.

At impact, the comet pieces of Shoemaker-Levy weighed about the same as packed snow (about 0.25 gram/cm^3). Similarly, NASA's Deep Impact mission reported that the comet Tempel-Tuttle "has a very fluffy structure that is weaker than a bank of powder snow." One of the scientists remarked in *Nature* (September 8, 2005), "You could probably dig from one side to the other with your hands, it's that weak." The astronomer Napier and his coworkers (2004) report that the calculations on twenty-seven known long-period comets suggest that they have an average mass similar to that of Shoemaker-Levy. This means that half of them are heavier than snow and half are lighter. You might logically ask how a fluffy dustball comet could be so dangerous; meteorites are rocky and more massive, so they seem more of a threat. Such comets may seem benign because they are so lightweight, but this is simply a matter of scale and velocity. For example, if a four-inch snowball hits you in the head, it may not hurt very much, but a four-foot-wide snowball certainly would. Furthermore, even a four-inch snowball traveling at thousands of miles per hour would be deadly. Comets may seem to be as benign as a fluffy snowball, but they are not.

SMALL COMETS

Those snowball comets hit other planets besides Jupiter, and they hit Earth as well. In the 1980s, physicists Louis Frank and John Sigwarth (1986,

1997) made the still controversial discovery that thousands of these small fluffy snowballs are colliding with Earth's atmosphere every day. Because these objects are usually no more than a few feet across, the scientists named them "small comets." Frank and Sigwarth provided photographic evidence that millions of them hit Earth every year, although most burn up before they reach the surface. But occasionally one gets through. Such lightweight objects should behave very differently from stone asteroids, so let's see what happens when they hit.

MORE ON THE DENSITY OF COMETS AND ASTEROIDS

Based on satellite measurements and photos, Frank calculated the mass of these small icy comets. When compared to ice (1.0 gram/cm^3), small comets are about the same density as newly fallen light powder snow (0.02 gram/cm^3), which is considerably less dense than Styrofoam (0.05 gram/cm^3). After studying the Tunguska impact in Russia in 1908, some scientists concluded that the original body also might have been lighter than snow. For comparison, an iron asteroid is about 800 times denser.

Figure 27.2 shows the comparative sizes of the two objects, which, incredibly enough, would weigh the same amount on Earth. They compare two objects of similar weight but different size. Now let's look at the case where two impacting objects are the same size. If a mile-wide comet hit the Earth, the effect should be considerably less destructive than that from an impact by a mile-wide iron asteroid that is 400 times denser. Because weight (or, more accurately, mass) helps determine crater size, the total impact energy of the large comet and small asteroid might be about the same—except for one thing. Because the comet is so much wider when it hits, it would create a wider, shallower crater than the meteorite would.

Comets are not the only objects that can be of very low density; rocky objects can also be of low density if they contain many voids. One such example is Hyperion, one of Saturn's moons, which is about 155 miles (250 km) across, as shown in figure 27.3. It has a bizarre appearance that indicates it does not have a typical rocky composition. Hyperion rotates chaotically, suggesting that it has a very low density. In fact, it may be more like a sponge than a rock, containing a vast system of caverns inside. If such an object ever hit Earth, it would behave much differently from a denser body and would leave a much wider, shallower crater.

Fig. 27.2. These two impactors would weigh the same on Earth, but the comet is 400 times larger than the rock. Because it is bigger and less dense, the snowball comet would create a wide, shallow crater, unlike the crater from the rock meteorite. *Adapted from NASA images*

Fig. 27.3. This image shows low-density Hyperion, a 155-mile-wide moon of Saturn. Its odd appearance and strange rotation indicate that it is a spongelike, rocky object. *Sources: Cassini Imaging Team, Space Science Institute, Jet Propulsion Laboratory, European Space Agency, NASA*

THE EFFECTS OF DIFFERENT TYPES OF IMPACTS

In 1982, Caltech scientists O'Keefe and Ahrens calculated what would happen if several types of impactors hit Earth.

Iron Meteorite

The researchers found that if a normal high-density iron meteorite impacts Earth, it transfers a lot of energy into the ground, fracturing the rock and blasting ejecta out in all directions, as seen in figure 27.4. Here's what they found about this type of impact.

- *Crater depth and width.* Usually the craters are very deep relative to the width.
- *Impactor shape.* The meteorite creates a bowl-shaped crater when it hits.

Lightweight Comet

The scientists also studied what happens when a low-density comet, weighing as much as powder snow, hits Earth, as in figure 27.5. The effects are very different.

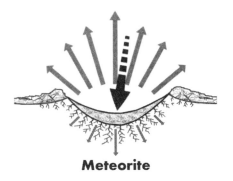

Meteorite

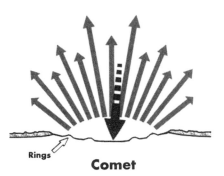

Rings

Comet

Fig. 27.4. A regular meteorite impact puts energy into the ground that fractures bedrock and forms a deep crater. *After Melosh (1989)*

Fig. 27.5. The impact of a lightweight comet sends a lot of energy upward into the atmosphere, has little effect on bedrock, and forms a shallow crater. *After O'Keefe and Ahrens (1982)*

- *Crater diameter.* The crater is about the same width as the comet.
- *Crater depth.* No matter what the diameter of the comet, even at hundreds of miles across, the crater always is exceptionally shallow. Like the Carolina Bays, a mile-wide crater might be only a few hundred feet deep or less.
- *Impactor shape.* When the parent comet enters the atmosphere, it frequently breaks up into an even lower-density cloud of smaller objects.

Carbon or Stone Dustball

A meteorite and a comet were not the only objects analyzed by O'Keefe and Ahrens. In another test, they calculated what would happen to a non-icy porous refractory body (fig. 27.6), meaning one composed of heat-resistant

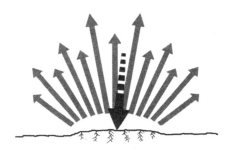

Dustball

Fig. 27.6. A porous impactor, made of black carbon and/or loosely bound dust and rock, is the strangest impactor of all. Rather than forming a crater, it can create an "impact mound." *After O'Keefe and Ahrens (1982)*

carbon compounds or lightweight stone dust. They calculated that the dustball would be about as massive as a Styrofoam cup. An impact involving one of these dustballs would look much like that of the lightweight comet.

- *Crater width*. The crater looks like that of a comet.
- *Crater depth*. In one of the oddest twists, O'Keefe and Ahrens found that such a dustball might produce no crater at all, and, in fact, the explosion would tend to fracture the ground slightly, causing the impact site to heave upward. The result is that, instead of creating a crater, the body produces an upside-down crater, or an "impact mound"! According to their figures, it would do this even if the dustball were hundreds of miles across, creating the baffling situation in which a 100-mile-wide dusty object would crash into Earth and produce no visible crater at all. This has important implications for the Event, as we will see later.

SOME MYSTERIES SOLVED

We have seen what happens when a cosmic object hits Earth, and now we are ready to add to the impact theory by offering some solutions to several long-standing questions.

QUESTION: How can the shallow basin in Lake Michigan be an impact crater?

The impactor that hit Lake Michigan was a lightweight comet or dustball arriving at a 5- to 15-degree angle that created a very shallow crater.

QUESTION: Why are bays and basins found only in sandy areas?

The lightweight ejecta that landed on loose sand left deeper, better-defined craters that have survived longer. Those that landed on hard rock or thick clay left no trace at all, or at most a faint crater that eroded quickly. Millions of secondary impacts whose marks are no longer visible most likely once occurred across North America, including every Canadian province, every Mexican state, and forty-nine U.S. states (not Hawaii). Parts of Europe and Asia probably suffered the same effects.

QUESTION: Why is there no impact debris around the many thousands of bays?

Unlike a typical stone or iron meteorite, the ice chunks contained very little except frozen water and carbon dioxide ice, which left no traces. Only the magnetic spherules, hollow spherules, and the glassy carbon remain detectable. Because this was a comet, the evidence looks very different from that for a meteorite.

QUESTION: If the comet was made of ice, how can there be streaks of meteorites across the Great Plains?

The original impactor was either an icy dustball or a dusty iceball. Either way, it contained a small percentage of the material that we define as meteorites, most of which are just small metal-bearing rocks and stones.

QUESTION: How could a dustball comet that landed in Lake Michigan or Canada throw streaks of debris all the way into Mexico and to the Pacific?

Fudali and Melson (1969) of the Smithsonian studied ejecta from an atomic bomb blast and from a volcano in Costa Rica and compared them with lunar craters. He concluded that in the nuclear explosion, ejecta traveled up to twenty times the bomb-crater radius, and in the volcanic eruption, some lava bombs traveled up to 100 times the volcano-crater radius. That gives us an estimated range of from 20 to 100 times the crater radius for impacts.

We found meteorite streaks extending about 2,000 miles from Lake Michigan. Dividing by the figures above, we find that the streaks could have come from an initial crater that was 20 to 100 miles wide, a range that includes our estimate of 65 miles for the diameter of the northern Lake Michigan crater.

QUESTION: How can we be sure that a supernova is related to the impacts? Maybe they were coincidental, or maybe something else caused them.

It is possible the impacts and the supernova are unrelated. There are a number of other theories about cosmic events capable of creating waves of impacts. In Richard Muller's book *Nemesis: The Death Star,* he suggests that the sun has a dark companion that travels in a 26-million-year orbit. He thinks Nemesis has kicked loose a barrage of comets in the past, as well as killed off the dinosaurs.

Walter Cruttenden, author of *Lost Star of Myth and Time,* presents the case for a dark companion of the sun that follows a 24,000-year orbit. When the orbit of the Lost Star brings it close to the sun, powerful gravitational forces shake loose a barrage of comets and asteroids that crash into all the planets, frequently causing extinctions on Earth.

In *Earth Under Fire: Humanity's Survival of the Ice Age,* author Paul

LaViolette proposes another theory: the center of the Milky Way galaxy periodically erupts in a massive explosion of galactic superwaves that have a similar effect to that of supernova radiation. When the pulse of gravity and particles arrives in our system, it causes impacts and extinctions just as we have found in the geologic record.

The question is which theory best fits the facts. At present, there are few facts to support any of them over the others, except for the potassium-40, which comes almost exclusively from supernovae. We have very high levels of it in 13,000-year-old sediment, but not above or below the Clovis-era layer. We accept that any or all of these alternative theories may be correct, but none of them contradicts the supernova theory. The universe is complex enough that all of them could have happened.

WHAT THE EVIDENCE SHOWS

- The impactor most likely was a dustball or icy comet.
- If so, it most likely would have been as dense as light powder snow.
- If the comet hit Lake Michigan, it landed on the ice sheet.
- Such an impact would have sent huge chunks of ice flying in all directions.
- That flying debris exploded upon impact to create the Carolina Bays and rainwater basins.
- Low-density fluffy ice and dusty debris explain the wide, shallow craters.
- Other similar theories can explain impacts, and all of them may be correct.

SUMMARY SO FAR

We see now that comet impacts, along with ejected glacial ice, might be responsible for most of the puzzling aspects of the bays and basins around North America. The evidence was there all along, but we missed it because . . . well . . . ice melts. It is very difficult to read clues that have evaporated. In addition, these types of comet impacts leave such odd craters that it has been difficult to relate them to something from outer space. Most scientists know what to look for in a meteorite crater, which is easier to detect, but this crater was not from a meteorite. To find dustball-comet craters, one has to look for different signatures in the rocks.

The story of the hidden craters is not over yet. There are more of them to the north in Canada, and they are immense.

28

POSSIBLE CANADIAN IMPACT CRATERS

QUESTION: The Event theory proposes multiple impacts. Are there any?

We have no conclusive evidence for any impact craters for the Event, and it may be that there are none, since according to the research by O'Keefe and Ahrens, some giant impactors do not leave any. But we have located some possibilities, and we propose them based solely on satellite evidence. If they are true craters, they will have to be confirmed by future on-site investigations, which are currently being planned.

The first potential crater is near Baffin Bay and is located underwater off the coast of Canada, northeast of Hudson Bay near Greenland. The crater averages about seventy-five miles in width. The subsea data in figure 28.1 come from Smith and Sandwell (1997), who used satellite information along with ocean-floor soundings by scientists from many countries, including Canada and the United States.

The deepest part of the crater is about 2,600 feet below the highest rim and about 800 feet below the break in the northeast rim. Such basins are rare on land or under the ocean, because they fill up rapidly by geologic standards. There are only a few along the entire coast of North America, and this suggests that the crater is relatively young, although its actual age is unclear. Unlike the other craters that we have checked, there is little information available with which to date it, so it may be very old. There is no evidence, however, that excludes it from having formed about 13,000 years ago.

POSSIBLE ARCTIC OCEAN CRATER

In the Northwest Territories of Canada, a possible 150-mile-wide underwater crater forms part of the Amundsen Gulf along the edge of the Arctic

280

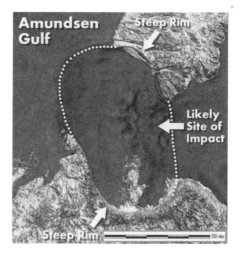

Fig. 28.1. This contrast-enhanced DEM image is of Home Bay on Baffin Island. The rough offshore crater is about seventy-five miles wide and a half-mile deep.

Fig. 28.2. In Canada's Amundsen Gulf, the northeast and southwest sides of one area show steep rims and a "cookie-cutter" shape, as if sliced from the terrain.

Ocean (fig. 28.2). Using underwater contour data from Smith and Sandwell (1997), we created a DEM file of the area, which has all the visual hallmarks of the type of comet impacts that are part of our theory.

The crater forms a shallow, rimmed ellipse that is oriented mostly northwest–southeast, just like the Carolina Bays. Part of the crater rim is on land and the rest is underwater, hidden from view. The scientific equipment used to survey it, however, can see through the water to show the shape of the basin. The image reveals an apparent impact site between the two peninsulas on either side. Oddly enough, the deepest part of the gulf is at the possible point of impact; it is nearly 2,100 feet deep just between the two closest points of land. Normally, one would expect such a spot to be very shallow, just the opposite of what it is. If it is an Event-related crater, it is deep enough to have been caused by a giant lightweight comet.

THE BIGGEST CRATER OF THEM ALL

The research showing the long axis lines leading from the Carolina Bays is the combined work of many scientists. As you may recall, some of the lines point to Lake Michigan, while the remaining lines point toward Canada. When we extended those bay lines, we found that they converge in or near

Fig. 28.3. These lines represent the long axes of Carolina Bays. Based on our research—along with that of Prouty, D. Johnson, and Eyton and Parkhurst—they converge near Hudson Bay.

another giant bay-shaped feature, Hudson Bay, as you can see in figure 28.3. Could it be that our comet has something to do with this giant 700-mile-wide inland sea in northern Canada?

BOTTOM OF THE BAY

Along the southeast shore of the bay, there is a uniformly curved shoreline known as the Hudson Arc. For years, many people have speculated that this is part of the rim of an old impact crater. Several field investigations have found no evidence that it is impact related, but the scientists could not rule out the possibility.

So if this is not a crater, where is the crater? The bay lines point more to the west under the waters of the bay. As we found for Lake Michigan, we propose that the crater is underwater in the center of the lake.

We created a view of the Hudson Bay lake bottom using highly detailed scientific data collected by teams of U.S. and Canadian scientists, who used satellite data and thousands of sonar soundings to map the bottom contours of the bay very precisely (fig. 28.4). What they found is that two deep holes in the lake form a depression that looks somewhat like an elongated Carolina Bay and is several times larger than the state of Massachusetts (inset). In addition, they found that the basin has a distinct curved rim that becomes a double or triple rim to the southeast. We think this may be the footprint of a comet.

A RING IN THE ROCKS

For more information about Hudson Bay, we need to look in detail at the rocks below the lake. In figure 28.4, you can see the bottom contours underneath the water; now we have to look beneath the soft bottom sediment to see the hidden rock layers. Scientists from the Geological Survey of Canada (GSC) did this, as shown in figure 28.5. Their data came from core samples, oil-well-drilling logs, and seismic survey work in the bay.

Their work revealed one very striking feature of Hudson Bay—a massive "ring," a highly unusual formation of rocks in the center of the bay that looks like a gigantic elongated bull's-eye. The ring is about 200 miles across and 400 miles long, and most importantly, it coincides perfectly with the raised underwater rim that we saw in the previous photo. That rock ring is the remnant of a geologic formation from the Cretaceous era: that is, from the time of the dinosaurs. That is a fitting coincidence, since most scientists now accept that a cosmic asteroid impact killed off the dinosaurs. Likewise, we believe that our more recent comet impact helped cause the extinction of mastodons, mammoths, and other megafauna.

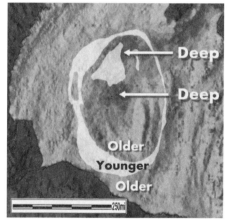

Fig. 28.4. In this contrast-enhanced DEM image of the bottom of Hudson Bay, notice the "Arc" that was once thought to be a crater but is likely only a geological feature. We think the 400-mile-long elliptical ringed depression near it is a genuine impact crater. Compare its size to that of Massachusetts. *Data source: Smith and Sandwell, Scripps, UCSD (1997)*

Fig. 28.5. In this close-up, we see a unique sediment ring that Canadian researchers have found on the bottom of Hudson Bay. The two deepest spots in the lake are in the middle of it. An extraordinary feature of the ring is that it is made of younger material than the lake sediment on each side of it, just as if it had been dumped on top of the lake bottom. *Based on Sanford and Grant (1990)*

Alan Grant, the GSC scientist who did many of the Hudson Bay studies, indicated that the ring is composed of very loose (unconsolidated) rock, sand, and gravel, unlike the formations underneath it, and that it is more than 300 feet thick in places. He considers the ring shape to have happened by coincidence, although he does not know of any other feature like it in the world (personal communication, 2004). To confirm this, we reviewed data from about half a dozen similar basins in North America and Africa, and we could find no other rings like it either.

One of the strangest features of the ring is the age of the rocks around it. According to GSC information, the ring is millions of years younger than the rocks on either side of it. In fact, there are entire stratigraphic sections of rock that should be under it, but they are missing, although they exist nearby in Hudson Bay. This is extraordinarily unusual. What happened to the missing strata, and how did the ring form?

Grant and others conducted hundreds of miles of seismic reflection work in the bay and produced a cross-section of the subsurface rocks. They concluded that the Cretaceous ring layer once covered the entire basin, but erosion removed most of it, so that now it exists only as a ring and in several isolated "islands" of loose sand and gravel in the center. The difficulty is in explaining how ordinary erosion could wear away the older formations on each side while leaving that thin perfect ring of younger rock on top of them. Some scientists suggest that the glacier created the ring, but the same objection applies. Since the massive weight of glacial ice is capable of carving deep grooves into hard granite bedrock, when heavy ice passed over the soft ring sediment, it should have scoured it away. Glaciers often leave curving lines of sand and gravel, but they do not leave complete rings. We are left with the puzzle that if the ring formed naturally, it is the only feature like it on the planet.

The ring can easily be explained, however, by a hypervelocity impact that blasted away the center rocks to form a ring-shaped sediment rim. There are dozens of examples of that on this planet, and there are millions on other worlds. In addition, there is other evidence around Hudson Bay that supports the impact theory. There are large raised banks forming a gigantic half-ring around the shore of the bay, the highest of which is about 250 feet above the ground level nearby. We believe these banks formed when the impact blew additional sediment out of Hudson Bay away from the impact site.

Here's what we think happened. Before impact, the undisturbed ring sediment in the center covered about 70,000 square miles beneath the ice sheet. Then the dustball-comet hit somewhere in Hudson Bay, blew an enormous hole in the ice, and blasted out the pulverized rock and ice to form a typical crater rim. The impact quickly altered Earth's climate, so that the ice sheet melted away without ever moving forward again. As the glacier retreated, it left the ring perfectly preserved beneath Hudson Bay.

If this scenario is correct, there is one important consequence. The impact occurred on top of nearly two miles of ice and blew a giant hole in the ice sheet. The ejected ice flew out in all directions, and, according to the flight lines from the Carolinas, some of it crashed into the coastal plain to create the bays. Other giant chunks landed across the Great Plains, gouging out the rainwater basins and the salinas. Most of the ice probably fell close to Hudson Bay, as is typical of most craters, forming a raised rim around the impact site. But that crater rim was very different from most impact rims; it was mostly ice with a lesser amount of rock. Over time, the ice melted away, removing nearly all traces of the impact. Only the shallow crater and ring remained visible in Hudson Bay, until the ocean level rose and the ring vanished from view. Scientists discovered it again only recently when they used high-tech equipment to peer beneath the water and sediment filling the bay.

DATING THE RING

Although we know the ring rock dates to the Cretaceous about 144 to 65 million years ago, the ring most likely formed much later. It is difficult to get an accurate date, since the ring is made of sand and gravel and contains almost no carbon-datable material, which makes it hard to link the rim directly to the impact. About the only certain thing is that a thin, unconsolidated open ring like that would not last long under the crushing, grinding weight of a two-mile-deep ice sheet. Therefore, the ring must have formed at the very end of the Ice Age, when the ice sheet was melting rather than advancing; this would have allowed the ring to remain relatively undisturbed until today. But is there any way to be more accurate about the collapse of the ice that filled Hudson Bay?

It is widely accepted that there was very thick ice over the bay for tens of thousands of years during the Ice Age. In fact, the bay was the location of an ice divide, or an ice center, meaning that the ice was much thicker there and flowed away from Hudson Bay in several directions, just as glaciers flow out of the mountains. Andrews and Peltier (1976) collected data about the Hudson Bay ice center and concluded that about 14,000 calendar years ago the ice over our proposed impact site in Hudson Bay was 10,000 feet (3,000 meters) thick, making it the thickest spot on the entire continental ice sheet. The ice was as deep as many mountains are tall.

Then one of the most puzzling aspects of the Ice Age occurred: within a short time, the thick ice over Hudson Bay nearly disappeared. Many researchers refer to it as the "catastrophic collapse" of the Hudson Bay ice center (USGS, Quaternary Geologic Map of Winnipeg, 2000; Mooers and Lehr, 1997). Andrews and Peltier state that by about 12,000 years ago, the entire center was gone, and ice no longer flowed away from it. Their range of dates includes the 13,000-year-old Event. Rather than melting away quietly, we think the Hudson Bay ice center exploded with a blast most likely heard around the world.

ICEBERGS SURGED INTO THE ATLANTIC

There is a specific date from another source—icebergs in the Atlantic Ocean. That may seem like an odd connection, but picture the following. When the comet hit Hudson Bay, the explosion blew a lot of ice into the air and across the continent. Then something else happened.

The immense force of the impact would have shattered the remaining ice sheet and shoved a gigantic pulse of broken ice into the Atlantic Ocean (fig. 28.6). This megasurge of icebergs near the end of the last Ice Age should show up in the glacial record—and it does, in hundreds of deep-sea cores drilled into the Atlantic Ocean from Canada to Africa.

Years ago, scientists recognized that icebergs periodically surged out of Hudson Strait into the Atlantic, carrying sand, silt, and gravel frozen into their icy layers. There were dozens of those surge events, and impacts had nothing to do with most of them; the natural cycle of ice sheets produced them about every 5,000 to 10,000 years. Then, as the newly formed icebergs melted, they dropped their load of rubble across the seafloor of

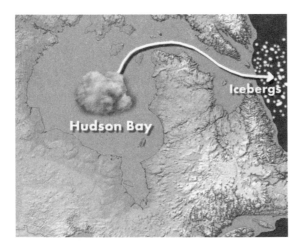

Fig. 28.6. When the comet exploded in Hudson Bay, the shock wave broke loose millions of icebergs that tumbled into the Atlantic along the path of the arrow. Within a short time they floated as far away as Ireland, Spain, and the northern coast of Africa.

the North Atlantic, and this debris, called IRDs (for ice-rafted detritus), shows up clearly as coarser sand-and-gravel bands in the ocean sediment. Scientists call these surges "Heinrich events" in honor of the scientist who discovered them. There was one important feature for us: scientists are able to date these events with radiocarbon dating.

DATING THE HEINRICH EVENTS

As shown in the next chart (fig. 28.7), Fronval (1997) retrieved an ocean core from east of Greenland, in which he found a major peak in IRDs from about 13,000 to 16,500 years ago, during the latest two Heinrich events, named H-0 and H-1. During that period, the IRDs increased substantially in a very short time and declined rapidly, maybe within less than 100 years, according to Hemming and colleagues (1998). The dotted line shows that part of the core was missing, but even so, that peak was the largest in the existing 160,000-year record.

QUESTION: Since there were many other Heinrich events, how can we be sure that the Event had anything to do with H-0 13,000 years ago or H-1 at 16,000 years ago?

We cannot be certain the Event was related to H-0 or H-1, but there is some evidence for a connection. Scientists such as Hemming and associates (1998) and Lagerklint (1999) conclude that these two events were different from all previous Heinrich events, and that they must have been caused by a uniquely different mechanism. Broecker and coworkers (1988) proposed that the ice dams of the huge glacial Lake Agassiz suddenly failed, sending frigid meltwater surging into the Atlantic at the time of H-0 13,000 years

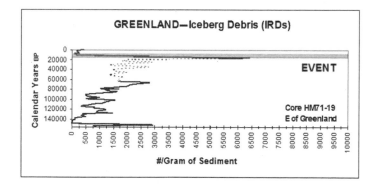

Fig. 28.7. This chart of iceberg debris shows that a pulse of icebergs occurred from 13,000 to 16,500 years ago. *Data from Fronval and Jansen (1997)*

ago, and possibly triggering the Younger Dryas cold spell. Andrews (1995) concluded that the pattern of H-0 IRDs indicated that H-0 happened just prior to massive meltwater flooding from Lake Agassiz. All this fits our picture, in which the comet impacts caused the collapse of the giant ice dams, which dumped cubic miles of ice and cold freshwater into the Atlantic and triggered the Younger Dryas climate change.

QUESTION: The Hudson Bay crater is 300 miles wide, but how large was the impactor?

There is not enough information to know how large the impactor was. Conservatively, it was ten to twenty miles across, but because the debris was so light, it could have been much larger. According to O'Keefe and Ahrens, the comet might have been about the same size as the crater. If so, as incredible as it sounds, the body would have been about 300 miles in diameter. It is difficult to envision such immense size, so let's try to visualize it.

The next image (fig. 28.8) is an actual NASA Shuttle photo of the darkened Florida peninsula facing west over the Gulf of Mexico. For comparison, we have superimposed a photo of a 300-mile-wide floating asteroid over the state. The original comet or cloud of supernova debris, which probably looked far blacker than this, would have extended nearly the full length of Florida, from Miami Beach almost to the Georgia border in the north. It would have been twice as wide as the entire state.

O'Keefe and Ahrens theorize that even if such an object was more or less solid, it would not have made it through Earth's atmosphere in one piece. Atmospheric friction, along with Earth's gravity, would most likely

Fig. 28.8. This image illustrates the possible size of the impactor in comparison to the state of Florida, the small sliver beneath it. The comet was nearly as long as the Florida peninsula. *Adapted from NASA*

have broken it up into a swarm of large pieces strung out as a stream of ice and dust. Either way, the impactor was presumably a deadly cloud of high-velocity debris, to which our usual notions of an icy-white comet do not apply. The comet-cloud was very dark or even black. NASA photos of Halley's comet confirm this, and other comets appear so dark that they reflect almost no sunlight. Scientists theorize that, over time, comets develop a thin shell of black hydrocarbon residue that does not boil off as the comet travels around the sun. This fact has an important connection to the black glassy carbon that we found at many Clovis-era sites, as we will see later.

Even though it was nothing more than a giant, dirty, porous snowball, or a loose cloud of supernova debris, and even though it was less massive than Styrofoam, that dustball-comet-cloud was big enough to have caused serious damage. Hitting Earth in multiple places at incredible speed, the cloud produced thousands of nearly instantaneous explosions that created shallow craters across the Northern Hemisphere. In Hudson Bay, the largest impact blasted away the Cretaceous rocks to form the large ring we see today.

LARGEST CRATER ON THE PLANET

If confirmed as an impact crater by further research, Hudson Bay would become the largest impact crater on the planet. It is about 300 miles across, and currently, the largest accepted crater is the Vredefort crater in South Africa. With a diameter of 186 miles (300 km), it is considerably smaller than Hudson Bay.

If Hudson Bay was a crater, however, it would differ from all others in one crucial manner: occurring 13,000 years ago, it had human eyewitnesses. The other large ones on the list range in age from 35 million to 2 billion years, long before human storytelling began to record such things. Because the explosions were so immensely powerful, it is understandable that Native American storytellers, as well as those in other ancient cultures all over the world, would tell of such huge catastrophic events.

It must have been a remarkable sight: a string of giant incandescent fireballs flashing across the noontime sky to smash with awesome power into the ice sheet. The eyewitnesses possibly saw them coming, and no one who saw them and survived would have forgotten the sight. For generations, such stories would have entertained wide-eyed listeners who had not been born until after the Great Event. Let's look at one of the stories that appears to have survived from that distant time.

THE SURVIVORS: THE WINTU

This tale from the Wintu, a California tribe, has many details in common with our scientific evidence. An allegory about the world immediately before this one, the story is about the adventures of two brothers called, appropriately enough, Shooting-Star and Fire-Starter. In this story, just as in our theory, Shooting-Star spreads over the landscape long streaks of fire that converge in the north. In fact, the comet debris, being high in hydrocarbons, may have burned like tar or pitch. Afterward, heavy rain and meltwater floods spread across the continent.

Fire-Starter and the World Fire

In the world before this one, the People were not like us. They had magical powers. One woman called Yonot and her husband, Fire-Starter, had an especially magical child, so that whatever the child touched burst into flames. They named it Fire-Child and were careful to keep its hands off anything that could burn.

One day at sunrise, Fire-Starter became very angry with another tribe that had stolen his fire-sticks. To punish them, he and his brother, Shooting-Star, touched Fire-Child with pine-pitch branches, which instantly burst into furious flames.

With these torches, Fire-Starter swiftly raced off to the southwest and Shooting-Star rushed away to the southeast, circling around the other tribe's lands. Magically covering miles with each stride, they dripped burning pitch from their torches, leaving long streaks of fiercely burning flames behind.

Soon exhausted, the men met in the north before noon. But when they looked back, the fire had raged out of control, so that long lines of fire stretched from horizon to horizon. Flames raced across the land in giant waves. All over the world, People rushed to escape, some jumping into streams, others hiding in caves, but most could not escape. The fierce heat was all around them.

Hearing the noise and feeling the temperature rise, the Creator looked down on the burning world. Through the dense smoke, the Creator saw that even the rocks and ground were on fire. Everything was burning. Working quickly, the Creator sent Hummingbird north to smash a hole in the sky. With the force of a hundred rivers, water poured down through the cracked sky to flow over the Earth below. The floodwaters surged off to the south in giant walls of muddy water, putting out the fires as they went. The waters rose until they reached as high as the mountains, finally quenching the last of the raging fires.

When it was over, the Creator spoke, and all the waters disappeared. But the Creator looked down on a devastated world. In every direction, the Creator saw no living thing. So the Creator set out to build a new world and to create new People to live in it. After this was done, the Creator told the People the story of the Wakpohas, the great world fire, that had been set by Fire-Starter and by Shooting-Star.

RETOLD FROM CURTIN, 1898

WHAT THE EVIDENCE SHOWS

- Underwater off Baffin Island in Canada is what may be a 75-mile-wide impact crater.
- There is another possible 150-mile-wide crater in the Amundsen Gulf in Canada.
- The Carolina Bay flight lines point to a 400-mile-long possible crater in Hudson Bay.
- Evidence for this includes an extraordinary giant ring of sediment on the bottom of Hudson Bay.
- The impacts caused massive iceberg surges into the Atlantic.
- Incoming cosmic debris caused the Heinrich 1 surge about 16,000 years ago.
- The comet impacts 13,000 years ago caused the last surge, known as Heinrich 0.
- The Hudson Bay dustball comet or cloud may have been 10 to 300 miles in diameter.
- If confirmed, the resulting crater in Hudson Bay would be the largest known on Earth.

Up until now, we have concentrated on North America. But what about Europe? When we reviewed the meteorite streaks across the Great Plains, we also saw apparent streaks pointing to northern Europe. Let's see if there are any impacts there.

29

IMPACTS ON OTHER CONTINENTS

QUESTION: If the shock wave was so wide, shouldn't we see evidence of impacts on other continents?

We would see the marks of other impacts only if they survived. Most of our evidence is for the Northern Hemisphere, suggesting that the barrage came from a northern direction in the sky, but it is possible that some impactors hit in the south as well. One good possibility exists in South America.

THE SOUTH AMERICAN CONNECTION: BOLIVIA

NASA discovered the Iturralde craterlike feature in 2002 from space shuttle photographs of the Amazon jungle in Bolivia. It is about five miles (8 km) wide and, remarkably, only about sixty-six feet (20 m) deep, dimensions

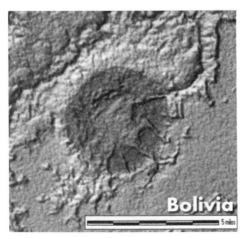

Fig. 29.1. DEM image of the five-mile-wide Iturralde crater. *Source: NASA*

similar to the shallow Carolina Bays (fig. 29.1). NASA sent a team of scientists into the dense, remote rain forest to research the crater. Not all the investigations are complete, but based on their on-site analyses of the rim, sediment, and other facts, the scientists are certain that it is an impact crater. Because the crater is so shallow, they suggest that the impactor was a very low-density object such as a comet. The pristine condition of the terrain—which looks as if a cookie cutter stamped the crater into the forest—leads them to think the impact happened between 30,000 and 11,000 years ago.

So could this wide, shallow crater be part of the shock-wave barrage? We cannot be sure, since there is only limited evidence at present. All the facts fit, however: the timing is right for the 16,000- to 13,000-year shock wave, the type of lightweight impactor is right, and the crater's depth-to-size ratio matches our northern impact sites. Maybe this impact was part of the barrage. If so, this Southern Hemisphere crater is the only one we have identified so far.

THE EUROPEAN CONNECTION

After we found long streaks of meteorites across the American Great Plains, we wondered whether they also existed in Europe, since our Event theory suggests that impacts or bolide explosions occurred in Europe at the same time as those in North America. To see if there is any evidence, we mapped the locations of 3,411 European, African, and Middle Eastern meteorites from the National History Museum (London) meteorite database (fig. 29.2), just as we did on the Great Plains.

As we suspected, there were long chains of old meteorite falls extending from northern Europe deep into Africa, even reaching as far as Oman on the Arabian Peninsula, a maximum distance of nearly 3,000 miles (4,800 km).

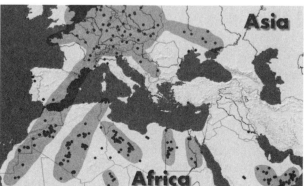

Fig. 29.2. Meteorite falls in Europe and Africa. The long streaks radiate away from northern Europe.

Very little radiocarbon dating is available, but the dates that are available support our theory. Jull (2002) reports that his studies found a large peak in meteorite dates between 20,000 and 15,000 years ago, a span that includes the arrival of the shock wave. When we plotted the ages of eighty-five African meteorites that have been reported in various scientific papers, we found a significant peak at the same time, indicating that more of them fell during the barrage.

We propose that those dates and that meteorite pattern are the footprint of a colossal explosion over northern Europe that blasted debris as far away as Africa. If it happened 16,000 to 13,000 years ago, as we believe, the impactor would have exploded over the Scandinavian ice sheet, blasting out immense chunks of ice and meteoritic debris in the same way that such impacts send rock debris flying into the air. When those ejected objects landed explosively across three continents, it would have seemed like a bomb barrage to the people and animals living there. If it happened that way, the crater might still be visible.

GIANT CRATER IN SCANDINAVIA

As you may recall from chapter 8, some of the largest drumlin fields outside Canada can be found in the countries near Scandinavia in northern Europe, which is exactly where our meteorite lines point. When we searched satellite photos of the area, we found something intriguing—a giant 150-mile-long arc, forming one side of a faint ellipse that spreads across most of southern Finland (see fig. 29.3). Such features are rare in nature.

European scientists who have studied this feature have concluded that it is glacial till: that is, great mounds of sand and gravel pushed up by the glaciers that covered nearly all northern Europe. The high mounds trapped water behind them, forming a region with hundreds of interconnected freshwater lakes and streams. The entire waterway network is called Lake Saimaa.

If a comet had hit this part of Scandinavia, it would have blasted a giant hole in the ice and blown mounds of glacial till out from under the edge of the ice sheet, just as they appear in the image. In fact, all that would be left would be the mounds, along with the huge shallow depression that is now filled with chains of lakes. The distinct curving lines visible in the satellite photo are exactly how the result might appear 13,000 years later. The timing is right since, according to European researchers, those lines of glacial moraine formed at the end of the Ice Age between 14,000 and 12,000 years ago.

QUESTION: So how do we know that all the proposed craters from Europe to North America are all the same age and from the same impact event?

Some evidence dates several of the proposed craters fairly well and other evidence suggests that all formed around the same time, but we do not know conclusively, because there just is not enough scientific data yet. We are planning expeditions to gather more evidence.

The next image, figure 29.4, shows one fact that suggests a connection. As you may recall, nearly all the largest craters are either slightly or very elliptical. When we align the long axes of the craters, we find that they all follow a nearly identical line, suggesting a common flight path roughly east of the North Pole. This is consistent with the Event having been limited to the Northern Hemisphere of Earth, and this direction is the same as for the radioactive bull's-eyes on the moon and Mars.

We can estimate the angle of impact, too, because the ring in the bottom of Hudson Bay is elliptical, as are those in Finland and Amundsen Gulf. To create all those elliptical craters, the impactors would have had to come in at a low angle of about 5 to 15 degrees above the horizon, as NASA found

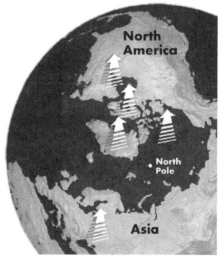

Fig. 29.3. We propose that the 150-mile-long arcing lines of glacial moraine around Lake Saimaa were created by a comet impact in Finland. *Contrast-enhanced composite satellite image © 2005 by Earth Satellite Corporation. Used with permission. Source: Terraserver.com*

Fig. 29.4. This image shows the alignment of the five large craters, suggesting that they may have formed from pieces of the same comet or wave of debris. *Adapted from NASA*

on Mars and the moon. Furthermore, all this correlates with the lines of meteorite debris strewn across the Great Plains. We already have seen such long lines following the direction of impact for craters on Mars and the moon. Everything we know suggests that at least five giant impactors came in from just above the northeastern horizon and slammed into Earth at a low angle 13,000 years ago.

TIME AND SEASON

The timing above allows us some more speculation. Let's see if we can estimate the time of day and season of the Event. Let's assume that Geminga is the supernova that caused all this trouble, even though that is still an unproven thesis. We can use star charts to look for that supernova remnant in the skies over Hudson Bay about 13,000 years ago. Geminga is in the constellation Gemini, which, along with the rest of the zodiac, has drifted around a little since the Event, but not very much. We want to correct for that, and then we want to find out when Geminga was just rising above the northeastern horizon at Hudson Bay.

Using star charts from NASA, we found that, amazingly, there are only a few weeks every year when Geminga would have risen over the northeastern horizon in that area 13,000 years ago. It happened only between about 11:00 AM and 1:00 PM for a few days before or after March 15 in late winter, as shown in figure 29.5.

Fig. 29.5. Moments before impact, the dustball-comet might have looked like this as it approached Earth from the direction of the supernova Geminga in Gemini, which is near Orion. This view is toward the northeastern horizon from Hudson Bay at about 1:00 PM on March 15, the year 13,000 BP. *Adapted from NOAA and NASA images*

TIMING FROM THE SURVIVORS

There are only a few stories describing the Fire and the Flood that mention details indicating the time of day or the season, but, surprisingly, all of them are consistent with March 15.

TRIBE	LOCATION	SEASON
Cheyenne	Kansas	"It was winter."
Pamarys	S. America	"It was cold."
Hopi	Arizona	"It was cold" and the world was "frozen."
Arawak	Caribbean	"The sun was out."
Toba	S. America	"The sun and moon were out."
Toltec	C. America	"The sun was out."
Wintu	California	"It happened just before sunrise and went until noon."
Apache	Texas	"It happened in the morning."
Inca	Peru, Bolivia	"It happened in the afternoon."
Greeks	Europe	"The sun was still up."

Given that information, let's see how the stories mesh with the star chart. Three stories say it was cold or was winter, and seven say it occurred at least partly during the day, so that fits with the rising of Gemini around March 15. Regarding the hour of the day, the key story is from the Wintu, who give a range of hours from just before sunrise to noontime. If the date was around March 15, when we check sunrise charts, we find that the sun rises around 6:00 AM on that date in California in modern times. Most likely, it rose near the same back then, so that gives us a range of from around 5:00 AM to noon Pacific time for the Event, or a total of seven hours from beginning to end.

Now, let's compare the times in the stories by looking at the time zones for each of those tribes. (We don't need to correct for daylight saving time, because the Clovis people never heard of it.) Consulting time-zone charts, we find that Greece is ten hours ahead of California, so the Event started after 3:00 PM there, or during the daytime, just as the story says. For the Apache tribe, the Event would have started several hours ahead of California at around 7:00 AM, in the morning, just as they tell us. Last, the Inca in Peru and Bolivia were a maximum of four hours ahead of the Wintu, meaning that the Event continued up until about 4:00 PM for them. Because the Inca were far away in the Southern Hemisphere, the long-range effects of the Event probably reached them some time later than it began in the north, so their report of "afternoon" is a reasonable match with the facts.

Obviously, that date is purely speculative and depends on many assumptions, but it is intriguing that every one of the cultural stories is consistent with the scientific evidence. Could it be that the Event actually happened between noon and 1:00, eastern standard time, on March 15? Probably not, since almost any day in the month of March would fit the evidence too, and, of course, there are too many uncertainties. The scientific evidence, however, shows that it could have occurred then.

OTHER CRATERS AROUND THE WORLD

Lake Saimaa is the last of five giant potential craters that we have found. We decided to see how all five compared to accepted craters around the world. The Earth Impact Database at the University of New Brunswick, Canada (www.unb.ca/passc/ImpactDatabase), had records for 171 widely recognized impact craters on the planet. If accepted as craters, those we have described would move to the top of the list as five of the eight largest impact craters on the planet. The new ones are in bold:

CRATER	LOCATION	DIAMETER
1. **Hudson Bay**	**Canada**	**300 miles (480 km)**
2. Vredefort	South Africa	186 miles (300 km)
3. **Lake Saimaa**	**Finland**	**180 miles (290 km)**
4. Sudbury	Canada	155 miles (250 km)
5. **Amundsen Bay**	**Canada**	**150 miles (241 km)**
6. Chicxulub	Mexico	106 miles (170 km)
7. **Baffin Island**	**Canada**	**75 miles (120 km)**
8. **Lake Michigan**	**United States**	**65 miles (105 km)**

If all these huge craters formed during the Event, then it is understandable that 40 million animals disappeared at the same time—the superheated shock wave and ejecta would have blanketed a sizable portion of the Northern Hemisphere.

THE SURVIVORS: THE GREEKS

Of the craters on the list above, the newest five are different in one crucial manner: they had human eyewitnesses. Unlike the others, which range in age from 35 million to 2 billion years, these new craters formed just 16,000 to 13,000 years ago. Because the explosions were so immensely powerful, it is understandable that ancient storytellers all over the world would weave dramatic stories about such catastrophic events. It was an unusual time and would have been highly memorable, to say the least. For the survivors' children who were born soon afterward and who grew up hearing these tales, the events must have seemed fantastic and unbelievable. Many people will continue to think the same thing today, in spite of all the scientific evidence to the contrary.

One such fantastical story, which comes from ancient Greece, tells of a great deluge with giant waves. The time of day fits, giving a global scope to the cultural evidence for the Event.

Zeus Ends the Golden Age

One day, the god Zeus wanted to see how the people of Greece were doing. He decided to visit them, as usual, as an impoverished and homeless itinerant. Upon arriving in the first city, he visited the king's great hall, only to find the court practicing human sacrifice; and in the second city, it was no better. The king served him a disgusting cannibal stew, made with the king's own brother.

Outraged, Zeus realized that the people were too evil to save, and he vowed to destroy the Earth with a mighty deluge and then rebuild it. Prometheus, who was close to the gods, found out about Zeus's vow and warned his son, Deucalion, to build an ark quickly and to stock it with food.

Just as Deucalion finished his boat, Zeus let loose a great rain that poured out over the Earth in sheets. Rivers swelled into huge torrents, and even the mighty ocean rose up out of its bed. Giant waves flooded the coastal plain of Greece, drenching the foothills in thick spray and lapping up against the mountains. The terrified people sought refuge in the mountains of Thessaly because they thought they would be safe, but Zeus split open the rugged mountains so that spouts of water gushed out to wash them away. Before long, the deluge swept the entire land clean of the evil influence of humankind. Only a few good people on the highest peaks survived, and by the time the deluge subsided, most of the land of Greece had washed into the sea.

Deucalion and his wife, Pyrrha, clung to their boat and rode the deluge for nine days, overwhelmed by the spectacle of giant waves and raging seas, until the waters gradually flowed back into the Earth again. At last, when they stepped out onto the muddy Earth, they were stunned by the spectacle and fearful about how to survive in the nearly empty world.

Deciding to search for survivors, they hurried off to find the Oracle of Themis, who had also been forewarned of the deluge. The Oracle advised Deucalion to throw the bones of his mother over his shoulder and told Pyrrha to do the same with her mother's bones. The two did not understand until they realized that the Earth was their true mother and the rocks were her bones. So, grabbing stones from the ground, they tossed the rocks behind them. The instant Deucalion's stones struck the Earth, they miraculously transformed into men, and Pyrrha's stones became women. In that way, they repopulated the Earth with people who sprang from the "bones" of Mother Earth, and who respected her laws and her bounty.

RETOLD FROM GASTER, 1969

WHAT THE EVIDENCE SHOWS

- NASA believes that 30,000 to 11,000 years ago, a giant comet crashed into the Amazon jungle.
- Meteorite streaks stretch 3,000 miles from northern Europe into Africa and the Middle East.
- There is a distinct peak in African meteorite bombardments from 20,000 to 15,000 years ago.
- There is a possible giant crater in Finland that dates to the end of the Ice Age.
- The long axes of all five of the craters we have studied suggest a common flight path out of the northern sky.
- The proposed craters would be at the top of the list as five of the eight largest on the planet.

Next, let's review some of the effects of massive impacts on various parts of the planet.

30

FIRES ACROSS THE LAND

QUESTION: Past impacts, like the one that killed the dinosaurs, happened at the same time as volcanic eruptions and global firestorms. Is there any evidence that a scenario that was similar in many regards unfolded 13,000 years ago?

Zielinski and Mershon (1997) uncovered some striking evidence for volcanic eruptions and global firestorms from the Greenland ice sheet. Cores drilled deep into the ice contain all the dust, debris, and gases from the atmosphere that the snow trapped when it fell eons ago, but the sulfur compounds were of special interest to the researchers. They believe that those provide a record of the eruptions of ancient volcanoes that were located mainly in Iceland, California, Alaska, the Pacific Northwest, and occasionally Europe. By measuring the sulfates, Zielinski and Mershon believed that they could create a continuous picture of eruptions stretching back for more than 100,000 years.

Prior to the work of Zielinski and Mershon and others, many scientists had assumed that the level of past volcanic eruptions was reasonably constant—maybe a few more or less in any given century, but mostly the same average over the millennia. What they found was not like that at all. There was a sudden evident jump in the number of eruptions in the period from 16,000 to 17,000 years ago, about 2,000 years before the climate began its radical switch to warmer conditions. The increase was startling. Suddenly, volcanoes near North America were erupting about eight times more often than normal, and then the frequency climbed even higher, eventually reaching a stunning sixteen times the previous low levels. According to Zielinski and Mershon, it was the greatest burst of volcanic activity in the entire 100,000-year ice-core record, and once it started, it continued for thousands of years.

Those were not just run-of-the-mill eruptions, however; they were major, climate-altering blasts that were severe enough to have injected millions of tons of dust, sulfur, and smoke high into the atmosphere, as shown in figure 30.1. In that graph, notice that before 16,000 years ago, when the shock wave arrived, volcanic activity was at very low levels. There were only occasional eruptions over nearly 15,000 years. Then all hell broke loose—volcanic hell, that is—when massive eruptions were occurring sixteen times more often than before. Compare that with recent times, when we can expect a major eruption larger than that of Mount St. Helens only every few hundred years. Incidentally, the Mount St. Helens eruption was tiny compared to most of the eruptions that Zielinski and his colleagues found.

There was evidence for massive volcanism, but the puzzle is this: what caused the sudden surge? It seems unlikely that so many widely separated volcanoes should suddenly become active by coincidence. Most scientists think that as the ice sheet built up, it forced the Earth's crust to sag under the weight. Then when the ice began to melt, it released the weight, allowing Earth's crust to rebound, and this uplift, in turn, caused earthquakes and eruptions.

But there are several problems with that theory. First, the sudden jump in eruptions began almost 2,000 years before there was significant melting of the ice sheet. Therefore, it is not true that the weight of the ice suddenly lifted—that happened thousands of years after the eruptions. In fact, the crust is still rising under Hudson Bay even today, as it has been since the Ice Age ended, but there are currently no major eruptions in the area.

There is evidence from New Zealand that the increase in eruptions was unrelated to the rebound of the crust. In figure 30.2, you can see that a major escalation in eruptions in New Zealand occurred at about the same time as those that show up in the Greenland ice cores. New Zealand is in the Southern Hemisphere, however, thousands of miles from the nearest ice sheet in Antarctica and even farther away from the North American ice sheets. Because of that, it is unlikely that crustal rebound was responsible for the eruptions there. But if the eruptions were caused by the arrival of the supernova shock wave and impacts, some such effects would be expected around New Zealand.

There's another problem with the eruptions. Heavy volcanic activity is known to cause global cooling, but that's not what happened around 16,000 years ago. Instead, the climate began to warm up, not cool down. What caused this discrepancy? There are several possibilities. First, it could be that the sulfur came from the supernova dust wave rather than volcanoes, and that the level of eruptions actually was normal rather than high.

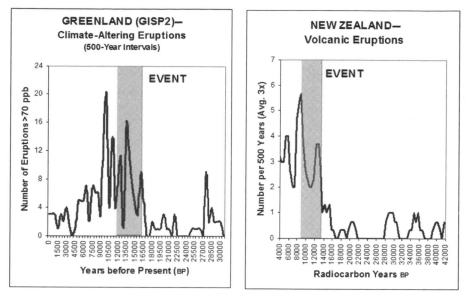

Fig. 30.1. Zielinski and Mershon used ice levels of sulfate, or SO₄, to pinpoint the times of climate-altering eruptions. *Data from Zielinski and Mershon (1997)*

Fig. 30.2. The frequency of eruptions rose suddenly about 13,000 to 16,000 years ago in New Zealand, which is far from the nearest ice sheets. *Data from Bryson (1988)*

If that was the case, the dust grains might have affected climate differently from volcanoes 16,000 years ago. In addition, while sulfate tends to cool things off, that wasn't the only thing brought into the atmosphere by the cosmic dust. Water vapor, carbon dioxide, and methane, three of the most potent greenhouse gases, make up about 50 to 90 percent of most comets and their dust. All these may have overwhelmed the effects of the sulfate to cause things to get warmer.

THE FIRESTORMS

Dust and ash were not the only problems from volcanoes. They also started forest fires, and we see evidence for this in the ice cores. When Legrand and De Angelis (1995), researchers at the European drill site in Greenland (known as GRIP), analyzed the ice core for ammonium, a compound produced by forest and grass fires, they expected there to be more fires as the climate warmed up, and there were. That's normal, since more vegetation and more thunderstorms create larger fires. But they were not prepared for what else they found.

The ice core went back nearly 400,000 years, and Legrand and De Angelis sampled various areas of all of it; but to their surprise, about 16,000 years ago ammonium levels in the ice began to rise slowly. Then, abruptly, around 13,000 years ago, the ammonium increased incredibly fast, as shown in figure 30.3. Levels had been only a few parts per billion for tens of thousands of years when suddenly they skyrocketed to sixty-three parts per billion, thirty times higher. Amazingly, in the samples they tested over the 400,000 years of the core, it was the highest reading up to that point. The fires rapidly became severe, widespread, and long lasting.

Once started, the fires continued at a higher-than-normal pace for thousands of years, but for them to start so suddenly, something unusual must have happened. The date of that massive spike is 12,340 years ago, and we think that may be the date of one of the impacts. In addition, the chart shows a distinct but smaller peak around 41,000 years ago, which we think may be the "footprint" of the supernova.

There is another source of ammonium besides the fires: the comets and comet dust. According to NASA, interstellar dust and comets contain ammonium in amounts ranging from about 0.6 to 10 percent. If our comets were hundreds of miles across, they might have dumped a huge load of ammonium into the atmosphere, enough for it to show up clearly in the ice core.

Also, do you recall the major nitrate peak in the same ice core back in chapter 16? That peak also occurred exactly 12,340 years ago. Legrand and De Angelis write in their paper that they did not always find nitrate and ammonium peaks together in the ice cores. To explain why, they cite a study by Lebel that showed a connection only when the fires were burn-

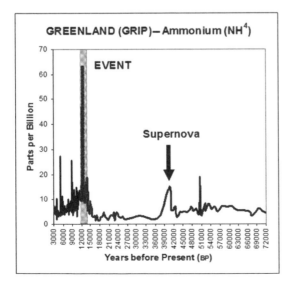

Fig. 30.3. There was a sudden peak in ammonium from fires at the time of the impact. *Data from Legrand and De Angelis (1995)*

ing fiercely rather than smoldering. In other words, both ammonium and nitrate appeared only after powerful firestorms. Because they were found together in this section of the ice core, the conclusion is that those were very hot fires, such as a dustball comet would have created.

Oxalate is another chemical that reaches a peak at the same time, and Legrand and De Angelis link it to fires as well. Like ammonium and nitrate, it peaks sharply at the impact time with a minor peak near the supernova. Scientists expect to see more fires near an interglacial, so some dismiss the oxalate and all the other peaks as just from normal burning events. But such high peaks are not necessarily linked to the interglacials, such as we now enjoy, because the scientists found no major peak in oxalate at the end of the previous ice age. The peak 12,340 years ago is the largest in the entire 400,000-year record. It is important to understand just how unusual that is: the researchers tested a lot of ice, and they did not find any event as severe as that one, which spanned about four ice-age cycles. We think that it was no ordinary fire; it was a cosmic one.

QUESTION: Okay, that shows evidence for fire from the ice cores, but Greenland is a long way from Hudson Bay. Is there evidence for fires anywhere closer to North America?

Paleoclimatologists have recovered many lake cores to study the pollen deposited there, and those cores tell a lot about the climate's effect on ecosystems. Sometimes, while looking for pollen, the researchers also came across charcoal that was deposited after forest fires. After looking through the data from hundreds of these scientific reports on lake-sediment cores, we found an unusual picture of the fire history of North America. Not many Ice Age lakes remain today, and most do not date back to the comet impacts. In addition, most were not tested for charcoal, just for pollen.

Out of hundreds of lake-core reports, we found only thirty-three sites with a charcoal record that spanned the Event (see appendix C for a list), including some of the Clovis-era sites that we visited. What we found is that prior to about 16,000 years ago, there is very little indication of large fires. During the coldest part of the Ice Age, the sparse vegetation and cold temperatures did not favor fires. Then, suddenly, fires began leaving an obvious black trail of charcoal in North American lakes about 16,000 years ago. Entire forests of spruce, fir, and pine disappeared in columns of dense smoke. During the span from 16,000 to 11,000 years ago, all across the North American landscape there were small fires, along with some of the largest fires ever recorded. Areas in every region of the continent began going up in smoke.

All thirty-three cores showed fires during the Event from about 16,000 to 11,000 years ago, and some showed distinct peaks around Clovis times, during the largest impacts. At many sites, the Event fires were the first ones to appear in the lake core, meaning that there was no indication of prior fires. In most cases, the Event fires were not as a large as later fires. Sometimes the peaks were small, because there was much less vegetation to burn during the Event. Dense, highly combustible vegetation did not appear until thousands of years later. Even so, the Event fires were often the largest fires for thousands of years, showing up as a distinct spike in the Clovis-era lake sediment.

QUESTION: Did the impact blasts and the fires kill all the megafauna?

No, we think the causes of the megafauna extinctions were much broader. The blast wave and burning ejecta, such as that which might have created the Carolina Bays, certainly would have caused some direct fires and would have been devastating to living things. The fires may have been isolated, however, because there was not a lot of vegetation around during the Ice Age, and cold temperatures tend to stifle fires, discouraging the kinds of wide-ranging forest fires we see today. In addition, there is an interesting negative connection between cosmic blasts and fires. Naturally, one would think they would go together, but when the Tunguska explosion occurred in Russia in 1908, the trees were knocked down and charred by the blast, but the fires went out quickly. The full reasons for that are unclear, but the explosion most likely consumed much of the oxygen necessary for combustion, or perhaps the rapidly moving blast winds actually blew out the fires.

We think that vegetation killed by the Event was possibly the most important cause of fires that occurred for hundreds of years following the impacts, although the shock wave was most likely equally lethal. Millions of trees and plants were killed by the combination of blast wave and rapid climate change, and so they would have provided ample fuel for forest fires long after the impact.

Climate change is one of the key points to understand about the impacts. Sudden climate change after the Event was catastrophic for plants and animals. Perhaps more important than the blast or the fires, the change in climate may have been primarily responsible for the extinctions. After the impact, global temperatures most likely plunged within hours, dropping to frigid levels that wiped out many plants. Their roots may have survived, as many do through severe northern winters, but the trunks and stalks may have been frozen back. Because of this and the fires, the food supply for many large animals, which had large appetites, vanished within days, and the animals that survived the blast would have starved. Survival favored smaller animals with smaller appetites and more varied diets, so the specialized foragers were in trouble. In fact, those animals and the animals that fed on them, such as tigers and bears, are the species that went extinct. Mammoths, horses, bison, and camels were all large animals that filled a very precisely defined grazing niche, and when the ample supply of grass and plants that they required burned or died, they were doomed.

Even worse, the climate change was not temporary; it persisted. Climate researchers know that many of the Ice Age ecosystems, especially the steppe vegetation favored by the large grazers, never returned to the areas that it once covered. New plants gradually replaced the former plants over centuries or decades. In the meantime, however, there was nothing for many of the highly specialized animals to eat.

Rapid, dramatic climate change has happened many times at the end of other ice ages, but we don't think past changes happened as abruptly as this one, and the suddenness is the key. At other times, the change happened gradually, giving animals the time to migrate or to adapt to a different food supply. This time, it wasn't simply a matter of walking somewhere else to get food—there was no food anywhere; it was dead or incinerated. It may have taken years or decades for the new ecosystems to establish themselves, and by that time, the large animals were gone. Smaller animals with broader diets and smaller appetites, including humans, survived even though their populations declined severely.

THE BLAST AND THE FIRESTORM

Biver and coworkers (2002) reported that comets might contain up to 6 percent methanol, or methyl alcohol, which is highly flammable. Being similar to lighter fluid or charcoal starter fluid, methanol would have added fuel to the fire, quite literally. The blast would have been unbearably intense, but that was not the only problem.

Comets are known to contain sizable amounts of cyanide, formaldehyde, and hydrogen sulfide, all highly toxic substances. Several breaths of any of the three can result in instant death, and formaldehyde, the deadly chemical used to kill and preserve biology specimens, peaked sharply during the Event. According to Cottin and colleagues (2001), the grains of Halley's comet contain about 7 percent of a parent compound of formaldehyde. If our impactor was similar, that means millions of tons of the deadly chemical plunged through Earth's atmosphere 13,000 years ago. We can only speculate about the consequences, but none of them would have been healthy for living things.

Millions of animals, including mammoths, mastodons, and people, were killed by the combination of the blast, shock waves, searing flames, the deadly brew of toxic cosmic chemicals, and the choking smoke. Afterward, many more died from plunging temperatures and from the lack of food.

When the fires finally burned out, vast stretches of the North American continent, especially in areas with baylike craters, may have looked like Mount St. Helens after its large eruption. That eruption devastated nearly 230 square miles (600 km²) of forest, leaving millions of tall trees either toppled or standing but dead.

The cosmic explosion at Tunguska was even larger than that at Mount St. Helens, flattening about 830 square miles (2,150 km²) of forest and leaving shallow holes in the peat bogs similar to the Carolina Bays. That area is about twice as large as the city of Los Angeles. The Tunguska comet is estimated to have been only a few hundred feet in diameter, far smaller than the Event impactors were. Even so, the devastation was extensive. If our impactors were thousands of times larger than the object at Tunguska, the devastation would have been almost beyond comprehension. The Paleo-Indians who witnessed the Event would not have forgotten it—that is, those who lived to tell about it.

THE SURVIVORS: THE INCA

Modern-day archaeological evidence suggests that people were living in South America, at Monte Verde, Chile, and other places, long before the Clovis people arrived. If so, they would have seen the Event unfolding in the northern skies. If the Iturralde crater (discussed in chapter 29) is part of the Event, they may well have had a ringside view of the impacts. The following story is from the Inca, who lived in Peru, and even though this story describes an apparent volcanic eruption, the comet impact could have caused all the events of the tale. Dust and smoke from the impact would have spread around the globe within one day's time, and the impact would have triggered earthquakes and volcanic eruptions. In addition, the toxic chemical stew may have fallen as the "blood-rain" described in this story.

✌

The Day the Inca City Sank

One day, several shabby-looking men came to the great city of the Inca on Lake Titicaca. Some people thought they were beggars, but later, some thought they were gods in disguise. They told everyone who would listen,

"Get ready. Leave the city. Death and trouble is coming, because the Earth is angry with you. The ground will shake and the sky will burn and the waters will rise up to strike you." But all these prophecies upset the rulers, who threw the men out of the city.

Soon afterward, someone saw a huge, rapidly expanding cloud on the horizon. First, it was red, then brown, and then as black as ashes. It headed toward the city, spreading out to cover the sky. After sunset, the clouds lit up the sky with an eerie reddish glow. Night never came.

Then, after sunrise, the Earth trembled violently. Some buildings shook but did not fall, although some homes crumbled, sending debris crashing into the streets. Almost before the shaking stopped, a thick, sticky blood-rain began to fall from the sky, covering everything. It stuck to all that it touched, piling up on roads and bridges and collapsing rooftops. Rubble choked the city.

While the blood-rain still fell, the ground shook again with an enormous rattle, far more violently than the last time. The remaining buildings collapsed with a great crash, throwing dust clouds into the air. The irrigation canals split open, pouring water into the city. Rivers changed course to send more water raging through the streets, and all the while, the shaking never stopped. Slowly, with great lurches and surges, the whole city sank slowly beneath the surface of the new lake. To this day, the cold blue waters of Lake Titicaca cover the grand city of the Inca.

RETOLD FROM GIFFORD, 1983

WHAT THE EVIDENCE SHOWS

- Sulfate levels in the ice cores suggest high volcanism during the Event.
- Ammonium, nitrate, and oxalate in the ice cores support firestorms during the Event.
- Comets also contain high levels of sulfur, ammonium, nitrate, and oxalate.
- Fire-related charcoal peaks were found at many Paleo-Indian sites in the Clovis-era layer.
- At a total of thirty-three sites, fire charcoal reached a peak during the Event 16,000 to 11,000 years ago.
- An impact would release millions of tons of lethal chemicals into the atmosphere.

According to the old stories, after the fires, the floods came. Let's look at the evidence for that.

31
FAST-MOVING ICE
AND WATER

QUESTION: Okay, so ice went flying across the continent to create the bays. But all the stories mention torrential rains and worldwide flooding. Is there scientific evidence for those?

The giant comets sent massive clouds of water vapor up into the atmosphere, but what goes up must come down, as rain and snow. The supersaturated atmosphere could not hold that much water for long, so soon afterward, steady rain began to fall and heavy snowfall started. Over the next few days, the airborne water and steam dispersed over the planet and fell at varying rates. Some areas got only a little and other areas got torrents, just as happens today.

Assuming that most of the 200,000 cubic miles of ice from Hudson Bay turned to water and vapor, if it had been spread evenly around the entire globe, it would have covered the Earth to an astounding 3.5 feet deep (about 1.1 meters). Coincidentally, that figure nearly equals the entire annual rainfall for the Earth. We don't know how long it took for it to fall, maybe a few weeks, probably less than a few months. But by the time the rain slowed down, the Earth had received, amazingly enough, an entire year's worth of rainfall in only a few weeks. The only comparable drenching in modern times would be the heavy rains from hurricanes and cyclones, except that such storms are localized, and they don't continue for weeks with almost no break. That, combined with warmer temperatures and more melting ice, contributed to the widespread flooding around the globe.

FLOODS ACROSS THE CONTINENT

When the comets hit, they also sent enormous surges of water racing toward the Gulf of Mexico, the Arctic Ocean, and the Atlantic. Many Canadian scientists have studied those meltwater flow paths across Canada very extensively, and Shaw (1999), Munro-Stasiak (2002), Russell and Arnott (2003), Baker (2002), Cutler and coworkers (2002), Fisher and associates (2002), and others have published papers describing what they found. They all agree that meltwater followed distinct paths under the ice sheet, forming drumlins as it went and melting the underside of the ice sheets, which added even more water to the flood surge. Shaw found evidence that the outburst came from a huge glacial lake that formed in Hudson Bay, just as would have happened in an impact.

When the water reached the edge of the sheet, it burst out into riverbeds to fill up immense glacial lakes along the ice margin. Once full, the lakes spilled water into other downstream rivers. Munro-Stasiak suggests that, to account for the massive erosion, as shown in figure 31.1, the giant flows lasted for days or weeks, carrying millions of cubic miles of water. If we look at how the flow traveled from Hudson Bay across Iowa and Oregon, the two main outlets for the surges, the total distance is much greater than the full length of the Mississippi River, and the total water flow also would have been far greater. In addition, there would have been many rivers running at catastrophic flood stage, including today's heartland rivers, such as the Arkansas, Colorado, Missouri, and Ohio. Those are the contemporary names for the rivers, but those were no modern-day rivers—they would have been roaring, churning, deadly monsters. Nowhere in the world today can we see one like them.

Licciardi and colleagues (1999) conducted a comprehensive study to

Fig. 31.1. This DEM image shows the western plains of Canada and the United States. Notice how the motion of the ice sheet and meltwater cut distinct grooves in the terrain that extend for several thousand miles across the continent.

determine the changes in outflow from the continental ice sheet, focusing on rain and snowfall and on the melting of the ice. The researchers acquired dozens of radiocarbon dates that indicated that the ice suddenly began to retreat from its most recent advance between 16,800 and 16,300 years ago (fig. 31.2), when it began to melt very rapidly. The combined flow of rivers, such as the St. Lawrence and the Hudson, dramatically increased to levels 100 times greater than their modern flow rates. At its peak, the Hudson River flowed at more than one-third the rate of the modern-day Amazon River—and the Mississippi flowed at more than half the rate of the mighty Amazon!

Incredibly, the total amount of water draining off the North American ice sheet equaled the combined massive flow rate of some of today's largest and most familiar rivers: the Amazon, Congo, Yangtze, Mississippi, Mekong, Danube, Nile, Rhine, Yellow, and Thames. Those Ice Age waterways were no languid and tranquil streams; they were dangerous, turbulent rivers in a state of nearly perennial flood.

THE IMPACT BLOCKS THE MISSISSIPPI

The dustball-comet impacts produced a surge in the glaciers that carved basins such as the Great Lakes and pushed out long glacial moraines that rimmed the ice sheet, damming up the water that once flowed to the Mississippi. At the same time, the impact blew out shattered ice, sand, and mud, which blocked the southern water-flow routes even more, so that the Mississippi suddenly dropped precipitously to a level an astounding ten times

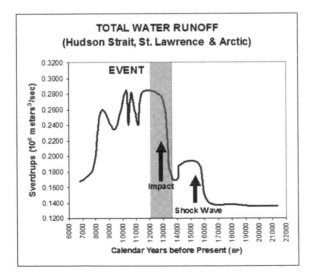

Fig. 31.2. This chart confirms that the impact melted the ice sheet and opened new drainage outlets. There was a small increase in the flow rate with the arrival of the shock wave, but the rate nearly doubled when the impacts occurred 13,000 years ago. *Data from Licciardi et al. (1999)*

lower than it had been. Licciardi's team found that comparatively little water flowed from the Great Lakes area to the gulf for the next 1,000 years, and instead it surged into the Atlantic, placing a near-freezing freshwater lid on the northern ocean and causing disastrous effects on the climate, which returned to nearly Ice Age conditions during the Younger Dryas.

There already were large glacial lakes, but with the ice sheet melting rapidly and retreating at one of the fastest rates of the Ice Age, even larger lakes formed. Acting as giant dams, the moraines and crushed ice were so high in some places that early lake levels were 200 feet above Lake Michigan today. Holding back an immense amount of frigid meltwater, some glacial lakes were larger than many states and provinces in North America. That water created a lethal hazard, because as more of it came surging in from the melting ice sheet, the lake levels rose higher to break through the moraine dams from time to time. When they did, skyscraper-high superfloods roared down the river channels, sweeping away everything in their paths.

THE MISSOULA FLOODS

It is difficult to imagine such flooding, especially considering how tranquil our rivers seem today. Because of that, the idea of catastrophic flooding has become widely accepted only recently. The proof for such floods came from the work of scientists, such as USGS researcher J. Harlan Bretz, who uncovered the superflood evidence along the Columbia River in the American Northwest. At first, no one believed that such events were possible, so Bretz suffered considerable criticism for his theory. But eventually he and other scientists amassed enough evidence to prove that it had happened as many as forty times along the Columbia.

The true scope of the floods became clear when USGS scientists found a narrow channel near Portland, Oregon, where floodwaters from glacial Lake Missoula reached a depth of at least 400 to 500 feet; some researchers think the floodwaters reached 1,000 feet in depth. The ice dam had been nearly a half-mile tall, making it more than twice as high as the current world's tallest dam in Asia and about 1,000 feet taller than the world's tallest building. When the dams broke, the sight must have been truly spectacular. Bretz called those awesome events the Missoula floods.

Beginning about 20,000 calendar years ago, the lake dams first began failing, but that had nothing to do with the shock wave, as far as we know. By about 16,000 years ago, with the arrival of the supernova shock wave, the dams were failing often, sending turbulent floods rushing down the

Columbia. At its largest, Lake Missoula was four times bigger than Lake Erie, and yet when its dams failed, in only eight hours' time it released as much water as there is in all of Lake Erie. People and animals hundreds of miles away would have heard the ground-shaking thunder of Lake Missoula's water rushing through the gap in the half-mile-high ice dam.

FLOODING FROM THE GLACIAL LAKES

There were dozens of other large ice-dammed glacial lakes at that time that would have been affected by a dustball-comet impact. As you may recall, in chapter 7 we discussed glacial Lake Hind, which Matthew Boyd showed had drained catastrophically in Clovis times. In the 13,000-year layer, we found high levels of magnetic grains, floating spherules, and charcoal from forest fires. In addition, we uncovered a black mat layer and abundant potassium-40, an isotope that typically comes from far beyond our solar system and is a telltale marker of supernovae. All this is direct evidence for a connection between the giant floods and the Event.

In addition, there is a strong impact connection with the largest glacial lake ever known, Lake Agassiz, which has dried up but once lay to the northeast in Manitoba (fig. 31.3). According to Rooth (1982), about 12,900 years ago (11,000 ^{14}C years), Lake Agassiz sent an immense pulse of meltwater surging downriver, and it was no run-of-the-mill flood. In a matter of days, the lake released enough meltwater to cover Canada, the United States, and Mexico fifteen inches deep in frigid glacial water, if it had been evenly distributed.

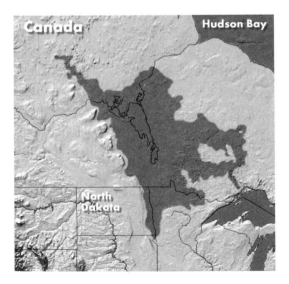

Fig. 31.3. Glacial Lake Agassiz covered most of Manitoba and parts of North Dakota and Minnesota. It was considerably larger than Lake Superior, shown at lower right.

Walter Broecker (1999) and others support the theory that the immense Agassiz flood initiated the return to near glacial conditions. Recently, however, Lowell and associates (2005) did extensive radiocarbon dating of the lake basin and called into question the idea of an Agassiz–Younger Dryas connection, since their dates show the failure occurred after the beginning of the cold spell. This dating issue is yet to be resolved, but we suggest that these researchers' dates may be too young, since there is a serious problem with radiocarbon dates near the Event. Due to the uneven influx of cosmic radiocarbon, some dates will appear considerably too young, whereas other dates nearby may be correct. It will take other kinds of dating besides radiocarbon to resolve the issue.

According to Broecker, water flow nearly ceased down the Mississippi, as we mentioned above, as the meltwater routes shifted to the north through the Hudson Strait and to the east through the St. Lawrence River and down New York's Hudson River. This situation caused the huge surge of frigid meltwater from Lake Agassiz to pour into the North Atlantic, where it initiated the return to Ice Age climate.

THE METEORITE THAT SURFED DOWN FROM CANADA

Along the Missoula flood routes, researchers found thousands of multiton boulders that the floodwaters had lifted and carried for miles. Some boulders, called erratics, are nearly house-sized, and they are difficult to move with the most modern excavating equipment. Yet the floodwaters carried them for hundreds of miles.

One of the strangest Ice Age boulders found turned out to be a 31,000-pound iron meteorite, one of the largest on the planet (fig. 31.4). Found in 1902 by a newly immigrated Welsh coal miner, the Willamette meteorite changed

Fig. 31.4. The massive 15.5-ton Willamette meteorite weighs as much as four SUVs but contains much more iron than they do. *Source: Lake Oswego Public Library*

ownership until the American Museum of Natural History acquired it for display in New York City.

Scientists who have studied the object concluded that the meteorite did not fall in Oregon, but rather on the ice sheet in Canada. They found it surrounded by erratics that were angular, suggesting that they had not been tumbled and smoothed by the floodwaters. The only way to explain that is if the erratics, along with the meteorite, had ice-rafted down the Columbia on one of the Missoula floods. Some theorize that the meteorite actually fell on top of the ice sheet and froze into the ice, until a flood carried the meteorite-laden ice block to Oregon, whereas others think the turbulent floods dug it out of the ground. What seems certain is that the Missoula floods carried it to Oregon.

There is no way to know precisely where it came from or when it fell, although it could have fallen less than about 40,000 years ago. That means it is possible the cosmic iron fell during the Event itself. Perhaps this 31,000-pound chunk of iron and nickel surfed the supernova wave to Canada and then surfed the flood waves to Oregon.

WHAT THE EVIDENCE SHOWS

The years 13,000 and 16,000 BP are "magic" numbers for the supernova/impact theory. Many key events we have discussed in this book occurred at those magic times:

- the disappearance of the ice center over Hudson Bay
- the dam failure on the largest glacial lake ever known
- the sudden increase in precipitation and meltwater discharges
- the abrupt shift from the Mississippi River to northern outlets
- the start of the Younger Dryas cold time
- the extinction of the mammoths
- the near disappearance of the Clovis-era culture
- the deposition of Clovis-era magnetic grains, magnetic spherules, and hollow spherules
- the appearance of high sediment levels of radioactivity and other markers

It is hard to see it as coincidence that all these events share a similar date.

Nearly all the cultural stories that we have read tell of world-altering rainfall and flooding. Matching the stories, most of the scientific research that we have seen provides compelling evidence for immense flood surges spanning thousands of years. Those megafloods disrupted life on land, and we find a revealing record in both the evidence and the moving stories of many native peoples who were overwhelmed by a disaster beyond their comprehension and experience. No matter how devastating the massive meltwater floods may have been, however, they always ended and the water eventually flowed away, but in this case, that did not happen very quickly.

32

THE WATERS COVER
THE LAND

QUESTION: If there was flooding caused by the comet impacts, the floods must have reached the oceans. Is there any evidence for that?

Searching for evidence of the Mississippi River floods, Aharon (2004) used seven seafloor sediment cores from the Gulf of Mexico to measure oxygen isotopes. Changes in the ratio of those isotopes reflect changes in ocean temperature caused by meltwater flooding. Aharon found that the gulf sediment clearly showed three major flood peaks (fig. 32.1).

All three of the peaks occurred within the period of possible bombardment of Earth by the shock wave and impacts. As we saw in chapter 1, two of the largest radiocarbon reversals took place around 18,000 and 13,000

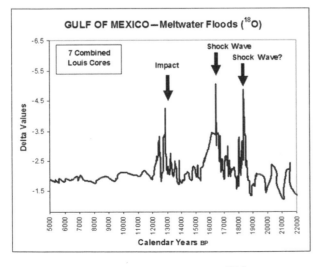

Fig. 32.1. In the ocean-floor sediment of the Gulf of Mexico, three major episodes of catastrophic flooding show up. All three fall almost exactly within the window of the shock wave and impacts. *Data from Aharon (2003)*

years ago, at the same time as two of these peaks. That timing could be merely coincidental, but that is hard to justify.

The first peak started more than 18,000 years ago, and the onset was sudden and dramatic. The second peak, the largest one in the record, reached maximum about 17,000 years ago, increasing at one of the fastest rates in the entire 22,000-year record. Clearly, massive flooding occurred just when other evidence supports the arrival of the shock wave from space. The third peak occurs at the time of the Hudson Bay impact 13,000 years ago. After that, the floodwater paths moved to northern routes, and meltwater pulses into the gulf became rarer.

EFFECTS ON SOUTH AMERICA

The consequences of the flooding into the oceans were not limited to North America; we also can find them near South America in the Cariaco basin, which is located off the coast of Venezuela. Piper and Dean (2002) conducted a study of seabed cores from the basin and found something very unusual: a very thick, fine-grained layer of gray sediment that dated to about 16,000 years ago, matching the date of one of the Gulf of Mexico meltwater events. Piper and Dean and other researchers concluded that the unusual layer represented a "freshwater event," or flood, and because radiocarbon dates at the top and bottom are nearly identical, Piper and Dean believed that it was sudden, brief, and very intense. Those trace minerals typically do not come from water sources, so the flood most likely originated on land in South America.

On analyzing the sediment, the researchers found that it contained high amounts of thorium, titanium, cobalt, and other rare earth elements, just like the KREEP meteorites from the moon and the magnetic grains found at all Clovis-era sites. The levels nearly doubled at 16,000 years ago within a very short time, and after the elemental levels rose, it took more than 2,000 years for them to return to normal levels, suggesting that large amounts of the elements were suddenly added to the local environment. Another peak appeared at 13,000 years, rising about 33 percent above its previous level. Something unusual had to have happened for these minerals to show up in such quantities.

QUESTION: The impacts should have raised sea levels. Is there any evidence for that during the Event?

The first evidence for a rise in sea levels comes from research by Fairbanks (1990) along the coral reefs of Barbados off the northern coast of South

America. He studied reefs that had been growing around the island for about 20,000 years, spanning the time we are discussing. Fairbanks knew that one particular type of common coral grows only in shallow water and dies if the water gets too deep, so that it would act as a marker to show the rate at which the sea level had changed over the course of the end of the Ice Age. Fairbanks and his fellow researchers drilled long cores into the reef, and then they looked for the marker shallow-water coral. By carbon-dating each sample, they were able to build up dates for various depths of the reefs, providing a record of the rise in sea level.

Up to that time, most scientists had assumed that the ice sheet had melted away very gradually and that sea levels rose steadily but slowly. To the astonishment of Fairbanks' team, they found large, rapid jumps in sea level, indicating that, at certain times, the ice sheet had melted at an amazingly fast pace, leading to the catastrophic global disappearance of low-lying coastal land.

The greatest and fastest rise began about 14,500 years ago. The Ice Age was ending, regardless of the supernova and impacts, and the ice was melting because of increasing sunlight and rising temperatures, so it is unclear how much of this climb was attributable to the shock wave, but it could have been a lot. Periodic heavy bombardments of small, hot, metallic grains and spherules landing on top of the ice could have produced some of the meltwater pulses.

RAPID RISE AT A RADIOCARBON REVERSAL

In the results from Barbados, Fairbanks uncovered radiocarbon plateaus or reversals right at the 13,000-year and 17,000-year depths (see fig. 32.2). The rest of the record was smooth with almost no reversals. This perfectly matches the influx of global radiocarbon found elsewhere. At those times, the cores showed impossible reversals of dates, indicating that something was wrong. Incredibly, some dates around 13,000 years ago reverse by more than 3,000 years. This can be explained only one of two ways: either older radiocarbon mixed with newer or new radiocarbon entered the system. For the latter, about the only way such a wild swing can happen is if radiocarbon rides in with a comet, asteroid, or cosmic dust or is produced by a sudden increase in cosmic rays.

At the time of this reversal, shown in figure 32.3, you can see from the slope of the two lines, one before the spike and one after it, that the ocean level rose by as much as fifty feet. Because it is not possible to determine the time interval accurately, we cannot say how fast this occurred, but it prob-

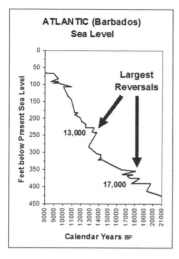

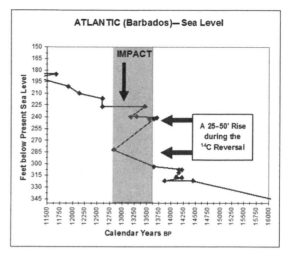

Fig. 32.2. Two large ¹⁴C reversals at 13,000 and 17,000 years ago. *Data from Fairbanks (1990)*

Fig. 32.3. Between the arrows, there is a 1,000-year reversal of dates, at which the oceans may have risen suddenly and catastrophically as much as fifty feet. *Data from Fairbanks (1990)*

ably was very rapid. In fact, Fairbanks's own records indicate that the ocean may have risen fifty feet in only a few weeks. Think about that from the perspective of millions of people living in coastal regions today. Fifty feet of water would cover their homes far above the rooftops—and the worst part is that it would not go down again.

QUESTION: How do we know that excess radiocarbon caused the spike? Maybe the measurement was in error, or maybe the sample was contaminated.

We wondered too whether excess radiation had in fact caused the spike, and to test the idea, we looked for other cores and found two other well-accepted coral-reef cores. Bard and colleagues (1996) recovered one from Tahiti in the South Pacific, and Hanebuth and associates (2000) extracted the other from the Sunda shelf in the South China Sea (see fig. 32.4). The Sunda is on nearly the opposite side of the Earth from Barbados, and Tahiti is thousands of miles southeast of it in the Southern Hemisphere, so they are good global test sites. If there was a real reversal at Barbados and if the sea-level rise is accurate, the effect should have been worldwide and should appear at these two other sites as well.

It does; both show almost identical results. Each site displays a rapid rise at 14,500 years ago following the shock wave, and both show large

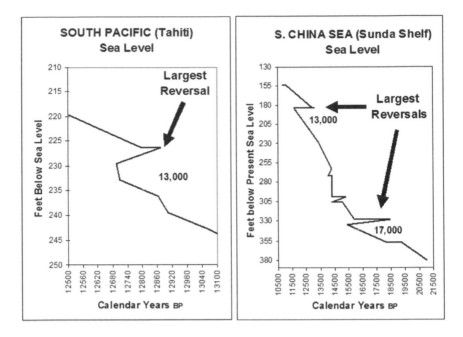

Fig. 32.4. On the left, the Tahiti core shows a reversal at 13,000 years; the core might have shown the reversal at 16,000 years, but it did not go back that far. *Data from Bard et al. (1996)*. To the right, the Sunda core shows two large reversals at 13,000 and 17,000 years ago. *Data from Hanebuth et al. (2000)*

radiocarbon reversals around 13,000 years ago. Also, while the Tahiti core does not go back to 16,000 years ago, the Sunda shelf has a dramatic 3,000-year reversal at 17,000 years. Altogether, these three sites provide compelling evidence for a global influx of radiocarbon during the Event.

RAGGED GAPS IN THE CORAL

In the research that Fairbanks published, there is striking evidence that relates to our theory. There are large gaps in the coral caused by rapid submergence of the reef or by physical damage. Of the coral reefs that he sampled, only four cores span the years from 17,000 to 13,000 years ago, and every one of them displayed extensive reef damage and gaps in the coral. No other interval in the entire 20,000-year-plus record showed anything like it, and in fact, Fairbanks reports no reef-damage rubble in any other section of any core, only during the years of the shock wave and impacts.

Figure 32.5 shows Fairbanks's record of the coral core, adapted from his paper. Parts of the 4,000-year section for each core were dead zones made of rubble and sand, indicating that the reefs were not growing or were suffering severe damage over a long period.

Because of the gaps, scientists cannot be certain how fast the sea level was rising during those 4,000 years. In analyzing the site later, Keigwin and coworkers (1991), as quoted by Aharon (2004), proposed an interpretation for the coral gaps. They suggested that it meant the sea level was rising so rapidly that the coral reefs could not grow fast enough to keep pace with it. We think there was more to it than that, as we will see in the next chapter. Later, after the impact Event was over, most of the coral reefs recovered and began to grow again, even though ocean levels continued to rise, occasionally at a rapid rate.

Weaver and associates (2003) interpret the Barbados data as indicating a rise in sea level of about sixty-six feet in 500 years, or thirteen feet per century. Likewise, Shaw and Gilbert (1990) use evidence of megafloods in Ontario, Canada, to conclude that the sea level rose forty to fifty feet around 17,000 to 16,000 years ago; and at one time during that span, they conclude that it rose an amazing six to ten feet in just a few years, and at least one foot in just a few weeks. We think ocean levels rose far faster, but even if we accept these researchers' estimates, the rise in sea level was incredibly fast when you think that our modern society is rightfully concerned about global warming producing a sea-level rise of a few feet in this century.

To put that amount in perspective, consider that a sixty-six-foot rise today would flood nearly all the cities, ocean ports, oil refineries, and farms that are within around fifty miles of all the world's oceans. This area is home to about half the world's current population, so such an event would be monumentally catastrophic. Today's oceanfront population percentage is probably similar to that of 16,000 years ago, when most groups depended

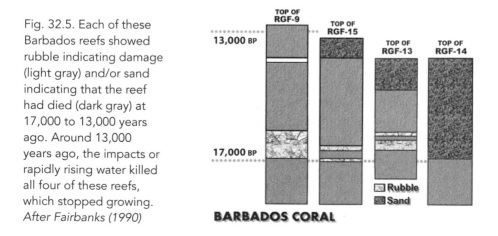

Fig. 32.5. Each of these Barbados reefs showed rubble indicating damage (light gray) and/or sand indicating that the reef had died (dark gray) at 17,000 to 13,000 years ago. Around 13,000 years ago, the impacts or rapidly rising water killed all four of these reefs, which stopped growing.
After Fairbanks (1990)

on the sea for food, and it was much warmer near the ocean during the Ice Age.

Altogether, the world's oceans rose an incredible 200 feet during the Event from 17,000 to 13,000 years ago, inundating millions of square miles of the world's most fertile land area, mostly along the shelf of each continent. This means that within a few thousand years, rich, abundant land twice the size of North America vanished under the waves. Knowing that, one can more easily explain the stories of catastrophic floods from all over the globe. Such worldwide flooding would have created extremely difficult problems for our ancestors.

THE SURVIVORS: THE KATO

Even though the cultural stories in this book are seemingly about disaster, they are really more about survival. Repeatedly, the stories delivered the same lesson to their people: to remain in touch with the Creator, to listen to the subtle warnings of trouble, to cooperate with each other when it came, and to take action no matter what others thought. Those who did this survived one of the greatest cataclysms ever to befall humankind.

The Kato tribe once lived in coastal California north of San Francisco. Along with many other ancient peoples, the Kato believed that the sky was made of stone. From that cosmological point of view, a rain of falling meteorites meant that the sky was falling, as in this story. The tale accurately describes a cosmic event with almost no embellishment or allegory.

The Day the Sky Fell

The Creator and the Creator's helper, called Thunder, originally made the sky out of sandstone. After a while, the sky became very old and began to break apart. Down below, the People of Earth heard a great, thunderous cracking noise, first from the east and then from all directions.

Before Creator and Thunder could fix the cracks, huge chunks of sky-stone began to fall into the ocean and onto the Earth, killing many People and animals, and flattening whole mountains. The ocean impacts created giant waves that crashed across the shorelines, sweeping trees and animals out to sea.

The People were terrified as sheets of water and heavy rain began to pour down through the cracks in the sky onto the Earth below. It rained for many days and nights, causing the waters of the great ocean to rise higher

Fig. 32.6. A Kato woman. *Source: Library of Congress*

and higher, until there was no land for a long way in every direction. The floodwaters swept away People, animals, and plants.

Some People jumped into their boats and some clung to logs and trees as the water swirled around them. Many perished, although a few People survived. Those who lived were the ones who listened to the Wind, the Rain, and the Rocks, and who knew how to live in harmony with them, who knew how to listen to the gentle whispers around them. The Kato tribe descended from those few People. For a while, all People lived in harmony because they were close to the troubled times. Before long, though, many forgot the lessons of those times and turned away from the Creator. Today, only a few still live that way.

RETOLD FROM GODDARD, 1929

WHAT THE EVIDENCE SHOWS

- Gulf cores show meltwater floods at 18,000, 16,000, and 13,000 years ago.
- Cariaco basin cores from Venezuela show floods at 16,000 and 13,000 years ago.
- During the floods, Cariaco shows high levels of radioactive thorium.
- Cariaco shows high levels of titanium, cobalt, and rare earth elements, just as at all Clovis-era sites.
- The ocean flood dates closely match supernova-related global radiocarbon reversals.

- Sea levels rapidly rose about 200 feet between 17,000 and 13,000 years ago.
- Barbados shows two large radiocarbon-reversal groups, one at 17,000 and another at 13,000 years.
- The reversal at 13,000 BP lasted for 1,000 years and the oceans may have rapidly risen twenty-five to fifty feet during the reversal.
- Between 17,000 and 13,000 years ago, something heavily damaged the reefs that were studied.
- On the opposite side of the planet, the Sunda shelf also shows reversals at 16,000 and 13,000 years ago.
- The largest reversal occurred 13,000 years ago.
- Tahiti reefs do not go back as far as 16,000 years, but they also show a reversal at 13,000 years ago.

The rising sea levels were not the only difficulty. The comets caused one more problem, something that you might not expect from a featherweight dustball comet. Yet it may have had the most powerful, far-reaching effects of all.

33

ENDLESS WAVES

✍

The Blessed Land of the Sea-God

Long before our time, a noble race lived peacefully along the edge of the Great Ocean in a fertile place they called the Blessed Land of the Sea-God. The sun bestowed upon the People extraordinary fruits and grains, watered by crystal-clear rivers. Having descended from the benevolent god of the sea, who took a beautiful mortal woman as his bride, the Blessed People were a handsome race, and, like their father, they favored the abundant sea.

Over many generations, however, the People became prideful and disdainful of their neighbors, who had not descended from the gods as they had. Determining to subjugate them, the Blessed People went to war with other lands, sparing only those who bowed down to them and killing the rest.

Fig. 33.1. The sea god Neptune.
Source: Library of Congress

Looking down from the heavens, the ruler of the gods became alarmed at how evil and belligerent humans had become, whereas once they had been good. The ruler of the gods decided to punish them and cause them to change their ways. This was not the first time that the gods had chastised humankind for its failings; it had happened many times before. Sometimes the gods shook loose bodies from space to cleanse Earth by fire, sometimes they purified the land with huge floods, and sometimes they did both.

With no warning, the gods turned loose their wrath on the Blessed People. Within a single day and night, violent earthquakes shook the land, destroying homes and fields and decimating the terrified People, who cursed the gods for their misfortune. Then, with a final shudder of the Earth, the Great Ocean rose to cover the entire island, and it sank from view under the turbulent waves.

Only a small remnant of the Blessed People survived, those who had lived in the high mountains, but the cataclysm shattered their lives along with their former land. For generations, the survivors struggled to meet their simplest needs, until, after a while, most forgot all about the Blessed Land of their ancestors. Even today, we know little of that abundant land, and we have forgotten that it perished because of its pride and evil ways.

RETOLD FROM PLATO

You probably already recognize that story. It is, of course, about Atlantis, and its key descriptions come directly from Plato's dialogues *Critias* and *Timaeus*. The legend of Atlantis fits very well with the known facts about the Event: earthquakes, tsunami waves, and rising sea levels, which destroyed a large island culture and all the people on it. Plato claimed to have taken some of the information for his writings from existing ancient Egyptian documents from around 11,600 years ago, not long after the Event, about an actual island.

Today, it is unclear just how much Plato embellished any facts that he had found in order to make the points in his dialogues. After all, he was a teacher. Apparently Aristotle, one of his disciples, thought Plato invented the part about an Egyptian papyrus describing Atlantis. He claimed that Plato, wanting to create a compelling teaching story, had made Atlantis sink beneath the waves all on his own, as a way of illustrating the evils of some governments. There is no way to be sure of the true source of Plato's story, although Crantor, a Greek commentator on the works of Plato, later claimed to have journeyed to Egypt where he saw ancient writings that confirmed the story of Atlantis.

In a related story from the Caribbean, Cuban geologist Manuel Iturralde told a *National Geographic* interviewer in May 2002 that the oral stories of the Maya and the native Yucatecos tribe describe a large island similar to Atlantis. According to them, the island vanished beneath the waves along with most of their ancestors. The stories place the island on the sunken continental shelf among the islands of the Caribbean.

RISING SEAS AND TOWERING WAVES

No one has ever uncovered indisputable scientific evidence for a civilization such as Plato described, but it is undeniable that there were many people living in groups and small villages along the ocean coasts 13,000 years ago when the Event happened. Let's see what happened to them. Our evidence for the Event can provide a striking and compelling view of what the inhabitants around the Atlantic Ocean would have gone through. Let's begin deep under the Atlantic, whose name may have been derived from Atlantis.

UNSTABLE CONTINENTAL SHELVES

As harrowing as it must have been to go through a bombardment of giant, exploding dustball comets, that may have been tame compared to what came next. For years prior to the impacts, glacial rivers had been dumping tons of sediment onto the steep continental shelf off all the world's continents, making the underwater cliffs unstable. Then, when the dustball-ice bombs exploded to create the Carolina Bays, they almost certainly exploded over the Atlantic too. If so, on impact with the ocean, the fast-moving icy chunks would have created immense steam explosions, much like submarine depth charges. That would have been disastrous for all the nearby fish, but there was an even bigger problem: the massive shock waves caused the cliffs to collapse in devastating underwater landslides.

Figure 33.2 shows the location of the well-documented slides, which stretched along the wall of the Atlantic shelf about 300 miles from Virginia to South Carolina. There were three major slides: the Cape Fear slide, which was the biggest, along with the Cape Lookout and the Black Shell slides. It is difficult to date these slides accurately, but Maslin (2004) reports that radiocarbon tests indicate the two largest happened around 17,000 to 16,000 years ago, although they may have occurred as recently as 12,000 years ago. That range puts them squarely within the time of the Event.

The small map in figure 33.2 does little to provide a feel for the immense size of these slides, so imagine this. If we dumped all the sand, silt, and

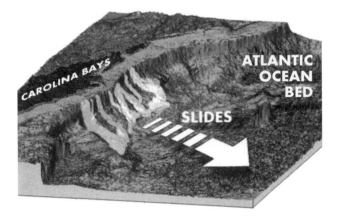

Fig. 33.2. The East Coast, showing the location of the largest U.S. underwater landslides just offshore from the Carolina Bays.

gravel from these slides over North and South Carolina, the layer would be more than twenty-one feet thick, deep enough to bury all the houses in those states along with all but the tallest buildings.

CASCADE OF SIMULTANEOUS SLIDES

The three slides along the Carolinas were not the only ones; Maslin (2004) describes nineteen extensively studied megaslides that occurred approximately between 40,000 and 9,000 years ago. Amazingly, thirteen of the nineteen slides, including three of the four largest ones, occurred in the small Event window between 16,000 and 12,000 years ago, as shown in figure 33.3. Three of the other six happened around 34,000 years ago, during the arrival of the hypervelocity particles that penetrated our mammoth tusks.

The greatest known slide during these times is the Storegga-1 slide off the coast of Norway, which researchers date to approximately 35,000 to 30,000 years ago. It is grouped with two other large older slides, all three of which may have occurred at the time of the first shock wave 32,000 to 34,000 years ago. As we saw in chapter 22, there was a magnetic field excursion at that time, which researchers link to the force of the shock wave, so it is reasonable to conclude that the shock could have triggered underwater landslides. That monster slide at Storegga released 935 cubic miles of sediment (3,900 km³), enough to bury the entire nation of Norway under about thirty-three feet of mud (10 meters).

There is one important point to note. Dating constraints for the slides are very poor, because the massive movement of debris obscures the actual dates. Many of the date-range uncertainties are for thousands of years, and

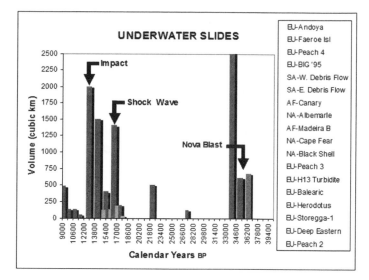

Fig. 33.3. Two major clusters of massive slides, one between 13,000 and 18,000 years ago, the other about 36,000 years ago, just when our mammoth tusks were bombarded. *Data source: Maslin (2004)*

even then, radiocarbon dates apply only to the items dated, rather than to the slide itself. For example, the date for a piece of coral from beneath the slide layer indicates that the coral was alive prior to the slide. There is no way to know, however, if the coral was alive exactly at the moment of the slide or whether it lived thousands of years earlier. That is why there is a great uncertainty.

In figure 33.4, we graphed the preferred slide date along with the uncertainty bars, which indicate that, according to Maslin and his colleagues, the

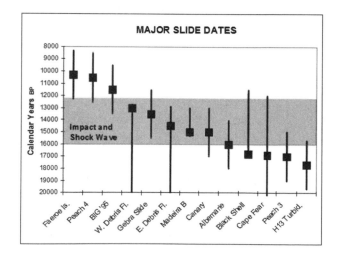

Fig. 33.4. Thirteen large underwater slides. Dating is uncertain, as represented by the bars, although all could have occurred during the shock wave and impacts between 13,000 and 16,000 years ago. *Data source: Maslin et al. (2004)*

slide could have happened anytime during the years covered by the bars. Where the uncertainty was unknown, we assumed it to be plus or minus 2,000 years. The graph shows that thirteen slides could have occurred between the arrival of the shock wave 16,000 years ago and the end of the impacts sometime later than 12,000 years ago.

Furthermore, since most of the uncertainty bars intersect the time from 12,000 to 14,000 calendar years ago, most of the thirteen slides might have occurred at exactly the same time. This is an important new addition to our theory. Let's sum it up: researchers who studied those massive slides independently provided a range of dates that allows us to conclude that each of those underwater landslides could have taken place *on exactly the day of the impacts in Clovis times.*

SHOCK WAVES AROUND THE ATLANTIC

Figure 33.5 shows the location of all eight main slides in the Atlantic basin. The three discussed above were off the Carolinas at #1, three were off Africa at #2, and the last two took place off the mouth of the Amazon River in South America at #3.

Most likely, the Carolina slides were the first to go at #1, because they were nearest the impacts. When they did, a huge volume of material moved down the underwater cliffs nearly as fast as the fastest speedboats, setting off mega-tsunamis. These giant tidal waves rushed away across the Atlantic

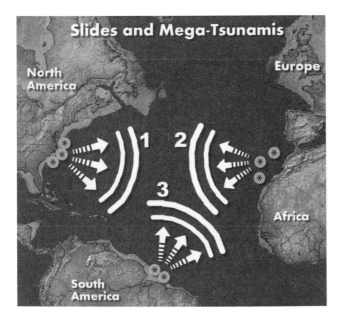

Fig. 33.5. Our proposed sequence of tsunami events. The #1 group of slides happened first, triggering tsunamis that caused the #2 slides to occur. Waves from this group triggered the #3 slides. All this probably happened within twenty-four hours.

at nearly 400 miles per hour to smash into Africa and Europe, triggering three slides there at #2.

TROUBLE IN AFRICA AND EUROPE

Ward and Day (2001), who studied one of the eight main events, the Canary Island slide, found that it tore away a huge notch in the northwestern side of the volcanic island, sending debris cascading nearly forty miles out across the Atlantic seafloor. The two scientists estimate that immediately after collapsing down the underwater slopes, the slides generated megawaves that were a staggering 3,000 feet high, more than twice the height of the world's tallest building.

Initially racing across the Atlantic at jet speed, the waves headed for North and South America. Along the way, the waves lost energy and became much smaller, but, according to Ward and Day's calculations, by the time they reached the North American coast, they were still about thirty-three feet (10 meters) high, and the one that approached South America was more than eighty feet (26 meters) from trough to crest.

When the giant waves hit the shallow, exposed shelves of those continents, they may have tripled in height in a process called runup. In North America, that means deadly waves up to 100 feet high crashed across the lowlands all along the Atlantic. The worst part was that, in nearly all cases, no one would have seen them coming. They just would have risen suddenly from the ocean to smash into the shorelines, obliterating all in their paths.

THE WORST WAS LAST

According to Ward and Day's estimate of ocean heights, when the waves reached South America, they suddenly increased to 240 feet or more to crash across the shallow lands along the continent. Giant waves like that may have traveled dozens or hundreds of miles inland over very shallow terrain, and they certainly would have surged far up the bed of the Amazon, reaching far inland.

These megawaves surging ashore in South America triggered the largest slides of all, which dumped mud and debris across more than 7,000 square miles of seafloor, an area greater than the state of Connecticut. If we could spread all that sediment over South America, it would bury the entire continent more than eight inches deep in mud and silt.

The tsunamis generated by the Amazon slides may have been the most massive of all, and they returned to roll across the coastlines of Europe,

Africa, and North America for the third time in less than twenty-four hours. Any one of them would have been bad enough, but those continents suffered from three successive onslaughts of monster waves.

ON THE REBOUND

Even after all those slides had occurred and sent tsunamis racing across the Atlantic, it was not over. Tsunamis are not one-way waves; they bounce off shorelines and return. With the slides triggering one another in a massive cascade, megawaves roared back and forth across the Atlantic for a full day, maybe longer. The evidence suggests that the mega-tsunamis inundated and devastated vast areas of the now underwater continental coastlines. The total area of destruction may have equaled more than the entire area of North America.

> QUESTION: There seems to be ample evidence for the slides. Is there any evidence for the destruction caused by the tsunamis?

Unlike slides, tsunamis leave much less evidence, and most of that evidence is now deep under the ocean. The reason for this is that the sea level has risen more than 400 feet since the end of the Ice Age, but even the largest of the waves that hit land were much less than 400 feet tall. At a maximum height of 100 feet, the waves could not have reached any land that is now above sea level, although they would have devastated some areas that are now below it.

Even so, there is evidence from cores along the coastal shelf of North America, as discovered by Fulthorpe and Austin (2004) off Long Island, New York, along the now submerged Hudson River drainage. They found thick deposits along the shelf that they think might be flood deposits, although they accept that a tsunami could have caused them. They date the deposits to the end of the Ice Age around 16,000 to 12,000 years ago. The large blocks that were ripped out of the existing sediment are so large that the authors conclude that only a very great water flow, such as a tsunami, could have moved them. Puzzled by what they found, they went on to conclude that it could have been caused by a second tsunami following the first within a very short time. We propose that this is exactly what happened as the tsunamis rebounded around the Atlantic.

In addition, do you recall the extensive reef damage that Fairbanks found in Barbados, detailed in chapter 32? The date range of that damage perfectly matches the date range for the tsunamis, suggesting that those megawaves were responsible for damaging or destroying all the reefs in that study.

BURPS OF METHANE

There is one particularly unusual connection of underwater slides to the Event. The Carolina Bay impacts include the area where the Cape Fear slide took place, and that huge slide is located, coincidentally, over the largest underwater deposit on the eastern seaboard of methane (fig. 33.6), a potent greenhouse gas that is highly flammable. Underwater, the gas exists in a frozen form that is known as gas hydrate or clathrate.

When sea plants and animals die, they sink down thousands of feet to the ocean bottom, where, over time, sediment buries them. When they decompose, they release methane gas, which freezes due to the extreme pressures and very low temperatures.

Jim Kennett and colleagues (2002) hypothesize that when massive underwater slides, such as the Cape Fear slide, took place, they released millions of tons of methane from the hydrate into the ocean and the air, enough to have had a sizable effect on global warming. If there had been such a sudden release of methane, the gas escaping from the ocean might have caught fire and burned for months or years. Other scientists argue that it had a minimal effect, but more studies are under way to find out whether there is a connection. If there is, once again we find a link to the Event.

TURNING OFF THE OCEAN CONVEYOR

There is another unexpected connection between the slides and climate through the ocean conveyor, that huge underwater current that flows around the world. The Gulf Stream is part of it. The location of the conveyor that loops through the Atlantic is shown in figure 33.7. Of the slides

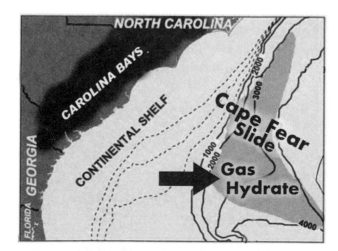

Fig. 33.6. The largest gas hydrate field is located just offshore from the highest concentration of Carolina Bays. Also, the huge Cape Fear slide lies directly above it. *Source: NOAA*

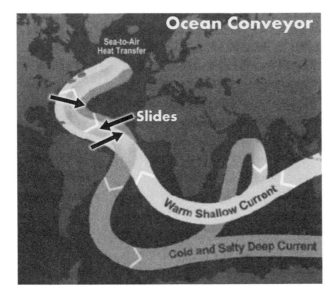

Fig. 33.7. The looping bands of the ocean conveyor pass directly through the major slide sites at the arrows. *Source: NOAA*

we reviewed above, all eight cross it, so when they occurred, they could have had an instant effect on the conveyor.

You might wonder how local slide events could have affected a gigantic ocean conveyor that is more than 10,000 miles long. To find out, let's do some quick math. Broecker (1997) provides figures for the conveyor flow rate as being equivalent to about 0.0036 cubic mile per second (0.015 km³/s). Now, compare that to the largest slide off the Amazon River in South America, for which, based on the slide size from Piper (1997), we estimate a flow rate of about 3.3 cubic miles per second (13.7 km³/s). Amazingly, just that one slide had a flow rate about 1,000 times greater than the conveyor, and all eight together may have been nearly 3,000 times larger. Such gigantic volumes of debris cutting right across the path of the conveyor could certainly have affected it. For example, Broecker and other scientists have found that in the past, the conveyor suddenly shifted its location sideways, producing radical changes in climate. Did such a thing happen 13,000 years ago when thousands of cubic miles of swirling, rapidly moving, muddy debris crossed paths with the conveyor? We think so.

THE SURVIVORS: THE YUROK

If such mega-tsunamis had occurred, there would have been eyewitnesses. One such account comes from the Yurok tribe, which lived on the Pacific

Coast, suggesting that the tsunamis were not limited to the Atlantic but occurred in the Pacific as well. This story is a highly accurate, nonallegorical description of an impact into the ocean, which claims that the "sky" fell, that mountain-high waves roared across the land, and that many people drowned.

The Day the Sky Fell

One day the sky fell. It crashed into the ocean, causing huge breakers that flooded inland far across the land, and it happened so suddenly that it caught the People by surprise.

Only two couples looked up in time to see the giant waves coming. Just as the first wave hit, they jumped into a tree-bark boat. Lifted high up into the air on the raging crest, they thought that the swirling waters would surely pull them under. But they held on for their lives and rode out the surging waves.

At last they landed on a mountaintop, but almost everyone else in the world drowned. These two couples ventured out after the world dried out, and their children became the Yurok tribe.

RETOLD FROM BELL, 1992

Fig. 33.8. A Yurok tribesman.
*Source: Library of Congress,
Edward Curtis Collection*

WHAT THE EVIDENCE SHOWS

We saw evidence that the impacts at 13,000 years BP set off a chain reaction of events that most likely:

- triggered a cascade of underwater slides off coasts of the Americas, Africa, and Europe
- exposed clathrate deposits that released huge amounts of methane, a greenhouse gas
- halted or rerouted the massive ocean conveyor in the Atlantic Ocean
- flooded the ocean with cold freshwater

All of which produced the return to Ice Age climatic conditions during the Younger Dryas.

34

DIAMONDS FROM
THE SKY

Now let's review our strongest lines of evidence for the Event.

IMPACT EVIDENCE #1: IRIDIUM

When the link was established between a meteorite impact and the extinction of the dinosaurs 65 million years ago, the crucial factor was iridium. That element is exceedingly rare on the surface of the Earth, yet very high in meteorites and cosmic-dust particles, although no one knows how much of it is in comets. Geologist Walter Alvarez and his father, Nobel Prize–winning physicist Luis Alvarez, knew they had a strong impact connection in 1980, when they discovered substantially elevated levels of iridium in the sediment layer immediately above the dinosaur extinction, or KT event. At first, they didn't know where the crater was located, but that didn't matter—they had iridium dust as evidence that the impact had occurred. Years later, scientists finally located the massive crater under the Yucatán Peninsula in Mexico.

To prove an impact connection with the mammoths, we looked through the scientific literature to see if anyone had found any iridium from around 13,000 years ago, and we found one recent study of the Greenland (GRIP) ice core by Gabrielli and colleagues (2004). Analyzing incredibly small particles of what they called "meteoritic smoke," which is all that remains of meteorites that exploded high in the atmosphere, the scientists found a significant peak in iridium at the end of the Younger Dryas (see fig. 34.1). It was two to three times higher than normal, indicating to us that there was an increase in small meteorites or that a large impact had occurred then.

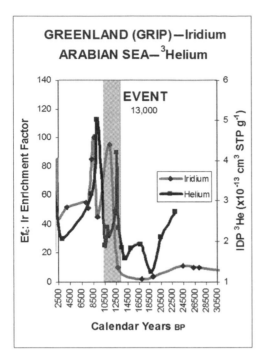

Fig. 34.1. Around 13,000 years ago, helium-3 and iridium show major peaks, which are among the highest in 23,000 years. *Data from: Ir, Gabrielli, 2004; He, Marcantonio, 2001*

IMPACT EVIDENCE #2: HELIUM-3

Another extraterrestrial impact marker is helium-3 (He-3), which is rare on this planet, unlike its common terrestrial cousin, helium-4 (He-4). Every year, millions of micrometeorites and cosmic-dust particles slam into Earth's atmosphere, producing a constant rain of tons of tiny particles of extraterrestrial dust that fall to Earth to become lodged in the polar ice sheets or in the ocean sediment. Because extraterrestrial chemistry can be very different from the chemistry of Earth, those particles usually contain higher levels of He-3. During an impact event, however, the amount of incoming He-3 can increase even more dramatically.

The impact–He-3 connection became clear when Becker and associates (2001, 2004) located a possible 125-mile-wide giant impact site near Australia, called the Bedout crater, that dates to the Permian extinction, 250 million years ago, when nearly 90 percent of all life on Earth disappeared. In the Permian extinction sediment, they discovered high levels of He-3.

Likewise, Marcantonio and coworkers (2001) tested ocean-sediment samples from the Arabian Sea and found elevated levels of He-3 that they concluded came from tiny interplanetary dust particles, or IDPs. One of the largest peaks in the 23,000-year record occurred 13,500 years ago, almost exactly at the date of the Event (fig. 34.1). Most research shows that comets contain more He-3 than meteorites do, supporting our idea that the impactor may have been a dustball comet. This may also mean that the Permian extinction was caused by a similar comet that produced the Bedout crater.

IMPACT EVIDENCE #3: FIRESTORM SOOT

You may recall from chapter 3 that we found a peak in charcoal at the Murray Springs site, as well as at all other Clovis-era sites and in the Carolina Bays. Charcoal and soot, classical markers for an impact event, also show up at the time of the dinosaur extinction, and the soot does not look as though it came from just any old wildfire. Wolbach (1985–1990) described the soot from the dinosaur extinction as appearing in "grape-bunch-like" clusters, and she and others have found it at dozens of sites around the world in the KT layer. She has looked for it at other known extinction layers but has not found any, indicating that the conditions for creating and preserving it are rare, and might be associated only with very large impacts.

Wolbach and Han Kloosterman (personal communication, 2006) together found the same "grape-bunch" soot in the black mat at Murray Springs—a striking connection to the Ice Age extinction. They plan to publish the information soon, and Kloosterman told us that Wolbach believes the amounts are "significant." Kloosterman indicates that it is nearly identical to the KT soot, validating our theory that massive firestorms occurred at the same time the mammoths disappeared. We have agreed to collaborate with the researchers to confirm this at other Clovis-era locations in Europe.

IMPACT EVIDENCE #4: HOLLOW FLOATING SPHERULES

In chapter 6, we reported finding hollow floating spherules at the Chobot site (see fig. 34.2). They were at nearly all Clovis-era sites except those in the Southwest, and we uncovered them in all the Carolina Bays. From Alberta to Manitoba, Michigan, and the Carolina Bays, the hollow spherules rise to a prominent peak only at the Clovis layer.

Incidentally, in the Carolina Bays, the peaks of those hollow spherules— along with magnetic grains, radioactivity, and all the other markers—are the strongest evidence we have that the bays are related to the impacts. Riggs and colleagues (2001) concluded that dozens of bays near Lake Waccamaw most likely formed between 16,000 and 13,000 years ago, and the peaks on our charts support that. If the bays formed long ago or at widely different times, these charts would not look so similar.

Our Northern Arizona University collaborators, Ted Bunch (retired

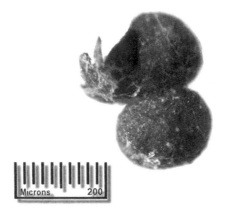

Fig. 34.2. These hollow, sometimes thin-shelled spherules occur only in the Clovis-era layers.

chief of exobiology at NASA Ames Research Center) and Jim Wittke, determined that these spherules contain a very high percentage of carbon. That is unusual for impact-related objects, which normally have a stonier composition with only a little bit of carbon.

So what are they? We do not yet have all the answers, but it is clear to us that the spherules are associated with the impact. Our investigations are continuing, although we are almost certain that they are one of two things.

First, the balls could be algal colonies that reached great size during a period of explosive growth following the impact. Hansen (2004), a Danish geologist, indicated that some hollow spherules have been found just after the KT extinction layer, having been formed by "disaster species" of algae that thrived after the extinction. He says that similar spherules are found just after the Permian extinction 250 million years ago. This algal interpretation is supported by Ted and Jim, who took scanning electron microscope (SEM) images (fig. 34.3) that reveal an apparent biological structure. Cell sizes are very small, ranging from two to twenty microns, or millionths of a meter. A human hair is only about 100 microns wide.

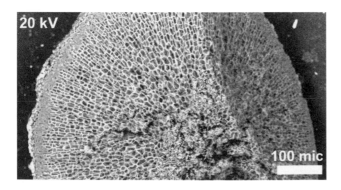

Fig. 34.3. An SEM image shows the foamlike texture of this broken spherule. The entire spherule is about the same width as a human hair.

Second, Ted and Jim also found some spherules that did not look biological and suggested that there may have been two processes at work. Two companion scientific presentations supporting a connection to impacts, by Hoffman and colleagues (2005) and Rosler and associates (2005), describe finding spherules like these around shallow craters that range from about 15 to 300 feet (5 to 100 meters) wide in southeast Bavaria. The researchers did not report an age, but they discovered the craters on top of glacial till, proving that they fell at or after the end of the Ice Age. This means that they might date to the time of the Event, and may represent the European evidence for it, or they may be part of a more recent, smaller cosmic event. The investigators also described finding high magnetic-susceptibility readings around the craters, indicating the increased presence of iron, just as we found at all Clovis-era sites. In addition, they found quartz grains that had been heated to very high temperatures, just as might have happened with the white sand of the Carolina Bays.

Most importantly, the scientists reported finding spheres "exhibiting foam-, sponge-, or cell-like internal structures with cell sizes of a few microns. Elemental compositions show a high portion of carbon." This description is almost identical to that of the Clovis-era hollow spherules shown above from the Chobot site (see again fig. 34.2). The spherules these researchers found may be biological but there are several facts that suggest otherwise, as discussed next.

IMPACT EVIDENCE #5: MICROSCOPIC DIAMONDS

In the most surprising discovery, the European scientists found diamonds. They occur in the hollow spherules in microscopic form, only one to five nanometers (billionths of a meter) wide. Such diamonds are very rare on Earth but common in meteorites, which sweep them up after they are blown away from the cores of enormous white dwarf carbon-stars. Those peculiar stars were proposed long ago, but the immense diamond core of the first such star was discovered in February 2004 by astronomers at the Harvard-Smithsonian Center for Astrophysics. About fifty light-years from Earth in the constellation Centaurus, the star is a single giant crystalline diamond, and the astronomers determined that it rings like an enormous crystal bell, giving off constant pulsations. The star measures only about 2,500 miles across, or about the distance from Los Angeles to New York City, making it one of the largest diamonds in our galaxy.

Since nanodiamonds form only in the hearts of those stars, the question arises as to how they ended up around the craters in Bavaria. Most likely, they rode in with an asteroid or comet, or on the supernova debris cloud, providing a possible link to the Event.

Nanodiamonds are not the only extraterrestrial things the researchers found. They also detected buckyballs, tiny soccer ball–like carbon cages. Also called fullerenes (after Buckminster Fuller, the inventor of the geodesic dome, which they resemble), these odd objects are believed to have formed in the heart of an exploding star, such as our supernova. They are common in meteorites but extraordinarily uncommon on Earth, being known to occur only occasionally from intense natural forces such as lightning strikes and, rarely, inside some lava flows. Throughout the multibillion-year history of our planet, only a few layers contain fullerenes, and most of those layers involve cosmic impacts.

Both the fullerenes and the nanodiamonds link the hollow spherules from Bavaria to some cosmic event, and their spherules sound just like the ones we have found in North America. In addition, there's another link.

IMPACT EVIDENCE #6: GLASSLIKE CARBON

In chapter 9, we described finding melted black glassy carbon at the Topper Clovis-era site, and later we found much bigger pieces in the Bladen County Carolina Bay. If carbon is pure, it melts at an incredible 6,400 degrees Fahrenheit, far higher than the melting point of iron, suggesting that something extraordinary happened to twist those pieces into their present forms. All the carbon glass from Alberta, Canada, to Michigan, to the Carolinas (see fig. 34.4) was similarly melted, and at every site, the glass showed a clear and distinct peak only in the Clovis-era boundary layer (see fig. 34.5). In addition, we uncovered the hollow spherules only at sites where we found black glass, suggesting that there is a connection between them. It may be that the spherules are just porous, ball-shaped forms of the black glass, all of which were created by the intense heat and pressure of impact.

In confirmation of the impact origin of the glass, Ernstson and colleagues (2004) found identical glasslike carbon lumps in the ejecta of the Azuara-Rubielos crater complex in Spain. These scientists propose that the glass formed during the impacts of multiple impactors across Spain that may have been part of a much larger comet swarm. The Azuara crater (nineteen miles wide), along with the Chesapeake Bay crater (fifty miles) and the Popigai crater in northern Siberia (sixty miles), is thought by some scientists to be one of several factors affecting the Eocene-Oligocene (EO) cluster of extinctions 40 million to 32 million years ago.

When we first showed the glass to Ted, who studied many aspects of

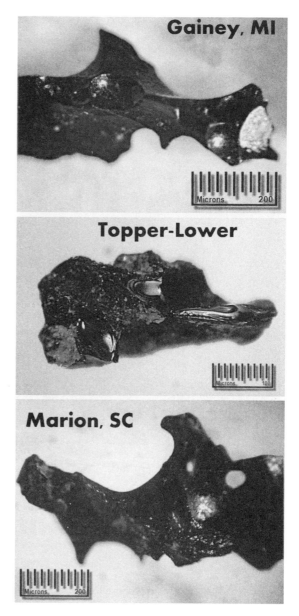

Fig. 34.4. The glassy carbon looks the same whether from Gainey, Michigan, the Topper site, or a South Carolina Bay. These sites are about 650 miles apart.

Earth impacts while at NASA, he was certain that it was in some way connected with an impact, and we planned a series of tests to determine its composition. First, microprobe testing revealed that it was nearly pure carbon, and other testing showed all of the lumps have numerous internal gas bubbles, another sign of extraordinarily high temperatures and sudden cooling. Next, we did a comprehensive NAA analysis on some of the lumps and discovered that both the magnetic grains and the glasslike carbon from

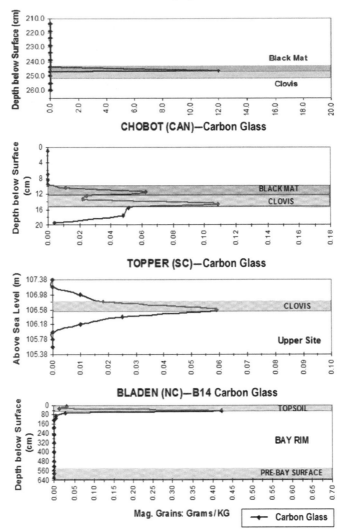

Fig. 34.5. This composite chart shows that the peaks in glasslike carbon are almost identical at the key Clovis-era sites of Murray Springs, Chobot, Topper, and a Bladen County Carolina Bay.

one Carolina Bay, just a short distance away from the Topper site, had elevated levels of radioactivity and KREEP, suggesting a connection to the radioactive impact scars on the moon.

The composition of the glass seems to match some of the facts astronomers know about comets. When the Giotto mission flew close to Halley's comet, scientists expected to see a giant dirty snowball, but instead they were stunned to see photos of an object so extraordinarily dark that it reflected almost no sunlight. Various analyses determined that the surface of Halley's is a black, tarlike, carbon-rich material, an almost exact description of the glassy carbon we have found.

Knowing that the glass is high in carbon, we decided to test it to see if there might be unusual forms of it, using a procedure called nuclear magnetic resonance (NMR) imaging, which is similar to modern medical imaging. We arranged for John Edwards of Process NMR Associates, LLC, in Connecticut, to do the testing. We ran two tests, from which John concluded that some of the carbon glass from a Carolina Bay near Marion, South Carolina, contained about 88 percent aromatic carbon. Such a high percentage of that particular type of carbon makes it unlike almost anything else on this planet, and John sent graphs of different types of coal and other common forms of carbon to prove the point. One tantalizing clue to the glassy carbon is the fact that scientists have found high levels of aromatic carbon in comets and in some meteorites. The NMR work does not prove an extraterrestrial origin, but it fits the known facts about comets.

The Bavarian results from the hollow spherules gave us several other things to look for, including nanodiamonds and fullerenes. We ran two tests to look for the diamonds. One test was negative, but another test showed a distinct peak for nanodiamonds. John calculated that the black glasslike carbon contained about 3 percent of them by weight, a remarkable result. That meant we had pulled enough black glass out of all the Clovis-era sites to form a three-carat black diamond! It would be only a low-grade diamond with scientific rather than monetary value, but the prospect of wearing a stunning black cosmic diamond interested everyone who heard the idea.

Does the black glass contain diamonds? More testing is planned, so we cannot say for certain, but it is possible. Similar tests on many meteorites have found diamonds, so it is plausible that the glasslike carbon contains them too.

IMPACT EVIDENCE #7: SPHERULES AND OTHER PEAKS IN THE CLOVIS-ERA LAYER

Powerful support for an impact comes from all the various markers found at the Clovis-era sites from Alberta, Canada, to Arizona, to South Carolina. Throughout the previous chapters, we have presented charts for magnetic grains, magnetic spherules, radioactivity, glassy carbon, hollow carbon

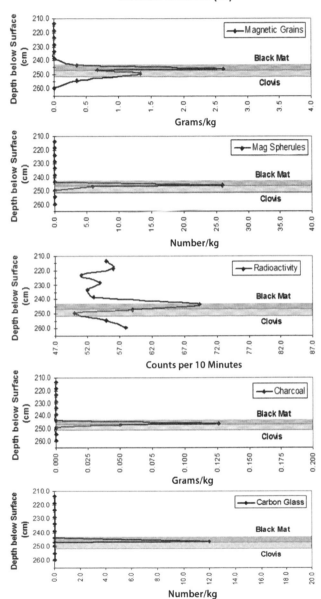

Fig. 34.6. This composite chart shows all the key markers from Murray Springs.

spherules, and charcoal, but here are all of them together in figure 34.6 for Murray Springs, which has every marker except for hollow spherules. In particular, magnetic spherules are widely accepted as indicators of an extraterrestrial event. Composite charts from other key sites are shown in appendix D.

IMPACT EVIDENCE #8: FULLERENES

We found one more stunning bit of evidence in the carbon glass. After the positive test for nanodiamonds, Ted contacted a colleague, Luann Becker, a leading researcher on fullerenes with the Institute for Crustal Studies at the University of California, Santa Barbara. She and Ted, along with a group of other researchers, have published many papers on fullerenes and other impact issues, and they have identified fullerenes in two of the five largest extinctions in Earth's history, the KT extinction of the dinosaurs and the Permian, or PT, extinction, when nearly 90 percent of life on Earth vanished. Her work and that of others has confirmed the existence of fullerenes in other impact craters, including, in 1994, the 1.85-billion-year-old Sudbury crater in Canada, which is 155 miles wide.

When Luann saw the black glass, she immediately suspected it contained fullerenes, so she went to work right away. Although the extraction process was laborious, it only took a few days over the weekend to get the first results. I remember vividly when she called with a report, saying simply, "There are lots of fullerenes, and some are *very* big." The oversized ones were among the largest she had ever seen, with some appearing as C-200, meaning they had 200 atoms of carbon, whereas the smaller C-60 fullerenes are more common in other impacts. That news gave us strong evidence of a powerful extraterrestrial signature—fullerenes form abundantly in the heart of a star, and she had millions of them in her lab.

With that result, we had fullerenes from the glasslike carbon that came from a Carolina Bay, but there was a problem with the other Clovis-era sites. We had only small quantities, too small for her to test. We were disappointed until Luann and I discussed the possibility that the carbon may have come down across North America as fine black dust, which is just what one would expect in a major impact that blew a lot of material into the atmosphere. If so, then it should be at other sites in a form different from the more obvious glass. It suddenly occurred to me that the black mat might be loaded with carbon-glass dust.

I immediately sent large samples of the mat from Murray Springs and Blackwater Draw to Luann, who promptly began to test them. The superfine mat was difficult to process, but within days Luann had results. First, she tested the Murray Springs sample and found fullerenes in it, but they were obscured somewhat by the thick algal debris, keeping her from getting the clearest signal. Then she turned to the Blackwater Draw black mat sample, and after a few days called to say that she could see them very clearly. That was exciting news, since now we had continental coverage, showing that the fullerenes were present in South Carolina, New Mexico, and Arizona, which are about 1,700 miles apart (2,800 km).

Although the fullerenes most likely arrived here with some huge impacting body, we knew that they can form naturally in very rare cases, so we set out to determine whether they had an extraterrestrial signature. One of Luann's colleagues, Bob Poreda of the University of Rochester, knew a way to determine their signature by using one of the unusual properties of the fullerene structure, which is that the carbon cage tends to trap gases such as helium, argon, and xenon. When heated, the fullerenes release the gases, allowing the gas levels to be measured. If the fullerenes came from this planet, they would have certain isotopic ratios, but if the fullerenes were cosmic in origin, the ratios would be very different. One of the key gases to find is helium-3, which exists at much higher levels beyond Earth.

The process was taking quite a while, so with our manuscript deadline only a few days away, I was resigned to not being able to report the results. That was a great disappointment, but there was little I could do about the deadline; the publisher wanted to get this story into print. On nearly the last day, Luann called to say, "We found the helium-3." Her voice was matter-of-fact, in contrast to the importance of her words. "The glassy carbon has a *very* high ratio of helium-3 that is equal to what we found in the Sudbury impact crater."

When her words registered, I felt a sudden thrill. There it was! We had solid evidence now that the fullerenes came from space and not from this planet. I had guessed as much on that day months before when I first dug a piece of glasslike carbon from the white-sand rim of a Carolina Bay, but it had taken months to prove it. Now we knew.

On that rainy day in South Carolina, I almost drove away and went home, but I decided to follow a hunch instead. Even though past hunches have led to good things, never in my wildest dreams did I imagine that the mysterious black glass contained such exotic stuff from a distant star. So it is that the strangest of clues has turned out to be the strongest proof so far.

It is also the tiniest evidence so far, as shown in figure 34.7, since fullerenes are measured in nanometers, or billionths of a meter. For a size comparison, the period shown here [.] is 1 million nanometers wide, meaning that, amazingly, about a million of our smallest fullerenes, laid side by side, would just stretch across that single dot! We think every one of those incredible hollow carbon balls, loaded with helium and argon, came to Earth

Fig. 34.7. The tiny ringlike structures are hollow fullerene spheres. The entire width captured by this image is so small that 4,000 views like it would fit across one human hair. *Source: Luann Becker, University of California at Santa Barbara, unpublished paper, 2006*

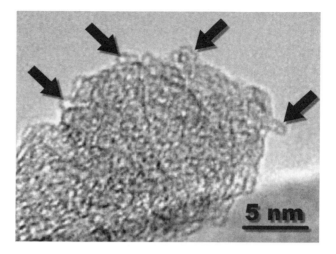

from hundreds of light-years away, surfing along on the shock wave of an exploding star, or riding in on the back of a dustball comet.

WHAT THE EVIDENCE SHOWS

Three lines of evidence support the occurrence of a supernova:

- *Radiocarbon.* Global ^{14}C suddenly nearly doubled 41,000 years ago.
- *Other isotopes.* Radioactive ^{10}Be, ^{26}Al, and ^{36}Cl peak in various 41,000-year-old cores.
- *Radioactivity.* Potassium-40 shows major increases in 13,000-year-old sediment.

Eight lines of evidence support a cosmic impact:

- *Iridium.* It also is found in the dinosaur extinction (KT) and in meteorites.
- *Helium-3.* It also peaks in the Permian extinction (PT) and in comets.
- *Soot.* Carbon soot from wildfires peaks in the KT extinction layers.
- *Hollow spherules.* These are also found at the KT and PT extinctions.
- *Nanodiamonds.* These occur at the KT extinctions and in meteorites.
- *Metallic spherules.* Spherules and other sediment markers all peak in the Clovis era.
- *Carbon glass.* Glassy carbon also occurs in Spain's Azuara crater in the EO extinction event.
- *Fullerenes.* These also are in the KT and PT layers, in other craters, and in meteorites.

LATE-BREAKING DISCOVERIES

Just before printing this book, we made a series of major breakthroughs, uncovering very high peaks of iridium (Ir) in the Clovis-era layers at Murray Springs, Blackwater Draw, Lake Hind in Manitoba, and the T13 Bay in South Carolina. The sediment peaks in Ir ranged up to 4 ppb, which is higher than many sites for the K/T dinosaur extinction, and some of the magnetic fractions contained as much as 24 ppb. At most sites, the Ir layer was only a few inches thick, but in the T13 Carolina Bay, the high levels of Ir appeared throughout most of the 10 feet (3 meters) of the rim, making a strong case that some bays formed 13,000 years ago during the Event.

We also extended the reach of the Event from the Atlantic Coast to offshore California in the Channel Islands. At the Daisy Cave Clovis site, Jim Kennett, his son Douglas, Jon Erlandson, and Luann Becker discovered glass-like carbon, fullerenes, carbon spherules, and magnetic particles. More testing is underway to look for iridium and other key extraterrestrial markers.

In addition, there is now a powerful connection between the Event and Europe. In our collaboration with Han Kloosterman, we tested sediment from the Usselo horizon near Lommel, Belgium, at a site occupied by the Magdalenian people. Contemporaries of Clovis, the Magdalenians also experienced a severe population decline during the Event 13,000 years ago. We found peaks in magnetic grains, metallic spherules, and charcoal, and there is a black mat just as Vance Haynes found at fifty sites in North America. In the magnetic fraction there, we found the highest Ir levels of all—an astounding 117 ppb.

FINAL THOUGHTS ON THE EVIDENCE

We have looked at a wide range of evidence, any one piece of which might not be enough on its own. When we step back to see the sum of it all, however, it becomes apparent how unusual it all is. There are only a few times in the last 500 million years when such an accumulation of evidence is found, and it is universally accepted as part of a cosmic impact. For example, similar evidence shows up from 65 million years ago, when the dinosaurs vanished, and the same comes from 250 million years ago, when 90 percent of all life disappeared in the largest extinction known to science, the Permian. Now we have another cosmic extinction event, with many lines of evidence indicating that it occurred just 13,000 years ago. Throughout the entire 500-million-year record up until today, we find this evidence associated only with times when there were cosmic catastrophes, and those cataclysms are linked to major extinction events.

35

WRAPPING IT UP

Something extraordinary and incredible happened between 41,000 and 13,000 years ago, when the Earth was suddenly blindsided from space, setting off a chain reaction of events that dramatically altered the planet and opened the way for the birth of modern civilization. Since that time, the skies have been mostly very calm—but the lack of danger is only an illusion. We have seen in this book that similar events have happened before, and they all occurred suddenly and without warning. They will happen again; the clock is still ticking.

SUPERNOVA DUE TO GO OFF?

We know that as many as fifteen supernovae occur each century in our galaxy, but most take place at safe distances from Earth. Eventually, one will happen close to us and toast one side of our planet. That is not likely to happen soon, since astronomers can see most close stars that are unstable, but it is possible.

In 2002, *New Scientist* reported that Harvard student Karin Sandstrom had found that a white dwarf star, designated HR 8210, in the Pegasus constellation is the best and closest supernova candidate discovered to date. Only 150 light-years away, the star is dangerously close, and if it does explode soon, the burst of radiation would touch off another round of extinctions.

Fortunately, astronomers calculate it may take millions of years before it explodes, but there is a great degree of uncertainty about that date, and it could happen much sooner. Unfortunately, they missed the importance of this star when it was discovered and ignored it for about fifteen years, even though it currently poses the greatest supernova threat to Earth. Have we missed others? The fact is that little money is spent assessing the threat of supernovae to Earth or for devising a survival strategy if one does occur. Such things are not very expensive. World governments eagerly spend a million dollars for each cruise missile to protect against shadowy enemies, when the amount spent on a few dozen cruise missiles would go a long way toward assessing a real but less recognized threat from space.

NOT-SO-HARMLESS SHOOTING STARS

If you want more evidence for what happened to the mammoths, you need only to look up at the clear night sky. In almost any month, you can see shooting stars from one of many meteor showers. Nearly every fiery streak you see is the tiny remnant of some giant comet that broke up into smaller pieces. Of course, most of those pieces are microscopic, but their parent comet was not—it was enormous. Astronomers know that, even today, hidden in those cosmic clouds of tiny remnants, there are some huge chunks of comet pieces. We pass through their clouds every year like clockwork, so eventually we will collide with some of bigger pieces.

In 1990, Victor Clube, an astrophysicist, and Bill Napier, an astronomer, published *The Cosmic Winter*, a book in which they describe performing orbital analyses of several of the meteor showers that hit Earth every year. Using sophisticated computer software, they carefully looked backward for thousands of years, tracing the orbits of comets, asteroids, and meteor showers until they uncovered something astounding. Many meteor showers are related to one another, such as the Taurids, Perseids, Piscids, and Orionids. In addition, some very large cosmic objects are related: the comets Encke and Rudnicki, the asteroids Oljato, Hephaistos, and about 100 others. Every one of those 100-plus cosmic bodies is at least a half-mile in diameter and some are miles wide. And what do they have in common? According to those scientists, every one is the offspring of the *same massive comet* that first entered our system less than 20,000 years ago! Clube and Napier calculated that, to account for all the debris they found strewn throughout our solar system, the original comet had to have been enormous.

So was this our megafauna killer? All the known facts fit. The comet may have ridden in on the supernova wave, then gone into orbit around the sun less than 20,000 years ago; or, if it was already here, the supernova debris wave may have knocked it into an Earth-crossing orbit. Either way, any time we look up into the night sky at a beautiful, dazzling display of shooting stars, there is an ominous side to that beauty. We are very likely seeing the leftover debris from a monster comet that finished off 40 million animals 13,000 years ago.

Clube and Napier also calculated that, because of subtle changes in the orbits of Earth and the remaining cosmic debris, Earth crosses through the densest part of the giant comet clouds about every 2,000 to 4,000 years. When we look at climate and ice-core records, we can see that pattern. For example, the iridium, helium-3, nitrate, ammonium, and other key measurements seem to rise and fall in tandem, producing noticeable peaks around 18,000, 16,000, 13,000, 9,000, 5,000, and 2,000 years ago. In that pattern of peaks every 2,000 to 4,000 years, we may be seeing the "calling cards" of the returning megacomet.

Fortunately, the oldest peaks were the heaviest bombardments, and things have been getting quieter since then, as the remains of the comet break up into even smaller pieces. The danger is not past, however. Some of the remaining miles-wide pieces are big enough to do serious damage to our cities, climate, and global economy. Clube and Napier (1984) predicted that in the year 2000 and continuing for 400 years, Earth would enter another dangerous time in which the planet's changing orbit would bring us into a potential collision course with the densest parts of the clouds containing some very large debris. Twenty years after their prediction, we have just now moved into the danger zone. It is a widely accepted fact that some of those large objects are in Earth-crossing orbits at this very moment, and the only uncertainty is whether they will miss us, as is most likely, or whether they will crash into some part of our planet.

That may seem like bad news, but there is a glimmer of good news too. For the first time in humankind's known history, we have ways to detect those objects and prevent them from hitting us again. One such effort is Project Spaceguard, a multinational cooperative attempting to locate those Earth-threatening objects, and other similar programs include the Near-Earth Asteroid Tracking (NEAT) telescope and the Spacewatch Project at the University of Arizona. Unfortunately, not one of them is funded nearly well enough to complete the job for many years, but they are working at it steadily.

No one knows exactly how many dangerous comets and asteroids are out there, but astronomers are certain that hundreds to thousands of them remain undiscovered. The worst part is that many of those space objects are so dark and difficult to see that they are nearly invisible until they come very close, and by then it is too late. It is certain that one of these monsters is on a collision course with Earth—we just do not know the details. Is it days from now or hundreds of years from now? Even if we were sure one was coming, there is just very little that we can do about it currently.

We are years away from being able to control our own destiny as it relates to supernovae and giant comets and asteroids, but scientists are working on solutions. This is not a high priority with the world's governments, however, which typically prefer to confront terrestrial threats rather than cosmic ones. To prevent one of those giant objects from smashing into us, collectively, we spend about $10 to $20 million annually, an amount less than the cost of one or two sophisticated fighter jets. Almost no money is spent trying to detect imminent supernovae.

Our politicians are seriously underestimating these severe threats, which are capable of ending our species, just as they snuffed out the mammoths a mere 13,000 years ago, only an eyeblink in cosmic terms. There are few threats of that magnitude facing us today. The survival of the human race is not seriously threatened by the avian flu, Al Qaeda attacks, the end of the Age of Oil, monster hurricanes, giant earthquakes, or enormous tsunamis; if any of those occur, most of us will continue with our lives. Furthermore, nothing on that list is broadly accepted as having caused worldwide extinctions in the past. The same cannot be said about supernovae and massive impacts. Those two cosmic events are implicated in many of the largest extinctions on our planet over the last millions of years. Fortunately, we survived them, but many of our fellow species did not. Humankind might not survive the next one. It seems reasonable to forgo several of our military fighter jets each year to decrease our chances of being "nuked" from space by a supernova or a comet.

THE LONG VIEW

We may ignore our spiraling overpopulation and all of mankind's related troubles, and we may pretend that supernovae and giant comets don't really threaten Earth, but pretending never alters the facts. Such massive impacts are Nature's way of cleaning house, causing hundreds of species to disappear and making room for others to step forward. Consider, for example, all the things that we value today—art, music, language, drama, writing, math, and technology. Evidence indicates that every one of those crucial aspects of human life arose or expanded greatly *after 41,000 years ago*. Can that be just coincidental? Maybe. But it is more likely that our modern way of life had its genesis in the brilliant flare of an exploding star and the thunderous flash of crashing comets.

ENDNOTE

Even as this book was going to press, our research continued and new discoveries were coming in. If you are interested in seeing what has happened since we wrote this book, there are free updates at www.cosmiccatastrophes.com.

Appendix A
FIND YOUR OWN STARDUST

If you want to look for your own stardust from supernovae, comets, and meteorites, all you need is a supermagnet, a few tools, some dirt, and some persistence. You can collect several gallons of dirt where you live. If you are lucky enough to live near a known Clovis site and can get some sediment from there, you can see exactly what we discovered throughout this book. Of course, not all that is magnetic comes from space, but some of it does. Here are the details on how to find it.

First, be sure you purchase a true rare earth, or neodymium, magnet, also called a supermagnet. Ask the dealer to make sure he or she knows the difference. If you get any other kind, it will not work as well. Ours cost only about $30 U.S. for one that is $1 \times 2 \times 1/2$ inch.

Although laboratories use high-tech equipment to separate magnetic grains from sediment, there are low-tech ways that work well, especially out in the field where lab equipment is not possible. Each of the methods in the list below works best with a different type of soil. All are easy to do, so

Fig. A.1. Neodymium supermagnet.

if you are curious, you can find magnetic grains on your own. Here's how we do it.

LOOSE OR SANDY SEDIMENT

- If you have sediment that crumbles easily in loose sand or silt, the best time to separate the grains is when it is dry. If it is not already dry, set it in the sun for a few hours to a few days.
- After it is thoroughly dehydrated, put it in a sturdy container and use some tool to break up the lumps, so that the sand flows easily. It is best to use nonmetallic containers and tools, because you don't want to add any foreign metal to the sample. A ceramic or stone mortar and pestle works well to break up the samples.
- Next, put the magnet in a plastic bag. This is essential, because anything that sticks to the magnet is hard to get off. The bag allows you to remove the magnet so that the grains fall off easily. Sandwich bags work, but thicker commercial ones last longer.
- Using at least three containers, pour a manageable sample of sand into one, leaving the other two empty.
- Next, pull the bag tight around the magnet. This is vital, because air spaces decrease the magnetic strength. Pour the sand from the container slowly over the bagged magnet, letting the sand fall into one of the empty containers. Anything that is magnetic will stick to the edges of the magnet.
- Once you have poured all the sand over the magnet, hold the magnet and bag over the other empty container. Next, while gripping the bag, slowly pull the magnet out of it. As you withdraw the magnet, the grains will drop freely into the container.

Fig. A.2. Magnet inside heavy-duty plastic bag.

- As strong as the magnet is, some grains are only weakly susceptible to a magnet, so you will need to repeat pouring the sand. It may take a few passes or as many as ten until you recover most of the grains. You can stop when only a few grains stick to the magnet each pass.
- If you plan to view the grains under a microscope, you will probably need to clean the dust off them. Distilled water works fine. Let them air-dry; don't use the microwave, because metal and microwaves often do not mix. Also, don't use an oven—some grains are hollow and may pop. That is not dangerous, just messy.
- When viewing with a microscope, a low-power scope with up to about 100× or 150× magnification works very well. Inexpensive used scopes are readily available on the Internet. Because the magnetic grains are opaque, it is best to get one that is lit from above and suitable for viewing rocks, rather than one that is lit from the bottom and used for viewing slides. The latter will still work well if you buy a separate high-intensity halogen lamp with an adjustable neck. Begin at about 50× and work up.

STICKY OR CLAYEY SEDIMENT

- If the sediment is sticky, the best way is to separate the grains is when it is wet. In most cases, it will be too sticky to separate when it is dry.
- Put the magnet in a plastic bag, as above.
- Using at least three containers, put a manageable sample of sand into one. Leave the second empty and fill the third with clear water.
- Put a sample in a sturdy container and add ample water to permit mixing. Use a nonmetallic tool or your hands to break up the lumps until you have fully liquefied the material. Use a screen or strainer to filter out the remaining lumps and break them up.
- There are two ways to separate the magnetic grains: (1) immerse the magnet while it is inside the plastic bag and swirl it around in the container of liquefied sediment; or (2) pour the slurry over the magnet into another container. Either way, anything that is magnetic will stick to the edges of the magnet.
- Once you have some grains stuck to the magnet, immerse the magnet in the other container with the clear water. Next, while gripping the bag, slowly pull the magnet out of it. As you withdraw the magnet, the grains will sink freely into the water-filled container.
- Repeat until most of the grains are out of the sediment.

- When you are through, you will need to dry the grains. To do so, you must remove most of the water. Hold up the container with the grains and place the magnet on the outside bottom of the container. Swirl the contents around until the grains are sticking to the magnet area. Slowly pour off the excess water. When most is gone, air-dry the still-wet grains.
- When viewing with a microscope, begin at about 50× and work up.

Only a tiny fraction of what you find will be extraterrestrial. If you find well-rounded, highly polished spherules, those are almost always from beyond the planet.

Happy grain hunting!

Appendix B

CHEMISTRY OF
THE COMET

Table B.1 shows the elemental concentrations (column 1) in magnetic particles from Clovis sites at Gainey (column 3) and Murray Springs (column 4). For comparison, it also shows the chemistry of clear nonmagnetic grains from Gainey (column 1). Notice that for the shaded elements, the Murray Springs grains look most like the KREEP values (column 5) and much less like values for the Earth's crust (column 6). All values are in parts per million unless indicated as percentages.

TABLE B.1

Element	Gainey Clear Nonmagnetic Particles	All Gainey Magnetic Particles	Murray Springs Magnetic Particles	KREEP from Lunar SaU 169[1] (% as oxides)	Earth's Crust
H	159	0.356%	0.579%	n/a	0.15%
B	—	29.5	41.3	n/a	8.7
O	53%	48%	45%	n/a	46%
Na	0.308%	2.20%	0.566%	1.18%	2.3%
Al	0.14%	5.60%	3.62%	1.634%	8.2%
Si	46.1%	28.1%	19.4%	n/a	27%
Mg	—	1.75%	1.2%	6.92%	2.9%
Cl	71	182	170	n/a	170
K	0.336%	1.67%	1.40%	0.88%	1.5%
Ca	930	1.59%	2.74%	10.6%	5%
Sc	0.2	25	49	18	26

Element	Gainey Clear Nonmagnetic Particles	All Gainey Magnetic Particles	Murray Springs Magnetic Particles	KREEP from Lunar SaU 169[1] (% as oxides)	Earth's Crust
Ti	173	0.941%	9.93%	1.47%	0.66%
V	—	302	1980	36	190
Cr	—	430	200	811	140
Mn	23	0.285%	1.44%	0.12%	.11%
Fe	532	9.85%	15.0%	8.8%	6.3%
Co	8.8	34	64	12	30
Ni	—	54	40	162	90
Zn	—	160	200	n/a	79
As	0.7	12	12	n/a	2.1
Br	—	2.1	4.2	n/a	3
Rb	—	61	100	20	60
Sb	—	0.7	3	n/a	0.2
Cs	—	1	6	0.9	1.9
Ba	—	340	—	1351	340
Ce	4	34	638	297	60
La	2.2	19	340	113	34
Nd	—	16	220	58	33
Sm	0.339	2.52	13.9	44.9	6
Eu	—	0.9	6.5	2.45	1.8
Gd	0.375	3.25	18.1	50.4	5.2
Tb	—	—	4.7	10.5	0.94
Yb	0.4	2.4	22.4	36	2.8
Lu	0.08	0.33	3.5	5.24	.56
Hf	—	5	17	34.7	3.3
Ta	—	2.3	54.8	4.16	1.7
W	—	2	56	2.5	1.1
Au	2	—	—	n/a	0.003
Th	0.6	8.8	105	21.7	6
U	—	2.0	17	5.83	1.8

[1]*Data from Gnos et al., in* Science 305 (2004): 657.

LAKES WITH FIRESTORM CHARCOAL

The table below shows all thirty-three sites with charcoal peaks indicating fires. All original third-party data are available from World Data Center at http://lwf.ncdc.noaa.gov/paleo/ftp-pollen.html.

FIRE SITES (33)	STATE	RESEARCHERS
Sithylemenkat Lake	AK	Earle
Bear Lake	AZ	Weng
Fracas Lake	AZ	Weng
Murray Springs	AZ	The authors
Bluff Lake	CA	Whitlock, Mohr
Siesta Lake	CA	Brunelle
Six (5) lakes	CAN, BC	Brown
Heal Lake	CAN, BC	Williams
Chobot	CAN, AB	The authors
Lake Hind	CAN, MB	Boyd and authors
Head Lake	CO	Jodry
Burnt Knob Lake	ID	Whitlock, Brunelle
Pittsburg Basin	IL	Teed
Gainey	MI	The authors
Kimble Pond	MN	Camill

FIRE SITES (cont'd)	STATE	RESEARCHERS
Sharkey Lake	MN	Camill
Baker Lake	MT	Whitlock
Pintar Lake	MT	Whitlock
Sheep Mtn. Bog	MT	Mehringer
Jerome Bog	NC	Buell
Salters Lake Bay	NC	The authors
Moon Lake	ND	Clark
Little Lake	OR	Whitlock
Marion Bay	SC	The authors
Browns Pond	VA	Kneller
Williams Lake Fen	WA	Mehringer
Cygnet Lake	WY	Whitlock
Slough Creek Lake	WY	Whitlock
Yellowstone	WY	Whitlock

Appendix D
KEY CLOVIS-ERA MARKERS

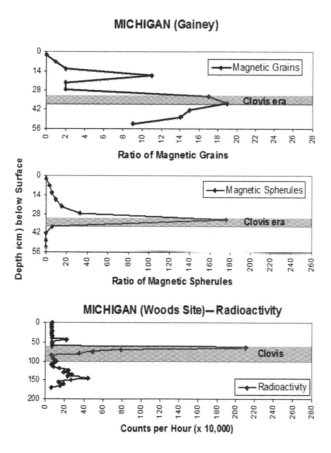

MICHIGAN (Gainey)

MICHIGAN (Woods Site)—Radioactivity

Fig. D.1. This composite chart shows all the key markers from Gainey, Michigan. Charcoal, carbon glass, and carbon spherules are abundant at Gainey, but because we had only one layer for testing, we did not graph them.

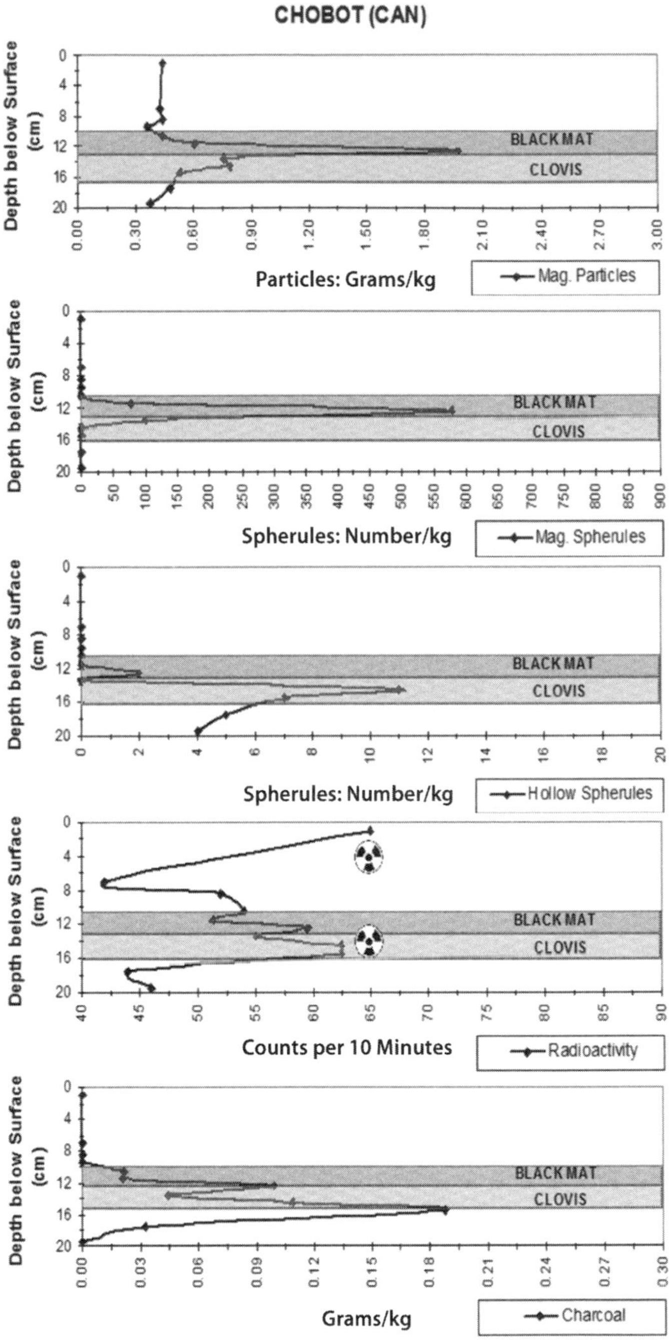

Fig. D.2. Key markers of the Chobot site in Alberta. Glasslike carbon is not shown but was there in small quantities.

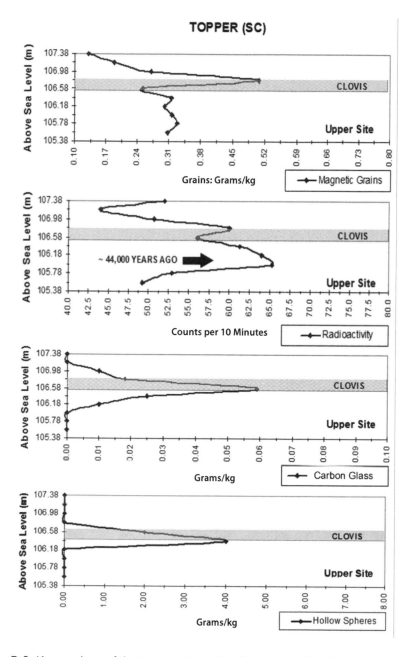

Fig. D.3. Key markers of the Topper site in South Carolina. Glasslike carbon is not shown but was there in small quantities.

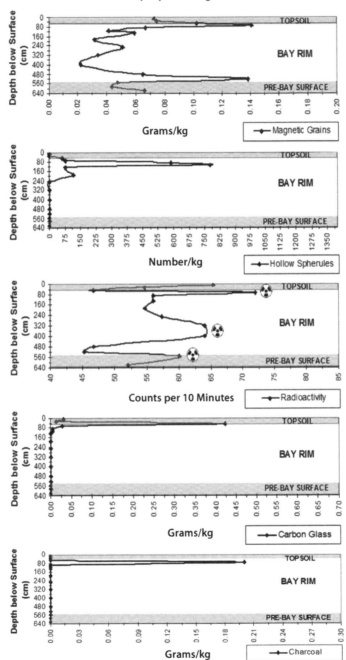

Fig. D.4. Key markers of a Carolina Bay in Bladen County, South Carolina. All markers except magnetic spherules were found. Note that the distribution looks very much like all the other sites, indicating that the bay rim dates back to Clovis times.

BIBLIOGRAPHY

Aharon, P. "Meltwater Flooding Events in the Gulf of Mexico Revisited: Implications for Rapid Climate Changes During the Last Deglaciation." *Paleoceanography* 18 (2003): 1079.

———. "Gulf of Mexico Deglacial Stable Isotope Data." IGBP PAGES/World Data Center for Paleoclimatology, Data Contribution Series # 2004-042.NOAA/NGDC Paleoclimatology Program, Boulder, Colo., 2004.

Akridge, G. "The Prehistoric Use of Meteorites in North America." *Meteorite!* 2 (no. 2, 1996): 20–22.

Allan, D. S., and Delair, J. B. *Cataclysm! Compelling Evidence of a Cosmic Catastrophe in 9,500 B.C.* Rochester, Vt.: Bear & Co., 1997.

Alpar, M. A.; Ogelman, H.; and Shaham, J. "Is Geminga a Glitching Pulsar?" *Astronomy and Astrophysics* 273 (1993): L35–L37.

Alvarez, L.; Alvarez, W.; and Klint, S. "Asteroid-Caused Extinctions." *Science News* 117 (1980): 22.

Anderson, D. G., and Faught, M. K. "The Distribution of Fluted Paleoindian Projectile Points: Update 1998." *Archaeology of Eastern North America* 26 (1998): 163–87.

Andrews, J. T. "Iceberg Rafted Detritus." Submitted to the NOAA Paleoclimatology Data Archive, 1987.

———. "A Heinrich-like Event, H-0 (DC-0): Source(s) for Detrital Carbonate in the North Atlantic During the Younger Dryas Chronozome." *Paleoceanography* 10 (1995): 943–52.

Andrews, J. T. and Peltier, W. R. "Collapse of the Hudson Bay Ice Center and Glacio-Isostatic Rebound." *Geology* 4 (1976): 73–75.

Baker, V. "The Study of Superfloods." *Science,* 29 March 2002.

Bard, E. "Tahiti Deglacial Relative Sea Level Reconstruction." IGBP PAGES/World Data Center for Paleoclimatology Data Contribution Series #2003-028. NOAA/NGDC Paleoclimatology Program, Boulder, Colo., 2003.

Bard, E., et al. "Sea Level Record from Tahiti Corals and the Timing of Deglacial Meltwater Discharge." *Nature* 382 (1996): 241–44.

————. "Hydrological Impact of Heinrich Events in the Subtropical Northeast Atlantic." *Science* 289 (2000): 1321.

Barefoot, Daniel W. *Touring the Backroads of North Carolina's Upper Coast.* Winston-Salem, N.C.: John F. Blair Publisher, 1995.

Baumgartner, S., et al. "36Cl Fallout in the Summit Greenland Ice Core Project Ice Core." *Journal of Geophysical Research* 102 (1997): 26659.

Beck, J. Warren, et al. "Extremely Large Variations of Atmospheric ^{14}C Concentration During the Last Glacial Period." *Science* 292 (2001): 2453.

Becker, Luann, et al. "Impact Event at the Permian-Triassic Boundary." *Science* 291 (2001): 1530–33.

————. "Bedout: A Possible End-Permian Impact Crater Offshore of Northwestern Australia." *Science* 304 (2004): 1469.

Bell, Rosemary. *Yurok Tales.* Etna, Calif.: Bell Books, 1992.

Bender, M., et al. "Climate Connections Between Greenland and Antarctica During the Last 100,000 Years." *Nature* 372 (1994): 663–66.

————. "On the Concentrations of O$_2$, N$_2$ and Ar in Trapped Gases from Ice Cores." *Journal of Geophysical Research* 100 (1995): 18651–60.

Benítez, Narcisco; Maíz-Apellániz, Jesús; and Canelles, Matilde. "Evidence for Nearby Supernova Explosions." *Physical Review Letters* 88 (2002).

Benito, G., and O'Connor, J. "Number and Size of Last-Glacial Missoula Floods in the Columbia River Valley Between the Pasco Basin, Washington, and Portland, Oregon." *Geological Society of America Bulletin,* May 2003.

Benson, L. V., "Timing of the Last Highstand of Lake Lahontan." *Journal of Paleoclimatology* 5 (1991): 115–26.

Biver, N., et al. "Chemical Composition Diversity Among 24 Comets Observed at Radio Wavelengths." *Earth, Moon and Planets* 90 (2002): 323–33.

Blunier, T., et al. "Biological Oxygen Productivity During the Last 60,000 Years from Triple Oxygen Isotope Measurements." *Global Biogeochemical Cycles* 16 (2002), art. no. 1029.

Boyd, M., et al. "Paleoecology and Geochronology of Glacial Lake Hind During the Pleistocene–Holocene Transition: A Context for Folsom Surface Finds on the Canadian Prairies." *Geoarchaeology: An International Journal* 18 (2003): 583–607.

Brackenridge, G. R. "Terrestrial Paleoenvironmental Effects of a Late Quaternary-Age Supernova." *Icarus* 46 (1981): 81.

Bradley, Bruce, and Stanford, Dennis. "The North Atlantic Ice-Edge Corridor: A Possible Palaeolithic Route to the New World." *World Archaeology* 36 (2004): 459–78.

Braun, A., and Pfeiffer, T. "Cyanobacterial Blooms as the Cause of a Pleistocene Large Mammal Assemblage." *Paleobiology* 28 (2002): 139–54.

Brett, William Henry. *Legends and Myths of the Aboriginal Indians of British Guiana.* London: Williams Wells Gardner, 1880.

Bretz, J. H., Smith, H. T. U., and Neff, G. E. "Channeled Scabland of Washington: New data and interpretations." *Geological Society of America Bulletin* 67 (1956): 957–1049.

Broecker, W. S. "The Ocean." *Scientific American* 249 (1983): 146.

———. "Thermohaline Circulation, the Achilles Heel of Our Climate System: Will Man-Made CO₂ Upset the Current Balance?" *Science* 278 (1997a).

———. "Will Our Ride into the Greenhouse Future Be a Smooth One?" *GSA Today* 7 (May 1997b): 1–7.

———. "What If the Conveyor Were to Shut Down? Reflections on a Possible Outcome of the Great Global Experiment." *GSA Today* 9 (1999): 1–7.

Broecker, W. S., et al. "The Chronology of the Last Deglaciation: Implications to the Cause of the Younger Dryas Event." *Paleoceanography* 3 (1988): 1–19.

Brook, E., et al. "Rapid Variations in Atmospheric Methane Concentration During the Past 110,000 Years." *Science* 273 (1996): 1087–1091.

Brook, E. J., et al. "Accretion of Interplanetary Dust in Polar Ice." *Geophysical Research Letters* 27 (2000): 3145.

Brooks, M., et al. "Carolina Bay Geoarchaeology and Holocene Landscape Evolution on the Upper Coastal Plain of South Carolina." *Geoarchaeology* 11 (1996): 481–504.

Brooks, M., et al. "Pleistocene Encroachment of the Wateree River Sand Sheet into Big Bay on the Middle Coastal Plain of South Carolina." *Southeastern Geology* 40 (2001): 241–57.

Brooks, M., and Taylor, B. "Age and Climate Correlates of Carolina Bays and Inland Dunes of the South Atlantic Coastal Plain." *Legacy* 6 (no. 2, 2001): 6–7.

Bruchac, Joseph. *Native American Stories.* Golden, Colo.: Fulcrum Publishing, 1991.

Bryson, R. A. "Late Quaternary Volcanic Modulation of Milankovitch Climate Forcing." *Theoretical and Applied Climatology* 39 (1998): 115–25.

———. "Volcanic Eruptions and Aerosol Optical Depth Data." IGBP PAGES/World Data Center for Paleoclimatology, Data Contribution Series # 2002-022. NOAA/ NGDC Paleoclimatology Program, Boulder, Colo., 2002.

Charles, T., and Michie, J. "South Carolina Paleo Point Database." In *PaleoIndian and Early Archaic Research in the Lower Southeast: A South Carolina Perspective,* edited by David G. Anderson, Chris Judge, and Kenneth E. Sassaman. Mount Pleasant, S.C.: Council of South Carolina Professional Archaeologists, 1992, 381–89.

Chylek, P., et al. "Biomass Burning Record and Black Carbon Concentration in the GISP2 Ice Core." *Geophysical Research Letters* 22 (no. 2, 1995): 89–92.

Clark, D. H.; McCrea, W. H.; and Stephenson, F. R. "Frequency of Nearby Supernovae and Climatic and Biological Catastrophes." *Nature* 265 (1977): 318–19.

Clark, Ella E. *Indian Legends of the Pacific Northwest.* University of California Press, 1953.

Clube, S. V. M., and Napier, W. M. "The Microstructure of Terrestrial Catastrophism." *Monthly Notices of the Royal Astronomical Society* 211 (1984): 953–68.

———. *Cosmic Winter.* New York: Universe Books, 1990.

Colgan, P. M., et al. "Glacial Landform-Sediment Assemblages Along the Southern Margin of the Laurentide Ice Sheet: Implications for Ice-Lobe Behavior and Subglacial Conditions." *Geological Society of America Abstracts with Programs* 32 (no.

7, 2000): A-20. Available online at: www.casdn.neu.edu/%7Egeology/department/staff/colgan/colgan00b.htm.

Colman, S., and Foster, D. "Stratigraphy, Descriptions and Physical Properties of Sediments Cored in Lake Michigan." OFR 90-478. Woods Hole, Mass.: USGS, 1990.

Conway, Thor. "The Conjurer's Lodge: Celestial Narratives from Algonkian Shamans," in Ray A. Williamson and Claire R. Farrer, *Earth and Sky*. Albuquerque: University of New Mexico Press, 1992.

Cortijo, E., et al. "Rapid Climatic Variability of the North Atlantic Ocean and Global Climate." *Quaternary Science Reviews* 19 (2000): 227–41.

Cottin, H., et al. "Polyoxymethylene as Parent Molecule for the Formaldehyde Extended Source in Comet Halley." *The Astrophysical Journal* 556 (no. 1, 2001): 417–20.

Cruttenden, W. *Lost Star of Myth and Time*. Pittsburgh, Pa.: St. Lynn's Press, 2005.

Culler, T. S.; Becker, T. A.; Muller, R. A.; et al. "Lunar Impact History from $^{40}Ar/^{39}Ar$ Dating of Glass Spherules." *Science* 287 (2000): 1785–88.

Curtin, Jeremiah. *Creation Myths of Primitive America*. Boston: Little, Brown, 1898.

Cutler, P., et al. "Sedimentologic Evidence for Outburst Floods from the Laurentide Ice Sheet Margin in Wisconsin, USA: Implications for Tunnel-Channel Formation." *Quaternary International* 90 (2002): 23–40.

Damon, P. E., et al. "Radiocarbon Production by the Gamma-ray Component of Supernova Explosions." *Radiocarbon* 37 (1995): 599.

Dar, A., et al. "Life Extinctions By Cosmic Ray Jets." *Physical Review Letters* 80 (1998): 5813.

Dey, W., et al. "Preliminary Geologic Cross Sections, Kane County, Illinois. *Illinois State Geological Survey*, Illinois Preliminary Geologic Map, IPGM Kane-CS, 1:100,000, 2004.

Dreschhoff, Gisella, and Zeller, Edward J. "Ultra-High Resolution Nitrate in Polar Ice as Indicator of Past Solar Activity." *Solar Physics* 177 (1998): 365–74.

Edmonds, M., and Clark, E. *Voices of the Winds: Native American Legends*. New York: Facts on File, Inc., 1989.

Ellis, J., and Schramm, D. "Could a Nearby Supernova Explosion Have Caused a Mass Extinction?" *Proceedings of the National Academy of Sciences USA* 92 (1995): 235–38.

Elmore, R. D., et al. "Black Shell Turbidite, Hatteras Abyssal Plain." *Geological Society of America Bulletin* 90 (2003): 1165–76.

Erdoes, R., and Ortiz, A. *American Indian Myths and Legends*. New York: Pantheon, 1984.

Ernstson, K., et al. "Unusual Melt Rocks from Meteorite Impact." 2004. Available on the Web at www.impact-structures.com/article/article_4.html.

Evans, P., et al. "Microcephalin, a Gene Regulating Brain Size, Continues to Evolve Adaptively." *Science* 309 (2005): 1717–20.

Eyton, J. R., and Parkhurst, J. I. "A Re-evaluation of the Extraterrestrial Origin of the Carolina Bays." Occasional Publication, Department of Geography Paper No. 9, University of Illinois at Urbana-Champaign, 1975.

Fairbanks, R. G. "The Age and Origin of the 'Younger Dryas' Climate Event in Greenland Ice Cores." *Paleoceanography* 5 (1990): 937–48.

Finkel, R., and Nishiizumi, K. "Beryllium 10 Concentrations in the Greenland Ice Sheet Project 2 Ice Core from 3-40 ka." *Journal of Geophysical Research* 102 (1997): 26699–26706.

Firestone, R. B., and Topping, W. "Terrestrial Evidence of a Nuclear Catastrophe in Paleoindian Times." *The Mammoth Trumpet* 16 (March 2001): 9.

Fisher, T. G.; Clague, J. J.; and Teller, J. T. "The Role of Outburst Floods and Glacial Meltwater in Subglacial and Proglacial Landform Genesis." *Quaternary International* 90 (2002): 1–4.

Fitting, J. "A Study of Natural Radioactivity in Osteological Materials from the Blackwater Draw, Locality Number 1." In *Studies in the Natural Radioactivity of Prehistoric Materials,* edited by A. Jelinek, et al. Ann Arbor: University of Michigan, 1963.

Foster, D., and Colman, S. "Preliminary Interpretation of the High-Resolution Seismic Stratigraphy Beneath Lake Michigan." OFR 91-21. Woods Hole, Mass.: USGS, 1991.

Frank, Louis A. *The Big Splash.* New York: Carol Publishing Group, 1990.

Frank, L. A.; Sigwarth, J. B.; and Craven, J. D. "On the Influx of Small Comets into the Earth's Upper Atmosphere." *Geophysical Research Letters* 13 (1986): 303–306.

Frank, L. A., and Sigwarth, J. B. "Influx of Small Comets into Earth's Upper Atmosphere." In *Instruments, Methods, and Missions for the Investigation of Extraterrestrial Microorganisms,* edited by Richard B. Hoover. *Proceedings of SPIE* 3111 (1997): 238–48.

Frazer, Sir James G. *Folk-Lore in the Old Testament,* vol. 1. London: Macmillan, 1919.

Frey, D. G. "Morphometry and Hydrography of Some Natural Lakes of the North Carolina Coastal Plain: The Bay Lake as a Morphometric Type." *J. Elisha Mitchell Scientific Society* 65 (1949): 1–37.

———. "Carolina Bays in Relation to the North Carolina Coastal Plain." *J. Elisha Mitchell Scientific Society* 66 (1950): 44–52.

———. "Pollen Succession in the Sediments of Singletary Lake, North Carolina." *Ecology* 32 (1951): 518–33.

———. "Regional Aspects of the Late-Glacial and Post-Glacial Pollen Succession of South-Eastern North Carolina." *Ecological Monographs* 23 (1953): 289–313.

———. "Evidence for Recent Enlargement of the 'Bay Lakes' of North Carolina." *Ecology* 35 (1954): 78–88.

———. "Stages in the Ontogeny of the Carolina Bays." *Proceedings of the International Association of Applied Limnology* 12 (1955): 660–68.

Fronval, T., and Jansen, E. "Eemian and Early Weichselian (140-60ka) Paleoceanography and Paleoclimate in the Nordic Seas with Comparisons to Holocene Conditions." *Paleoceanography* 12 (1997): 443–62.

———. "Nordic Seas Eemian Paleoceanography Data." IGBP PAGES/World Data Center-A for Paleoclimatology Data, Contribution Series # 97-029. NOAA/NGDC Paleoclimatology Program, Boulder, Colo., 1997.

Fudali, R., and Melson, W. "Secondary Craters as a Clue to Primary Crater Origin on the Moon." *Meteoritics and Planetary Science* 4 (1969): 273.

Fuhrer, K., and Legrand, M. R. "Continental Biogenic Species in the Greenland Ice Core Project Ice Core: Tracing Back the Biomass History of the North American Continent." *Journal of Geophysical Research* 102 (1997): 26735.

Fulthorpe, C., and Austin, J. "Shallowly Buried, Enigmatic Seismic Stratigraphy on the New Jersey Outer Shelf: Evidence for Latest Pleistocene Catastrophic Erosion?" *Geology* 32 (2004): 1013–1016.

Gabrielli, P., et al. "Meteoric Smoke Fallout over the Holocene Epoch Revealed by Iridium and Platinum in Greenland Ice." *Nature* 30 (2004): 43223.

Gardner, J. V., et al. "A Climate-Related Oxidizing Event in Deep-Sea Sediment from the Bering Sea." *Quaternary Research* 18 (1982): 91–107.

Gaster, Theodor H. *Myth, Legend, and Custom in the Old Testament.* New York: Harper & Row, 1969.

Gifford, Douglas. *Warriors, Gods and Spirits from Central & South American Mythology.* Glasgow: William Collins, 1983.

Gillis, J. J.; Jolliff, B. L.; and Korotev, R. L. "Lunar Surface Geochemistry: Global Concentrations of Th, K, and FeO as Derived from Lunar Prospector and Clementine Data." *Geochimica et Cosmochimica Acta* 68 (2004): 3791–3805.

Gnos, E., et al. "Pinpointing the Source of a Lunar Meteorite: Implications for the Evolution of the Moon." *Science* 305 (2004): 657–59.

Goddard, P. E. "Kato Texts." University of California Publications in American Archaeology and Ethnology 184 (no. 2, 1929). From Stith Thompson, *Tales of the North American Indians.* Bloomington: Indiana University Press, 1929.

Gong, G., et al. "Association Between Bone Mineral Density and Candidate Genes in Different Races and Its Implications." *Chinese Medical Journal* 115 (2002): 116–121.

Goodyear, A. "Results of the 1999 Allendale Paleoindian Expedition." *Legacy: Newsletter of the South Carolina Institute of Archaeology and Anthropology* 4 (nos. 1–3, 1999): 8–13.

———. "The Topper Site 2000: Results of the 2000 Allendale Paleoindian Expedition." *Legacy: Newsletter of the South Carolina Institute of Archaeology and Anthropology* 5 (no. 2, 2000): 18–25.

———. "Evidence for Pre-Clovis Sites in the Eastern United States." An expanded version of a paper presented at the "Clovis and Beyond" Conference in Santa Fe, N.M., 29 October 1999; final version, 8 February 2001.

Goodyear, A., et al. "The Earliest South Carolinians: The Paleoindian Occupation of South Carolina." *Occasional Papers* 2. Columbia: Archaeological Society of South Carolina, 1990.

———. "Archaeology of the Pleistocene-Holocene Transition in Eastern North America." *Quaternary International* 49/50 (1998): 151–66.

———. "Evidence of Pre-Clovis Lithic Remains in Allendale County, SC." Paper presented at the Annual Meeting of the Southeastern Archaeological Conference, Greenville, S.C., 1998.

———. "Evidence of Pre-Clovis in the Savannah River Basin, Allendale County, South Carolina." Paper presented at the 64th Annual Meeting of the Society for American Archaeology, Chicago, 24–28 March 1999.

———. "The Early Holocene Occupation of the Southeastern United States: A Geoarchaeological Summary." In *Ice Age Peoples of North America,* edited by R. Bonnichsen and K. Turnmire. Corvallis, Or.: Oregon State University Press, 1999, 432-81.

Goodyear, A., and Steffy, K. "Evidence for a Clovis Occupation at the Topper Site, 38AL23, Allendale County, South Carolina." *Current Research in the Pleistocene* 20 (2003): 23–25.

Goodyear, A., et al. "Evidence of Pre-Clovis Sites in the Eastern United States." In *Paleoamerican Origins: Beyond Clovis,* edited by R. Bonnichsen, et al. Texas A&M University Press (in press for 2006).

Grant, J.; Brooks, Mark J.; and Taylor, Barbara E. "New Constraints on the Evolution of Carolina Bays from Ground-Penetrating Radar." *Geomorphology* 22 (1998): 325–45.

Green, David A. "Galactic Supernova Remnants: An Updated Catalogue and Some Statistics." *Bulletin of the Astronomical Society of India* 32 (2004): 335.

Grimley, D. "Glacial and Nonglacial Sediment Contributions to Wisconsin Episode Loess in the Central United States." *Geological Society Bulletin* 112 (2000): 1475–95.

Grinnell, George Bird. *Pawnee Hero Stories and Folk-Tales.* Lincoln: University of Nebraska Press, 1961. Reprinted from Forest and Stream Publishing Company, New York, 1889.

Hagiwara, K., et al. "Review of Particle Physics." Particle Data Group. *Physical Review D 66,* 2001.

Hanebuth, T., et al. "Rapid Flooding of the Sunda Shelf: A Late-Glacial Sea-Level Record." *Science* 288 (2000): 1033.

Hansen, H. J. "Was There or Was There Not a Meteoritic Impact at the K/T Boundary 65 Million Years Ago?" Presentation to the Micropalaeontological Society, 2004.

Haskin, L. A., et al. "The Nature of Mare Basalts in the Procellarum KREEP Terrane." Abstract number 1661, 31st Lunar and Planetary Science Conference, Houston, March 2000.

Haynes, C. V., Jr. "Stratigraphy and Late Pleistocene Extinction in the United States." In *Quaternary Extinctions: A Prehistoric Revolution,* edited by P. S. Martin and R. G. Klein. Tucson: University of Arizona Press, 1984, 353–65.

———. "Clovis Origin Update." *The Kiva* 52 (no. 2, 1987): 83–93.

———. "Curry Draw, Cochise County, Arizona: A Late Quaternary Stratigraphic Record of Pleistocene Extinction and Paleo-Indian Activities." Geological Society of America, Centennial Field Guide, Cordilleran Section, 1987.

―――. "Geoarchaeological and Paleohydrological Evidence for a Clovis-Age Drought in North America and Its Bearing on Extinction." *Quaternary Research* 35 (1991): 438–50.

―――. "Contributions of Radiocarbon Dating to the Geochronology of the Peopling of the New World." In *Radiocarbon After Four Decades*, edited by R. E. Taylor, A. Long, and R. S. Kra. New York: Springer-Verlag, 1992, 355–74.

―――. "Clovis-Folsom Geochronology and Climatic Change." In *From Kostenki to Clovis: Upper Paleolithic Paleo-Indian Adaptations*, edited by Olga Soffer and N. D. Praslov, 219–36. New York: Plenum Press, 1993.

―――. "Investigator Describes Site Formation." *Mammoth Trumpet* 13 (no. 2, 1998).

―――. "Geochronology of the Stratigraphic Manifestation of Paleoclimatic Events at Paleoindian Sites." Paper presented at the 63rd Annual Meeting of the Society for American Archaeology, Seattle, 1998.

―――. "Younger Dryas 'Black Mats' and Other Stratigraphic Manifestations of Climate Change in North America." Abstract and presentation to XVI INQUA Congress, Reno, Nev., Geological Society of America, 2003.

―――. "Nature and Origin of the Black Mat, Stratum F2. Appendix B." In *Murray Springs: A Clovis Site with Multiple Activity Areas in the San Pedro Valley, Arizona*, edited by C. V. Haynes Jr. and Bruce B. Huckell. Tucson: University of Arizona Press (in press).

Haynes, C. V., Jr.; Stanford, Dennis J.; Jodry, Margaret; et al. "A Clovis Well at the Type Site 11,500 B.C.: The Oldest Prehistoric Well in America." *Geoarchaeology* 14 (1999): 455–70.

Hemming, S. R., et al. "Provenance of Heinrich Layers in Core V28-82, Northeastern Atlantic: 40Ar=39Ar Ages of Ice-Rafted Hornblende, Pb Isotopes in Feldspar Grains, and Nd–Sr–Pb Isotopes in the Fine Sediment Fraction." *Earth and Planetary Science Letters* 164 (1998): 317–33.

Henbest, Nigel, and Couper, Heather. *The Guide to the Galaxy.* Cambridge University Press, 1994.

Hoffman, V. et al. "Characterization of a Small Crater-like Structure in SE Bavaria, Germany." Presentation to the 68th Annual Meteoritical Society Meeting, Gatlinburg, Tenn., 2005.

Holcombe, T., et al. "Small Rimmed Depression in Lake Ontario: An Impact Crater?" *Journal of Great Lakes Research* 27 (2001): 510–17.

Holland, J. G., and Lampert, R. St. J. "Major Element Chemical Composition of Shields and the Continental Crust." *Geochimica et Cosmochimica Acta* 36 (1972): 673–83.

Holliday, V., et al. "Lithostratigraphy and Geochronology of Fills in Small Playa Basins on the Southern High Plains, United States." *Geological Society of America Bulletin* 108 (1996): 953–65.

Horcasitas, Fernando. "An Analysis of the Deluge Myth in Mesoamerica." In *The Flood Myth*, edited by Alan Dundes. Berkeley and London: University of California Press, 1988.

Hoyle, F., and Wickramasinghe, C. "Cometary Impacts and Ice-Ages." *Astrophysics and Space Science* 275 (2001): 367–76.

Hughen, K., et al. "A New ^{14}C Calibration Data Set for the Last Deglaciation Based on Marine Varves." *Radiocarbon* 40 (1998): 483.

Hughen, K., et al. "^{14}C-Activity and Global Carbon Cycle Changes over the Past 50,000 Years." *Science* 203 (2004): 202–207.

Hughen, K. A., et al. "Synchronous Radiocarbon and Climate Shift During the Last Deglaciation." *Science* 290 (2000): 1951–54.

Hussey, T. C. "A 20,000-year History of Vegetation and Climate at Clear Pond, Northeastern South Carolina." Master's thesis, University of Maine, Orono, 1993.

Isphording, W. C., and Flowers, G. C. "Karst Development in Coastal Plain Sands: A 'New' Problem in Foundation Engineering." *Bulletin of the Association of Engineering Geologists* 25 (1988): 95–104.

Ivester, A., et al. "Carolina Bays and Inland Dunes of the Southern Atlantic Coastal Plain Yield New Evidence for Regional Paleoclimate." *Geological Society of America Abstracts with Programs* 34 (no. 6, 2002): 273.

———. "Concentric Sand Rims Document the Evolution of a Carolina Bay in the Middle Coastal Plain of South Carolina." *Geological Society of America Abstracts with Programs* 35 (no. 6, 2003): 169.

Ivester, A. H., et al. "Chronology of Carolina Bay Sand Rims and Inland Dunes on the Atlantic Coastal Plain, USA." Third New World Luminescence Dating Workshop. Department of Earth Science, Dalhousie University, Halifax, Nova Scotia, July 4–7, 2004a, p. 23.

———. "The Timing of Carolina Bay and Inland Activity on the Atlantic Coastal Plain of Georgia and South Carolina." *Geological Society of America Abstracts with Programs* 36 (no. 5, 2004b): 69.

Johnson, D. W. *The Origin of the Carolina Bays.* New York: Columbia University Press, 1942.

Johnson, Elias. *Legends, Traditions, and Laws of the Iroquois, or Six Nations, and History of the Tuscarora Indians.* New York: Union Printing and Publishing, 1881.

Johnson, G. H., and Goodwin, B. K. "Elliptical Depressions on Undissected Highland Gravels in Northern Chesterfield County, Virginia" (abstract). *Virginia Journal of Science* 18 (1967): 186.

Jull, A. J. T. "Carbon-14 Terrestrial Ages and Weathering of 27 Meteorites from the Southern High Plains and Adjacent Areas (USA)." *Meteoritics and Planetary Science* 28 (1993): 188–95.

———. "Terrestrial Ages of Some Meteorites from Oman." *Meteoritics and Planetary Science* 37 (suppl., 2002): A74.

Jull, A. J. T., et al. "^{14}C Depth Profiles in Apollo 15 and 17 Cores and Lunar Rock 68815." *Geochimica et Cosmochimica Acta* 62 (1989): 3025-3036.

Karner, D. B., et al. "Constructing a Stacked Benthic d^{18}O Record." *Paleoceanography* 17 (2002): 1–11.

Katz, L. *The History of Blackwater Draw.* Portales: Eastern New Mexico University Printing Services, 1997.

Kauffman, George B. "Martin D. Kamen: An Interview with a Nuclear and Biochemical Pioneer." *The Chemical Educator* 5 (2002): 252–62.

Keigwin, L. D., et al. "Deglacial Meltwater Discharge, North Atlantic Deep Circulation, and Abrupt Climate Change." *Journal of Geophysical Research* 96 (1991): 16811–26.

Kennett, James P., et al. *Methane Hydrates in Quaternary Climate Change the Clathrate Gun Hypothesis.* American Geophysical Union, Washington, D.C., (2002): 224.

Kitagawa, H., and van der Plicht, J. "A 40,000-Year Varve Chronology from Lake Suigetsu, Japan: Extension of the ^{14}C Calibration Curve." *Radiocarbon* 40 (1998): 505.

Kobres, Bob. "More Carolina Bay Information." 2005. Available on the Web at abob.libs.uga.edu/bobk/cbaymenu.html.

Kooyman, B., et al. "Identification of Horse Exploitation by Clovis Hunters Based on Protein Analysis." *American Antiquity* 66, no. 4 (2001): 686–91.

Korotev, R. L.; Jolliff, B. L.; and Ziegler, R. A. "The KREEP Components of the APOLLO 12 Regolith" (abstract no. 1363). Thirty-first Lunar and Planetary Science Conference, Houston, March 2000.

Kruse, S. "Uranium Minerals Deposited on Mammoth and Bison Bones at Blackwater Draw Archaeological Site." Portales: Eastern New Mexico University, thesis, 2000.

Lagerklint, I. M. "Late Glacial Warming Prior to Heinrich Event 1: The Influence of Ice Rafting and Large Ice Sheets on the Timing of Initial Warming." *Geology* 27 (1999): 1099–1102.

Lal, D.; Jull, A. J. T.; Burr, G. S.; et al. "On the Characteristics of Cosmogenic In Situ ^{14}C in Some GISP2 Holocene and Late Glacial Ice Samples." *Nuclear Instruments and Methods in Physics Research, Section B: Beam Interactions with Materials and Atoms* 172 (2000): 623–31.

Lallement, R.; Welsh, B. Y; Vergely, J. L.; et al. "3D Mapping of the Dense Interstellar Gas Around the Local Bubble." *Astronomy and Astrophysics* 411 (2003): 447–64.

LaViolette, P. *Earth Under Fire: Humanity's Survival of the Ice Age.* Rochester, Vt.: Bear & Company, 2005.

———. *Genesis of the Cosmos: The Ancient Science of Continuous Creation.* Rochester, Vt.: Bear and Co., 2004.

Leakey, R., and Lewin, R. *The Sixth Extinction.* New York: Doubleday, 1995.

Legrand, H. E. "Streamlining of the Carolina Bays." *Journal of Geology* 61 (1953): 263–74.

Legrand, M. R., and De Angelis, M. "Origins and Variations of Light Carboxylic Acids in Polar Precipitation." *Journal of Geophysical Research* 100 (no. D1, 1995): 1445–62.

Legrand, M. R., et al. "Large Perturbations of Ammonium and Organic Acids Content in the Summit-Greenland Ice Core: Fingerprint from Forest Fires?" *Geophysical Research Letters* 19 (1992): 473–75.

Leonard, J. A., et al. "Ancient DNA Evidence for Old World Origin of the New World Dog." *Science* 298 (2002): 1613–1616.

Leon-Portilla, Miguel. "Mythology of Ancient Mexico." In *Mythologies of the Ancient World*, edited by Samual Noah Kramer. Garden City, N.Y.: Anchor Books, 1961.

Libby, W. F.; Anderson, E. C.; and Arnold, J. R. "Age Determination by Radiocarbon Content: World-Wide Assay of Natural Radiocarbon." *Science* 109 (1949): 227–28.

Licciardi, J. "Variable Responses of Western U.S. Glaciers During the Last Deglaciation." *Geology* 32 (2004): 81–84.

Licciardi, J., et al. "Freshwater Routing by the Laurentide Ice Sheet During The Last Deglaciation." In *Mechanisms of Global Climate Change at Millennial Time Scales*. Edited by P. U. Clark, R. S. Webb, and L. D. Keigwin. AGU Geophysical Monograph 112 (1999): 177–201, 1999. Available on the Web at www.unh.edu/esci/licciardi_et_al_1999_agu.pdf.

Lin, H.-L., et al. "Late Quaternary Climate Change from delta O-18 Records of Multiple Species of Planktonic Foraminifera: High-resolution Records from the Anoxic Cariaco Basin, Venezuela." *Paleoceanography* 12 (199): 415–27.

Lowell, T. V., et al. "Testing the Lake Agassiz Meltwater Trigger for the Younger Dryas." *EOS, American Geophysical Union* 86 (no. 40, 2005).

Marcantonio, F., et al. "Abrupt Intensification of the SW Indian Ocean Monsoon During the Last Deglaciation: Constraints from Th, Pa, and He Isotopes." *Earth and Planetary Science Letters* 184 (2001): 505–14.

Markewich, H. W., and Markewich, W. "An Overview of Pleistocene and Holocene Inland Dunes in Georgia and the Carolinas: Morphology, Distribution, Age, and Paleoclimate." *U.S. Geological Survey Bulletin* 2069 (1994).

Markman, Roberta H., and Markman, Peter T. *The Flayed God*. New York: HarperCollins, 1992.

Marsh, Nigel D., and Svensmark, Henrik. "Low Cloud Properties Influenced by Cosmic Rays." *Physical Review Letters* 85 (2000): 5004–5007.

Martin, P. S. "Prehistoric Overkill: The Global Model." In *Quaternary Extinctions*. Edited by P. S. Martin and R. G. Klein. Tucson: University of Arizona Press, 1984.

Maslin, M., et al. "Linking Continental-Slope Failures and Climate Change: Testing the Clathrate Gun Hypothesis." *Geology* 32 (2004): 53–56.

May, J. H., and Warne, A. G. "Hydrogeologic and Geochemical Factors Required for the Development of Carolina Bays Along the Atlantic and Gulf of Mexico, Coastal Plain, USA." *Environmental and Engineering Geoscience* 5 (1999): 261–70.

Mayewski, Paul A., and Legrand, Michel R. "Recent Increase in Nitrate Concentration of Antarctic Snow." *Nature* 346 (1990): 258–60.

Mayewski, P. A., et al. "An Ice Core Record of Atmospheric Response to Anthropogenic Sulphate and Nitrate." *Nature* 346 (1990): 554–56.

Mayewski, et al. "Major Features and Forcing of High-Latitude Northern Hemisphere Atmospheric Circulation Using a 110,000-Year-Long Glaciochemical Series." *Journal of Geophysical Research* 102 (1997): 26345–66.

Mazur, M. J., et al. "The Seismic Signature of Meteorite Impact Craters." *CSEG Recorder* (2000).

McDonald, J. "The Reordered North American Selection Regime and Late Quaternary Megafaunal Extinctions." In *Quaternary Extinctions*. Edited by P. S. Martin and R. G. Klein. Tucson: University of Arizona Press, 1984.

McHargue, L. R.; Damon, P. E.; and Donahue, D. J. "Enhanced Cosmic-ray Production of Be-10 Coincident with the Mono Lake and Laschamp Geomagnetic Excursions." *Geophysical Research Letters* 22 (1995): 659.

Mead, J., and Meltzer, D. "North American Late Quaternary Extinctions and the Radiocarbon Record." In *Quaternary Extinctions*. Edited by P. S. Martin and R. G. Klein. Tucson: University of Arizona Press, 1984.

Melosh, H. J. *Impact Cratering: A Geologic Process*. New York: Oxford University Press, 1989.

Melott, A. L., et al. "Did a Gamma-ray Burst Initiate the Late Ordovician Mass Extinction?" *International Journal of Astrobiology* 3 (2004): 55–61.

Melton, F. A., and Schriever, W. "The Carolina 'Bays': Are They Meteorite Scars?" *Journal of Geology* 58 (1933): 128–34.

Mooers, H., and Lehr, J. D. "Terrestrial Record of Laurentide Ice Sheet Reorganization During Heinrich Events." *Geology* 25 (1997): 987–90.

Mourant, A. E.; Kopec, A. C.; and Domaniewska-Sobczak, K. *The Distribution of the Human Blood Groups and Other Polymorphisms*, 2nd ed. Oxford: Oxford University Press, 1976.

Muller, Richard. *Nemesis: the Death Star.* London: Mandarin Press, 1990.

Munro-Stasiak, M. "The Blackspring Ridge Flute Field, South-Central Alberta, Canada: Evidence for Subglacial Sheetflow Erosion." *Quaternary International* 90 (2002): 75–86.

Napier, W. M., et al. "Extreme Albedo Comets and the Impact Hazard." *Monthly Notices of the Royal Astronomical Society*, 1365–2966, 2004.

Nei, M. "Evolution of Human Races at the Gene Level." *Progress in Clinical and Biological Research* 103 (1982): 167–81.

Nelson, Byron C. *The Deluge Story in Stone.* Minneapolis: Augsburg, 1931.

Nishiizumi, Kunihiko; Finkel, Robert C.; and Welten, Kees C. "26Al in GISP2 Ice Core." Tenth International Conference on Accelerator Mass Spectrometry, Berkeley, Calif., 2005.

NOAA. National Climactic Data Center, Paleoclimatology Data Search. 2005. Available on the Web at lwf.ncdc.noaa.gov/paleo/ftp-search.html.

———. National Geophysical Data Center. 2005. Available on the Web at www.ngdc.noaa.gov.

———. World Data Center. 2005. Available on the Web at www.ngdc.noaa.gov/wdc.

O'Keefe, J. D., and Ahrens, T. J. "Cometary and Meteorite Swarm Impact On Planetary Surfaces." *Journal of Geophysical Research* 87 (1982): 6668–80.

Pellizza, L. J., et al. "On the Local Birth Place of Geminga." *Astronomy and Astrophysics* 435 (2005): 625–30.

Perry, D. L.; Firestone, R. B.; Molnar, G. L.; et al. "Neutron-Induced Prompt Gamma Activation Analysis (PGAA) of Metals and Non-metals in Ocean Floor Geothermal Vent-generated Samples." *Journal of Analytical and Atomic Spectrometry* 16 (2001): 1–7.

Peterson, L. C., et al. "A High-Resolution Late Quarternary Upwelling Record from the Anoxic Cariaco Basin, Venezuela." *Paleoceanography* 6 (1991): 99–119.

Petit, J. R., et al. "Climate and Atmospheric History of the Past 420,000 Years from the Vostok Ice Core, Antarctica." *Nature* 399 (1999): 429–36.

———. "Vostok Ice Core Data for 420,000 Years." IGBP PAGES/World Data Center for Paleoclimatology Data Contribution Series #2001-076. NOAA/NGDC Paleoclimatology Program, Boulder, Colo., 2001.

Phillips, J. D.; Duval, J. S.; and Ambrosiak, R. A. "National Geophysical Data Grids: Gamma-Ray, Magnetic, and Topographic Data for the Conterminous United States." *United States Geological Survey Digital Data Series DDS-9,* 1993.

Piper, D., and Dean, W. "Trace-element Deposition in the Cariaco Basin, Venezuela Shelf, under Sulfate-Reducing Conditions: A History of the Local Hydrography and Global Climate, 20 ka to the Present." U.S. Geological Survey Professional Paper 1670. Denver: USGS Information Services, 1997.

Plato. *The Dialogues of Plato* translated into English with Analyses and Introductions, vol. 3. Translated by B. Jowett. Oxford: Oxford University Press, 1892.

Powars, D. S., and Bruce, T. S. "The Effects of the Chesapeake Bay Impact Crater on the Geological Framework and Correlation of Hydrogeologic Units of the Lower York-James Peninsula." U.S. Geological Survey Professional Paper 1612. Reston, Va., 1999.

Price, G., and Sobbe, I. "Pleistocene Palaeoecology and Environmental Change on the Darling Downs, South Eastern Queensland, Australia." *Memoirs of the Queensland Museum,* 2005.

Prouty, W. "Carolina Bays and Their Origin." *Geological Society of America Bulletin* 63 (1952): 167–224.

Raisbeck, G. M., et al. "Evidence for Two Intervals of Enhanced ^{10}Be Deposition in Antarctic Ice During the Last Glacial Period." *Nature* 326 (1987): 273.

Ramadurai, S. "Geminga as a Cosmic Ray Source." *Astronomical Society of India Bulletin* 21 (1993); 391–93.

Raup, D. M., and Sepkoski, J. J. "Periodicity of Extinctions in the Geologic Past." *Proceedings of the National Academy of Sciences USA* 81 (1984): 801–805.

Rauscher, T.; Heger, A.; Hoffman, R. D.; et al. "Nucleosynthesis in Massive Stars with Improved Nuclear and Stellar Physics." *The Astrophysical Journal* 576 (2002): 323–48.

Reimer, P., et al. "IntCal04 Terrestrial Radiocarbon Age Calibration, 0–26 cal Kyr BP." *Radiocarbon* 46 (2004): 1029–58.

Riggs, S., et al. "The Waccamaw Drainage System: Geology and Dynamics of a Coastal Wetland, Southeastern North Carolina." Department of Geology, East Carolina University, for North Carolina Department of Environment and Natural Resources, 2001.

Roberts, Richard G., et al. "New Ages for the Last Australian Megafauna: Continent-Wide Extinction About 46,000 Years Ago." *Science* 292 (2001): 1888–92.

Rooth, C. "Hydrology and Ocean Circulation." *Progress in Oceanography* 11 (1982): 131–49.

Rosler, W., et al. "Diamonds in Carbon Spherules: Evidence for a Cosmic Impact?" 68th Annual Meteoritical Society Meeting, Gatlinburg, Tenn., 2005.

Russell, D. A. *A Vanished World: The Dinosaurs of Western Canada.* Ottawa: National Museums of Canada, 1977.

———. "The Cretaceous-Tertiary Boundary Problem." *Episodes* (1979): 21–24.

———. "The Enigma of the Extinction of the Dinosaurs." *Annual Review of Earth and Planetary Sciences* 7: (1979b): 163–82.

Russell, H., and Arnott, R. "Hydraulic-Jump and Hyperconcentrated-Flow Deposits of a Glaciogenic Subaqueous Fan: Oak Ridges Moraine, Southern Ontario, Canada." *Journal of Sedimentary Research* 73 (2003): 887–905.

Sandstrom, K., as reported in "Supernova Poised to Go Off Near Earth." *New Scientist*, 23 May 2002.

Sanford, B. V., and Grant, A. C. "New Findings Relating to the Stratigraphy and Structure of the Hudson Platform." In *Current Research, Part D: Geological Survey of Canada* Paper 90–1D. 1990, 17–30.

Savage, H., Jr. *The Mysterious Carolina Bays.* Columbia: University of South Carolina Press, 1982.

Savolainen, P.; Zhang, Y.; Luo, J.; et al. "Genetic Evidence for an East Asian Origin of Domestic Dogs." *Science* 298 (2002): 1610–13.

Schimel, David S. "Terrestrial Ecosystems and the Carbon Cycle." *Global Change Biology* 1 (1995): 77–91.

Schramm, A.; Stein, M.; and Goldstein, S. "Calibration of the ^{14}C Time Scale to >40 ka by ^{234}U-^{230}Th Dating of Lake Lisan Sediments (Last Glacial Dead Sea)." *Earth Planet Science Letters* 175 (2000): 27.

Severinghaus, J. P., et al. "Timing of Abrupt Climate Change at the End of the Younger Dryas Interval from the Thermally Fractionated Gases in Polar Ice." *Nature* 391 (1998): 41–146.

Severinghaus, J. P., and Brook, E. J. "Abrupt Climate Change at the End of the Last Glacial Period Inferred from Trapped Air in Polar Ice." *Science* 286 (1999): 930–34.

Shara, M., et al. "HST Imagery of the Non-Expanding, Clumped 'Shell' of the Recurrent Nova T Pyxidis." *Astronomical Journal* 114 (1997): 258.

Sharitz, R. "Carolina Bay Wetlands: Unique Habitats of the Southeastern United States." *Wetlands* 23 (2003): 550–62.

Shaviv, Nir J., and Veizer, Ján. "Celestial Driver of the Phanerozoic Climate." *GSA Today* 13 (2003): 4.

Shaw, J. "Kinematic Indicators in Fault Gouge: Tectonic Analog for Soft-Bedded Ice Sheets—Comment." *Sedimentary Geology* 123 (1999): 153–55.

Shaw, J., and Gilbert, R. "Evidence for Large-Scale Subglacial Meltwater Flood Events in Southern Ontario and Northern New York State." *Geology* 18 (1990): 1169–72.

Shelton, Ian. "Supernova 1987A in the Large Magellanic Cloud." *International Astronomical Union (IAU)* Circular No. 4316, 1987.

Sherrell, R. M., et al. "Trace Metals in GISP2 Ice Core and Recent Snow at Summit, Greenland." *Journal of Geophysical Research,* Special Issue on the GISP2/GRIP Ice Cores, 1997.

Smith, H. J.; Wahlen, M.; and Mastroianni, D. "The CO_2 Concentration of Air Trapped in GISP2 Ice from the Last Glacial Maximum-Holocene Transition." *Geophysical Research Letters* 24 (1997): 1–4.

Smith, W. H. F., and Sandwell, D. T. "Global Sea Floor Topography from Satellite Altimetry and Ship Depth Soundings." *Science* 277 (1997): 1956–1962.

Stanford, D., and Day, Jane, eds. *Ice Age Hunters of the Rockies.* Boulder: University Press of Colorado, 1992.

Stanford, D., and Bradley, B. "The Solutrean Solution." *Discovering Archaeology* 2 (2000): 54–55.

———. "Ocean Trails and Prairie Paths? Thoughts on Clovis Origins." In *The First Americans: The Pleistocene Colonization of the New World.* Edited by Nina G. Jablonski. *Memoirs of the California Academy of Sciences,* no. 27. San Francisco, 2002, 255–72.

Steig, E. J., et al. "Synchronous Climate Changes in Antarctica and the North Atlantic." *Science* 282 (1998): 92–95.

———. "Wisconsinan and Holocene Climate History from an Ice Core at Taylor Dome, Western Ross Embayment, Antarctica." *Geografiska Annaler* 82A (2000): 213.

Stephenson, F. Richard, and Green, David A. *Historical Supernovae and Their Remnants.* Gloucestershire, U.K.: Clarendon Press, 2002.

STScI, Press Release. "Blobs in Space: The Legacy of a Nova." Space Telescope Science Institute, September 8, 1997. Available on the Web at www.xtec.es/recursos/astronom/hst/hst3/9729e.html.

Stuiver, M., et al., eds. "INTCAL98 Radiocarbon Age Calibration, 24,000-0 cal BP." *Radiocarbon* 40 (no. 3, 1998),

Taylor, K. C., et al. "Biomass Burning Recorded in the GISP2 Ice Core: A Record from Eastern Canada?" *The Holocene* 6 (1996): 1–6.

Taylor, R. E.; Haynes, C. Vance; and Stuiver, M. "Clovis and Folsom Age Estimates: Stratigraphic Context and Radiocarbon Calibration." *Antiquity* 70 (1996): 515–25.

Taylor, Walter K. *Wild Shores: Exploring the Wilderness Areas of Eastern North Carolina.* Winston-Salem, N.C.: Down Home Press, 1993.

Thom, B. G. "Carolina Bays in Horry and Marion Counties, South Carolina." *Geological Society of America Bulletin* 81 (1970): 783–814.

Thompson, R. S., et al. "Climatic Changes in Western United States Since 18,000 B.P." In *Global Climates Since the Last Glacial Maximum.* Edited by H. E. Wright Jr., et al. Minneapolis: University of Minnesota Press, 1993, 468–513.

Thurston, M. *The Lost History of the Canine Race.* Kansas City, Kans.: Andrews and McMeel, 1996.

Vecsey, Christopher. *Imagine Ourselves Richly.* San Francisco: HarperCollins, 1991.

Vereshchagin, N. K., and Baryshnikov, G. F. "Quaternary Mammalian Extinctions in Northern Eurasia." In *Quaternary Extinctions*. Edited by P. S. Martin and R. G. Klein. Tucson: University of Arizona Press, 1984.

Voelker, A. H. L., et al. "Correlation of Maine ^{14}C Ages from the Nordic Seas with the GISP2 Isotope Record: Implications for ^{14}C Calibration Beyond 25 ka BP." *Radiocarbon* 40 (1998): 517–34.

Vogel, J. "^{14}C Variations during the Upper Pleistocene." *Radiocarbon* 25 (1983): 213.

Wade, Nicholas. "Brain May Still Be Evolving, Studies Hint." The *New York Times*, 9 September 2005.

Ward, S., and Day, S. "Potential Collapse and Tsunami at La Palma, Canary Islands." *Geophysical Review Letters* 26 (2001): 3141–44.

Waters, Frank. *Book of the Hopi*. New York: Penguin Books, 1963.

Waters, Michael R., and Haynes, C. Vance. "Late Quaternary Arroyo Formation and Climate Change in the American Southwest." *Geology* 29 (2001): 399–402.

Watts, W. A. "Late Quaternary Vegetation History of White Pond on the Inner Coastal Plain of South Carolina." *Quaternary Research* 13 (1980): 187–99.

Wdowczyk, J., and Wolfendale, A. W. "Cosmic Rays and Ancient Catastrophes." *Nature* 268 (1977): 510.

Weaver, A., et al. "Meltwater Pulse 1A from Antarctica as a Trigger of the Bølling-Allerød Warm Interval." *Science* 299, 2003: 1709–13.

Wells, R. S., et al. "The Eurasian Heartland: A Continental Perspective on Y-chromosome Diversity." *Proceedings of the National Academy of Sciences USA* 98 (2001): 10244–49.

Wells, S. *The Journey of Man: A Genetic Odyssey*. New York: Random, 2004.

Welsh, C. "Quaternary Geologic Mapping of the Barrington Quadrangle, NE Illinois." *Geological Society of America Abstracts with Programs* 35 (no. 6, 2003): 67.

Whitehead, D. R. "Late-Wisconsin Vegetational Changes in Unglaciated Eastern North America." *Quaternary Research* 3 (1953): 621–31.

——. "Palynology and Pleistocene Phytogeography of Unglaciated Eastern North America." In *The Quaternary of the United States*. Edited by W. E. Wright Jr. and D. G. Fre. Princeton, N.J.: Princeton University Press, 1965, 417–32.

——. "Studies of Full-Glacial Vegetation and Climate in Southeastern United States." In *Quaternary Paleoecology*. Edited by E. J. Cushing and H. E. Wright Jr. New Haven, Conn.: Yale University Press, 1967.

——. "Developmental and Environmental History of the Dismal Swamp." *Ecological Monographs* 4 (1972): 301–315.

——. "Late-Pleistocene Vegetational Changes in Northeastern North Carolina." *Ecological Monographs* 51 (1981): 451–71.

Whitehead, D. R., and Barghoorn, E. "Pollen Analytical Investigation of Pleistocene." *Ecological Monographs* 32 (1962): 347–69.

Whitlow, S. I., et al. "An Ice Core Based Record of Biomass Burning in North America." *Tellus* 46B (1994): 239–42.

Wilbert, J., and Simoneau, K. *Folk Literature of South American Indians.* Los Angeles: UCLA Latin American Center Publications, University of California, 1975.

Willard, D., and Korejwo, D. "Holocene Palynology from Marion-Dufresne Cores MD99-2209 and 2207 from Chesapeake Bay: Impacts of Climate and Historic Land-Use Change." *U.S. Geological Survey Open-File Report* 00-306 (2001), chapter 7.

Williams, I. "The Evolution of Meteoroid Streams." Proceedings of the 150th colloquium of the International Astronomical Union held in Gainesville, Fla.; Astronomical Society of the Pacific (ASP 104), 1996.

Wolbach, W., et al. "Cretaceous Extinctions: Evidence for Wildfires and Search for Meteorite Material." *Science* 230 (1985): 167–70.

———. "Global Fire at the Cretaceous/Tertiary Boundary." *Nature* 334 (1988): 665–69.

———. "Fires at the K-T boundary: Carbon at the Sumbar, Turkmenia Site." *Geochimica et Cosmochimica Acta* 54 (1990): 1133–46.

———. "Major Wildfires at the K-T Boundary." *Geological Society of America* Special Paper 247 (1990): 391–400.

Woosley, S. E., and Weaver, T. A. "The Evolution and Explosion of Massive Stars. II. Explosive Hydrodynamics and Nucleosynthesis." *The Astrophysical Journal Supplement Series* 101 (1995): 181–235.

Wright, E., et al. "Geomorphology and Stratigraphy of Two Overlapping Carolina Bays in Northeastern South Carolina." Presentation at the Geological Society of America Annual Meeting, Reno, Nev., 2000.

Yiou, F., et al. "Beryllium 10 in the Greenland Ice Core Project Ice Core at Summit, Greenland." *Journal of Geophysical Research* 102 (1997): 26783–94.

Zanner, C. "Nebraska's Carolina Bays." Presentation at the Geological Society of America Annual Meeting, Boston, November 5–8, 2001.

Zielinski, G. A., and Mershon, G. R. "Paleoenvironmental Implications Insoluble Microparticle Record in the GISP2 (Greenland) Ice Core During the Rapidly Changing Climate of the Pleistocene-Holocene Transition." *Geological Society of America Bulletin* 109 (1997): 547–59.

Zielinski, G. A., et al. "GISP2 Sulfate Data." IGBP PAGES/World Data Center for Paleoclimatology, Data Contribution Series, NOAA/NGDC Paleoclimatology Program, Boulder, Colo., 2004.

Zook, Herbert A. "On Lunar Evidence for a Possible Large Increase in Solar Flare Activity ~23×10⁴ Years Ago." In *Proceedings of the Conference on the Ancient Sun.* Edited by R. Peppin, J. Eddy, and R. Merrill. New York and Oxford: Pergamon Press (1980): 245–66.

INDEX